AF734405

TRANSGENIC PLANTS: NEW RESEARCH

TRANSGENIC PLANTS: NEW RESEARCH

OLIVER T. CHAN
EDITOR

Nova Science Publishers, Inc.
New York

Copyright © 2009 by Nova Science Publishers, Inc.

All rights reserved. No part of this book may be reproduced, stored in a retrieval system or transmitted in any form or by any means: electronic, electrostatic, magnetic, tape, mechanical photocopying, recording or otherwise without the written permission of the Publisher.

For permission to use material from this book please contact us:
Telephone 631-231-7269; Fax 631-231-8175
Web Site: http://www.novapublishers.com

NOTICE TO THE READER

The Publisher has taken reasonable care in the preparation of this book, but makes no expressed or implied warranty of any kind and assumes no responsibility for any errors or omissions. No liability is assumed for incidental or consequential damages in connection with or arising out of information contained in this book. The Publisher shall not be liable for any special, consequential, or exemplary damages resulting, in whole or in part, from the readers' use of, or reliance upon, this material. Any parts of this book based on government reports are so indicated and copyright is claimed for those parts to the extent applicable to compilations of such works.

Independent verification should be sought for any data, advice or recommendations contained in this book. In addition, no responsibility is assumed by the publisher for any injury and/or damage to persons or property arising from any methods, products, instructions, ideas or otherwise contained in this publication.

This publication is designed to provide accurate and authoritative information with regard to the subject matter covered herein. It is sold with the clear understanding that the Publisher is not engaged in rendering legal or any other professional services. If legal or any other expert assistance is required, the services of a competent person should be sought. FROM A DECLARATION OF PARTICIPANTS JOINTLY ADOPTED BY A COMMITTEE OF THE AMERICAN BAR ASSOCIATION AND A COMMITTEE OF PUBLISHERS.

LIBRARY OF CONGRESS CATALOGING-IN-PUBLICATION DATA

Transgenic plants : new research / Oliver T. Chan (editor).
p. cm.
ISBN 978-1-60692-017-6 (hardcover)
1. Transgenic plants. 2. Crops--Genetic engineering. I. Chan, Oliver T.
SB123.57.T734 2008
631.5'233--dc22

2008033200

Published by Nova Science Publishers, Inc. New York

CONTENTS

PREFACE

Short Communication A - For the 21st century a large increase in human population is expected. In fact, in the year 2050, estimated population will reach 9.1 billion, mainly in developing countries (USCB, 2008).

To sustain only this growth, agricultural outputs are expected to increase significantly. Wheat production for instance is expected to rise from the actual 600 million tons per year to one billion tons in 2020, which would imply doubling the current mean productivity of such crop (Rajaram, 2001).

Short Communication B - Improvements of gene delivery systems are always consequential works for transgenic plants and crop recombination. However, cell walls and plasma membranes are natural barriers to prevent most large and exogenous molecules from entering cells, thereby raising the difficulty of DNA delivery. Several DNA transfection methods have been developed recently, and they can be grouped into viral and nonviral delivery systems. In spite of amelioration of transfection efficiencies, applications of these methods were limited because of either immunogenetic problems or cell injuries. Protein transduction domains, also called cell-penetrating peptides, are small peptides with less than twenty amino acids highly rich in basic residues. These peptides have been used for intracellular delivery of various cargoes. In our previous research, AID (arginine-rich intracellular delivery) peptides, protein transduction domains composed only of arginine residues, could interact with DNA optimally in a specific nitrogen/phosphate (N/P) ratio, and cause retardation of DNA complexes. Plasmid DNAs not only entered the plant cells but also expressed in plant cells during two-day incubation. Lots of studies indicated that AID peptide-mediated delivery was in a multi-tentacled pattern and functioned in different organisms, including plants, mammals, and other species. The most notable property of AID peptides used in DNA delivery was its lack of cytotoxicity. Besides, simple and convenient protocols were being developed and indicated that AID peptides have the potential to be a new category of studies in cell biology and molecular biology in addition to traditional methods. Not surprisingly, more comprehensive applications of DNA delivery via AID peptides will be seen in the future.

Chapter 1 - Although wheat transformation has been succeeded by several research groups using particle bombardment and *Agrobacterium* infection methods, wheat cultivars suitable for transformation are limited and the transformation efficiency still remains to be low. Wheat transformation is effectively applied to molecular breeding of improved wheat varieties and functional analysis of newly identified wheat genes. Alternatively, heterologous plant species have been used for functional studies revealed by overexpression of transgenes from wheat. The authors' recent studies of abiotic-stress responsive genes showed that the heterologous tobacco transformation system was efficient for the evaluation of wheat gene function. GUS reporter assay using promoter regions of wheat abiotic stress-responsive genes reflected to their gene expression patterns in transgenic dicotyledonous plant species such as tobacco and *Arabidopsis*. Transgenic dicot plants overexpressing wheat transcription factor genes associated with abiotic stress responses significantly increased in tolerance levels against abiotic stress including freezing stress. Crossing among the transformants clearly proved the *cis-trans* relationship between the transcription factors and their direct target genes. Therefore, the heterologous transformation system gives us valuable and helpful information for molecular basis of the wheat signal networks in abiotic stress responses.

Chapter 2 - Ornamental plants (OPs) play important roles in society, religion and science. A wealth of species are considered to be OPs, and this makes this group of plants an increasingly important group as model plants for a range of physiological, biochemical, genetic and biotechnological studies. The introduction of horticulturally and agronomically useful traits into OPs through well-developed and fairly well-established transgenic systems would be one way in which to overcome some of the barriers of conventional breeding programmes and to shorten what might easily take years or decades to achieve, especially in some OPs whose vegetative cycles or vegetative-transition stage takes months or even years. The success of a transformation protocol invariably depends on the inherent success of the tissue culture protocol. Genetic engineering to incorporate selector, marker or useful genes into OPs has been primarily achieved by gene transfer mediated by *Agrobacterium* and particle bombardment, although other methods have been employed. In this chapter,te author provides a brief overview of how the main technologies have been used to develop genetically modified OPs. The desire for new traits (color, improved scent, cut-flower longevity and post-harvest robustness, modified architecture, acclimatization efficiency, and tolerance to biotic and abiotic stresses) and a greater overall acceptance of GM OPs, more than GM fruits or vegetables, make this group of plants important for every fundamental type of plant science.

Chapter 3 - Lignin is complex, three-dimensional polymer existing in plant cell wall. Lignin biosynthesis has become increasingly highlighted because it plays the important roles in plant growth and development as well as for applications in the human life. In this brief review, the progress in the lignin biosynthesis has been discussed. The genetic engineering approaches on lignin biosynthesis are reviewed. The transgenic plants with alternation on lignin biosynthesis and their applications in agriculture and industry productions are presented, particularly in pulping process, forage digestibility, and bio-fuel. Based on the information generated to date, implications and future research directions are discussed.

Chapter 4 - *Agrobacterium tumefaciens*-mediated plant transformation represents a powerful biotechnological technique for introducing desirable traits into commercially important plant species. Multiple approaches were undertaken during the last decade in order to improve the efficiency of plant transformation. Unfortunately, a relatively low frequency of stable integration events, high-copy-number insertions, a need for using antibiotic based selection markers, and a number of plant species recalcitrant to *Agrobacterium* transformation vote for continuous search for improved transformation procedures. The work of the authors' group was directed on elucidating novel factors that influence the efficiency of tissue regeneration and transgenesis. The integration of foreign DNA into the plant genome is believed to be strongly dependent on the activity of the host DNA repair machinery and, therefore, can be stimulated via upregulation of homologous recombination rate (RR) in host cells just before transformation. This should enable the manipulation of a frequency and perhaps, quality of integration events. The group was able to develop several unique growth medium compositions that allow a significant increase in RR without causing a negative effect on plants. The medium composition that exhibited no genotoxic influence was further tested for its effect on stable transformation in tobacco. The results obtained demonstrated that the application of a novel medium composition allows them to achieve higher transformation rates, as compared to those obtained while using conventional media. Plants grown on the modified medium generally looked healthier and produced more transgenic calli per amount of tissue transformed. The authors demonstrated that the application of this novel medium resulted in callus induction and plant regeneration as well as an increase in transformation rates by 2–3-fold, as compared to the application a conventional medium. Moreover, it resulted in the quality of integration events similar to that for the conventional medium. Currently, the authors continue searching for other factors that can be successfully used to improve plant transformation efficiency. Several chemical agents displaying a positive effect on *Agro*-transformation have been successfully identified. The possibility of using these factors alone and in combination is currently under the scope of our study. Overall, they believe that their approach represents a low-cost and highly efficient strategy for achieving a high number of stable transformants in a single round of plant transformation with *Agrobacterium*.

Chapter 5 - Genetic engineering of higher plants typically involves stable introduction of antibiotic or herbicide resistance genes as selective markers in order to obtain transgenic plants recovered from rarely occurred transformation events against a background of a large number of non-transformants. Over the past several years, concerns over the environmental and human health risks associated with antibiotic- and herbicide-resistance genes present in transgenic crops have been raised. Developing technologies to eliminate the marker genes and other superfluous T-DNA fragments from transgenic plants would be helpful not only for public acceptance of transgenic crops but also for studying plant gene functions and for generation of 'super' transgenic crop varieties with improved complex traits. In this paper, the authors reviewed the emerged techniques for removing marker genes from transgenic plants in the following aspects: (1) intra-genomic relocation of transgenes via transposable elements, (2) site-specific recombination systems, (3) intra-chromosomal recombination; and (4) co-transformation segregation approach. They put our emphasis on the application of site-specific recombination systems for use in plant gene transformation and discuss some future directions and improvements of this technology.

Chapter 6 – Genetic transformation has played a key role in gaining and applying knowledge of the roles of HMW-GS in wheat end-use properties. Reliable and stable expression of transgenes as well as the characterization and field adaptation of transgenic lines are prerequisites for the successful application of gene technology.

Loci that appear to be stably expressed initially can become progressively silenced over generations. The stability and the behaviour of transgenes are influenced by several factors, such as chromosomal location, transgene copy number and their interaction with the host genotype.

Traditionally such factors are characterized using Southern analysis which can be time consuming and laborious. Recent results obtained in various crops indicate that Real-Time

PCR could be a powerful tool for the detection and characterization of transgene locus structures. The determination of transgenic locus number through Real-Time PCR overcomes the problems linked to phenotypic segregation analysis and can analyze hundreds of samples in a day, making it an efficient method for estimating copy number integrated in a transgenic line.

This study was conducted to determine transgene copy number in transgenic lines and to investigate potential differences in sensitivity, resolution and variability between two different Real-Time chemistries (SYBR Green I dye and TaqMan probes). The authors have applied Real-Time PCR to a set of four transgenic durum wheat lines previously obtained. A total of four experiments (two experiments for each gene) were conducted and standard curves were obtained from serial dilutions of the plasmids containing the genes of interest. The correlation coefficients of the standard curves were rather good, being in the range between 0.95 and 0.97. By using TaqMan quantitative Real-time PCR we were able to achieve estimates of 1 to 42 copies of transgenes per haploid genome in T_4 homozygous transformants. Conversely, SYBR Green dye method revealed unable to accurately quantify transgene copy number as it failed in detecting the inserted genes when integrated in few copies.

In their study, the authors assessed Real-Time PCR as a fast, sensitive and reliable method for the detection of transgene copies in durum wheat, which can be a valid alternative to Southern analysis.

Chapter 7 - Viruses are particularly fast evolving pathogens, so they are capable of overcoming natural and engineered virus resistance. This is why engineering of virus resistance has become a testing area for new transgenic approaches. Nowadays pathogen derived resistance is the most common strategy to protect crop plants against viruses through transgenesis. Generally, pathogen derived resistance provides moderate protection against a narrow range of related viruses (protein-mediated resistance) or high levels of protection only against closely related strains of a virus (RNA-mediated resistance). So far this strategy has not offered protection against a broad spectrum of heterologous viruses. At the same time, several alternative strategies were developed that have conferred protection against multiple viruses. It was shown that transgenic plants expressing human ά- and β-interferon; interferon-induced 2'-5' oligoadenylate synthetase, pokeweed antiviral proteins (PAP) and PAP analogues, intracellular ds-RNA specific RNases and extracellular bovine pancreatic RNase were broadly resistant to multiple viruses. Recently some modifications of pathogen derived strategy such as transgenic production of artificial microRNA or chimeric peptide "aptamers" were made to provide broad-spectrum resistance. All these approaches offer different degrees of protection, specificity and biosafety that will be reviewed in this article.

Chapter 8 - Genetically modified crops are an effective management tool for worldwide growers. However, there is a concern that pests may develop resistance to Bt toxins produced by genetically-modified Bt plants. The authors study the impact of invasion of Bt-resistant pests on Bt crops. They show that the invasion leads to a fall in Bt crop production. The biomass of the Bt plants is shown to have only a weak dependence on the initial biomass of the insect invaders. Bt crop production is found to be essentially dependent on the initial Bt plant biomass. Furthermore, variations of the crop production under invasion of the Bt-resistant insects are shown to undergo badly predictable irregular oscillations. The complex character of the response of the Bt crop-pest system to the invasion of Bt-resistant insects can lead to significant complications in attempts to regulate the system dynamics.

Chapter 9 - Plant cell wall contains various components such as polysaccharides, lignin and proteins whose composition and concentration change during plant development and growth. Polysaccharides are major components in plants cell walls and contain three major classes: cellulose, hemicelluloses, and pectins. During plant development a series of coordinated biochemical processes occurred in plant including biosynthesis and degradation of cell wall polymers. Biosynthesis and remodulation of cell wall polysaccharides are regulated by complex enzymatic machinery including various enzymes belonging to families of glycoside hydrolases and glycoside transferases. These enzymes play crucial roles in the structure and function of the plant cell wall polysaccharides. Various molecular, genetic and biochemical approaches have been used to study of these enzymes. Transgenic plants modified in the expression of biosynthetic or modifying enzymes have been intensively used to investigate the role of the particular activities and to determine alteration in cell wall composition. The chapter describes utilisation of mutant and transgenic plants related to glycosyl transferases and glucoside hydrolases in order to determine the molecular basis of the synthesis and remodulation of plant cell walls polysaccharides. The modulation of plant cell wall polysaccharides using transgenic plants could be important not only for fundamental research but also in plant biotechnology.

In: Transgenic Plants: New Research
Editor: Oliver T. Chan

ISBN 978-1-60692-017-6
© 2009 Nova Science Publishers, Inc.

Short Communication A

TRANSGENIC PLANTS AND DEVELOPING NATIONS: A 21ST CENTURY CHALLENGE

André Martinho de Almeida*
IICT – Instituto de Investigação Científica e Tropical, Lisboa, Portugal

AN OVERVIEW ON AGRICULTURAL TRENDS FOR THE 21ST CENTURY WITH EMPHASIS ON DEVELOPING NATIONS

For the 21st century a large increase in human population is expected. In fact, in the year 2050, estimated population will reach 9.1 billion, mainly in developing countries (USCB, 2008).

To sustain only this growth, agricultural outputs are expected to increase significantly. Wheat production for instance is expected to rise from the actual 600 million tons per year to one billion tons in 2020, which would imply doubling the current mean productivity of such crop (Rajaram, 2001).

In such a context, the pressure to enlarge the areas used for crop production is continuously increasing. Such area enlargement will necessarily be noted at the level of soils that until now have been used for other purposes (pastures or woodlands for example), which would bring increasingly higher environmental costs, or at the use of marginal soils (desert or semi-desert, with high salt concentrations, marshes or cold regions) where productivity is extremely reduced. Also the irrigated areas are expected to rise although water pollution, exhaustion of existing fossil aquifers, growing contamination with salt water and competition with industrial or urban usage are major setbacks to such increase (Carruthers *et al.*, 1997).

The enlargement of agricultural areas will be specially noticed on developing nations: if today only China, Egypt, Kenya, Bangladesh and Vietnam have "per capita" crop land availability below 0.25 ha, in 2025 other countries such as Peru, Pakistan, Tanzania, Indonesia and the Philippines are expected to join this group (Sharma *et al.*, 2002).

* Correspondence should be addressed to: André M. de Almeida, IICT-CVZ, Fac. Med. Veterinária., Av. da Universidade Técnica, 1300-477 Lisboa, Portugal. amalmeid@itqb.unl.pt

In addition to the above mentioned situation, three factors are believed to further increase the worldwide shortage for crops, especially cereals:

1) The need of plant material (with special relevance to corn and sugar cane) to be used in the production of bio-ethanol in order to cut down greenhouse effect gases in the developed nations. This aspect is of particular relevance as the demand for corn for bio-ethanol productions is expected to occur in countries such as the United States of America, Canada, Argentina, Australia and the European Union that are major exporters of corn (both in the commercial factor and in hunger relief actions) and under constant pressure to decrease the production of carbon dioxide to the atmosphere. It is likely that the importance of sugar cane as an "export cash crop" will increase rendering it very interesting to developing nations to become exporters of this crop, leaving the production of staple food crops only marginal areas hence contributing to the food security problems in these nations.
2) An increasing demand for cereals and cereal-based animal products (pork, poultry, milk and eggs) in emerging economies, with special reference to China and India, that are presently relevant importers of agricultural products with a tendency to increase their expenditures on these commodities.
3) The rise in oil prices are also extremely important as this affects agricultural inputs and transportation costs and hence the need to continuously increase the production.

These factors combined contribute to general context of food scarcity, lowering living standards in most of the developing nations and increasing or creating social problems that affect severely such countries, with special reference to Southeast Asia, Latin America, Sub-Saharan Africa and some of the Caribbean and the Pacific island nations. An example of the above-stated can be found in the recent riots in countries, such as Haiti or Burma, that are under particularly fragile economic and political conditions.

Water Is the Major Constraint to Agricultural Output Increase in Developing and the Developed World

Plants in the field are subjected to several external factors that limit their growth and development and therefore their productivity. Such factors can be divided in two large groups: biotic and abiotic. Biotic stresses are dependent of biological factors and include the action of disease-causing pathogens, pests, mechanical injury or competition with weeds for access to water, nutrients or light. Abiotic stress refers to facts of physical nature such as intense light, herbicides or polluted and flooded soils, and also extremely important factors such as temperature stress (heat, chilling or freezing), drought and salinity. These last three factors (temperature, drought and salinity) are grouped in what is generally referred to as Water Deficit Stress (Slater *et al.*, 2003). Water deficit is created by a situation in which the plant's demand for water exceeds the supply available. Such situation is evident during drought, as water content in the soil decrease, but also during high-salt conditions (osmotic potential of salt solution is lower than that of water), high environmental temperatures (increasing water loss through evaporation) and freezing temperatures (formation of ice crystals in the extra-cellular space reduces water potential and results in the efflux of intracellular water) (Slater *et al.*, 2003).

Abiotic stresses have a very significant effect on crop productivity. Slater and collaborators (2003), presented the average yield of several major crops (wheat, barley, soybean, maize, potato and sugar beet) with yields previously registered, where limiting factors were considered minimal. These authors estimated that losses due to abiotic and biotic stresses range between 87% and 65%. Of such losses, the majority are due to abiotic stress (82% for wheat, 75% for barley, 69% for soybean, 66% for maize, 54% for potato and 51% for sugar beet). In productivity terms, such percentages represent losses of almost 12.000 kg/ha in vital plants for human consumption, such as the cereals wheat and maize. In contrast, the effect of biotic stresses is much less important: losses are estimated by the above mentioned authors as 5% for wheat, 7% for barley, 9% for soybean, 10% for maize, 19% for potato and 14% for sugar beet.

It can therefore be concluded that abiotic stresses are the major factors affecting crop productivity, limiting not only agricultural output and farmer's benefits but in a global perspective, the issue of food security in the 21st century. Consequently, it is possible, to assume that the selection of plants towards tolerance to abiotic stress factors is a major goal of both scientists and farmers in the 21st century. It has however to be said that, although measures to counter abiotic stresses would lead to better yields, it is clear that the improvement of pest-tolerant crops is also of extreme importance.

Towards Increasing Crop Productivity: Conventional Breeding or Genetic Engineering Approaches?

Over the last century, conventional plant breeding has been the methodology used to obtain increased production in crops that sustain humankind. Such methodologies allowed, for instance, the development of high yielding varieties in the 1960's-70's that launched the green revolution in Asia and averted mass starvation (Jauhar, 2001). In the last 20 years, novel techniques of gene transfer have been developed. Generally defined as Genetic Engineering, they involve the insertion of a well-characterized gene(s) into plant cells able to multiply and differentiate and the subsequent recovery of fully fertile plants with the inserted gene stably integrated in their genome (Jauhar, 2001). Such plants are defined as transgenic (or genetically modified) and their development resulted in numerous applications: bioremediation, fiber synthesis, biodegradable plastics, industrial enzymes or ornamental crops (for non-food applications); several food and feed applications; and recombinant proteins for therapeutics or diagnostics, (Dunwell, 1999; Dunwell, 2002).

Conventional plant breeding is a slow and laborious process: transferring a desired gene from a landrace or a closely related species to a crop variety or species takes from five to twelve years. When dealing with a less related species, if not impossible, that period could be of 15 years or longer (Jauhar, 2001). Genetic engineering allows a reduction of the time needed to transfer qualitative characteristics overcoming specie's barriers.

Together with other measures such as the use of biological fertilizers, improved pest control, soil and water conservation and the use of conventionally improved varieties, transgenic crops are considered to be a key component in answering questions of feeding the increase human population and agricultural output need for the 21st century (Herrera-Estrella, 2000).

Although threatened by difficulties involving technology transfer and food safety issues (Herrera-Estrella and Alvarez-Morales, 2001), advances in plant biotechnology have benefited farmers of developing nations in concrete situations, such as *Bacillus thurigiensis* cotton and cereals resistant to the European corn borer in Latin America or "witchweed" (*Striga*) control in sub-Saharan Africa (Toenniessen *et al.*, 2003). Although still only at experimental level, obtaining water deficit tolerance through methods of genetic engineering has been perceived as an important goal to agriculture in developing nations (Toenniessen *et al.*, 2003).

In a context of rising world population and escalating food prices, it is expectable that a fast response has necessarily to be given by both farmers and agricultural scientists in order to avoid severe famine situations and rupturing pressure on scarce natural resources. As transgenic plants hasten the whole process of selection and hence the development of necessary novel more productive and adaptable crop varieties they are therefore a key issue in the response of the challenges likely to be posed by agriculture in the 21st century.

CONCLUDING REMARKS

Figure 1 summarizes what was previously stated and what measures can be taken in order to answer challenges posed by the current global agricultural situation.

From what was stated above, the following conclusions could therefore be ascertained:

1. In the 21st century, conditions are created for a general scarcity of staple food sources.
2. To answer such rise in crop demands, an energetic response from farmers, researchers, public authorities is needed.
3. Regarding plant scientists and institutions that finance agricultural research, a strong emphasis should be put on the development of varieties of major crops capable of tolerating water deficit conditions as this will be a strong constraint to agricultural production in both the developed and the developing world.
4. Plant Genetic Engineering (Biotechnology) with the development of transgenic plants is considered the fastest way of increasing a certain trait of a determined plant variety, such as, for instance, tolerance to water deficit, enabling a more capable response by farmers worldwide.
5. The development of such transgenic crop varieties would have to have in consideration the most important setback of transgenic crops: food safety, especially regarding allergen ability of genetically engineered plants.
6. Once such varieties are created and adequately tested, it is imperative for farmers in developing nations to have access and hence be able to grow them.

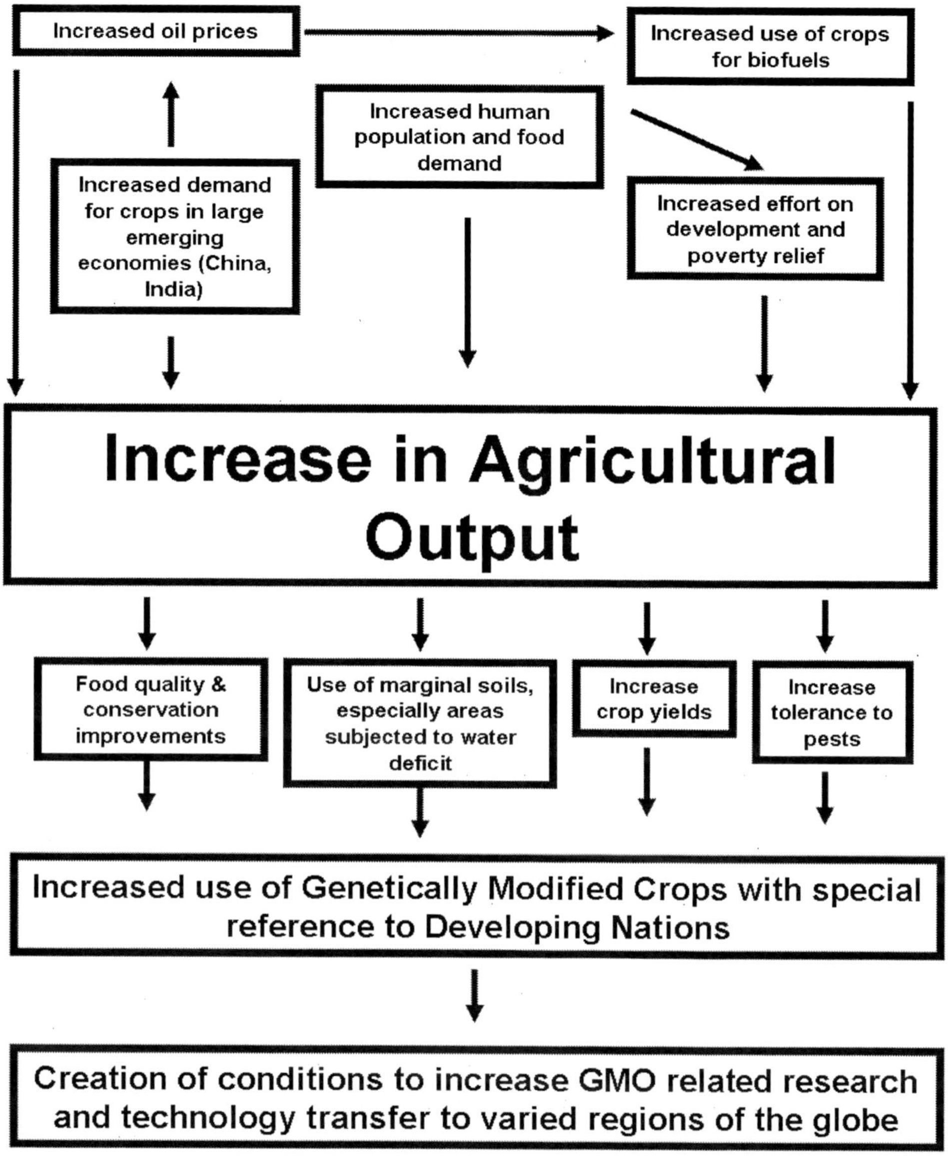

Figure 1. Economic context and agricultural trends for the 21st century overviews, possible responses and the importance of the use of genetically modified crops to answer such problems, with special reference to developing nations.

References

Carruthers, I., M.W. Rosegrant & D. Seckler (1997). *Irrigation and food security in the 21st century. Irrigation and Drainage Systems* 11: 83-101.

Dunwell, J.M. (2001). Transgenic crops: the next generation or an example of 2020 vision. *Annals of Botany* 84: 269-277.

Dunwell, J.M. (2002). Future prospects for transgenic crops. *Phytochemistry Reviews* 1: 1-12.

Herrera-Estrella, L. & A. Alvarez-Moralez (2001). Genetically modified crops: hope for developing countries? *EMBO reports* 21(4): 256-258.

Herrera-Estrella, L. (2000). Genetically modified crops and developing countries. *Plant Physiology* 124: 923-925.

Jauhar, P.P. (2001). Genetic Engineering and accelerated plant improvement: opportunities and challenges. *Plant Cell, Tissue and Organ Culture* 64: 87-91.

Rajaram, S. (2001). Prospects and promise of wheat breeding in the 21st century. *Euphytica* 119: 3-15.

Sharma, H.C., J.H. Crouch, K.K. Sharma, N. Seetharama & C.T. Hash (2002). Applications of biotechnology for crop improvement: prospects and constraints. *Plant Science* 163: 381-395.

Slater, A., N. Scott & M. Fowler (2003). *Plant Biotechnology – The genetic manipulation of plants.* Oxford University Press, Oxford, England, 350pp.

Toenniessen, G.H., J.C. O'Toole & J. DeVries (2003). Advances in plant biotechnology and its adoption in developing countries. *Current Opinion in Plant Biology* 6: 191-198.

USCB, 2008. *World population information.* United States Census Bureau. Available Online at: http://www.census.gov/ipc/www/idb/

In: Transgenic Plants: New Research
Editor: Oliver T. Chan

ISBN 978-1-60692-017-6
© 2009 Nova Science Publishers, Inc.

Short Communication B

DNA Transfection and Expression Mediated by Arginine-rich Intracellular Delivery Peptides in Plant Cells

Han-Jung Lee*, Jyh-Ching Chou, Chung-Pin Chen, Betty Revon Liu

Department of Life Science, National Dong Hwa University, Hualien 97401, Taiwan

Abstract

Improvements of gene delivery systems are always consequential works for transgenic plants and crop recombination. However, cell walls and plasma membranes are natural barriers to prevent most large and exogenous molecules from entering cells, thereby raising the difficulty of DNA delivery. Several DNA transfection methods have been developed recently, and they can be grouped into viral and nonviral delivery systems. In spite of amelioration of transfection efficiencies, applications of these methods were limited because of either immunogenetic problems or cell injuries. Protein transduction domains, also called cell-penetrating peptides, are small peptides with less than twenty amino acids highly rich in basic residues. These peptides have been used for intracellular delivery of various cargoes. In our previous research, AID (arginine-rich intracellular delivery) peptides, protein transduction domains composed only of arginine residues, could interact with DNA optimally in a specific nitrogen/phosphate (N/P) ratio, and cause retardation of DNA complexes. Plasmid DNAs not only entered the plant cells but also expressed in plant cells during two-day incubation. Lots of studies indicated that AID peptide-mediated delivery was in a multi-tentacled pattern and functioned in different organisms, including plants, mammals, and other species. The most notable property of AID peptides used in DNA delivery was its lack of cytotoxicity. Besides, simple and convenient protocols were being developed and indicated that AID peptides have the potential to be a new category of studies in cell biology and molecular biology

* E-mail: hjlee@mail.ndhu.edu.tw

in addition to traditional methods. Not surprisingly, more comprehensive applications of DNA delivery via AID peptides will be seen in the future.

ABBREVIATIONS

AID	arginine-rich intracellular delivery;
CPP	cell-penetrating peptide;
GFP	green fluorescent protein;
N/P	nitrogen/phosphate;
PBS	phosphate buffered saline;
PTD	protein transduction domain;
SR9	synthetic nona-arginine.

INTRODUCTION

Plasma membranes act as protective barriers for cells that are impermeable to biomacromolecules, such as DNA, RNA, peptides, and proteins. Typically, delivery of macromolecules can be mediated by lipids, electroporation, or through viral vectors. However, these manners have several limitations, including the inability to deliver into primary and non-dividing cells, the requirement for optimization with each cell type, low transfection levels, cell injury and cytotoxicity [1].

Recent studies identified several basic peptide sequences, namely protein transduction domains (PTDs), cell-penetrating peptides (CPPs), or arginine-rich intracellular delivery (AID) peptides in our group, appearing to rapidly translocate into cells [2]. These short AID peptides can also deliver biologically active molecules across membrane barriers and into animal cells, when synthesized as recombinant fusion proteins or covalently cross-linked to a wide variety of cargoes up to 200 nm in diameter [3, 4]. However, the mechanism of AID peptides entering cells is still unknown. The internalization of AID peptides was originally described as being independent of energy, classical endocytosis, and receptors [5]. It was hypothesized that AID peptides enter cells by lipid raft-dependent macropinocytosis, a specialized form of fluid phase endocytosis [6-8]. Recent reviews proposed that the internalization mechanisms of AID peptides may involve multiple pathways simultaneously [9, 10].

Our group recently demonstrated that AID peptides were able to deliver protein in plants. We found that AID peptides could covalently [11, 12] or non-covalently [13-16] deliver green fluorescent protein (GFP) and other proteins into different plant cells in fully biologically active forms. In animal system, the DNA delivery via AID peptides was demonstrated in 2003 [17-19]. However, very few reports described AID peptide mediated DNA delivery in plant cells. In this chapter, we showed that AID peptides could transfer plasmid DNA into plant cells, and the transient expression of the transferred plasmid DNA was performed [20]. Based on this study, we may develop a novel method for transgenesis in plants.

MATERIALS AND METHODS

Gel Retardation Assay of Peptide/plasmid Complex

To test if AID peptides could interact with plasmid DNA, pEGFP-C1 plasmid (Clontech, Mountain View, CA, USA) was pre-mixed with various amounts of SR9 peptide in PBS to form different N/P ratios (0, 0.2, 0.4, 0.5, 0.6, 0.8, 1, and 2). N/P ration was defined by the molar ratio between the cation group ($-NH_3^+$) and the anion group ($-PO_4^-$) of peptide/DNA molecules. This SR9 peptide was used in the absence of any plasmid as the maximal N/P ratio (∞). In order to form peptides/plasmid complexes, these mixtures were incubated at 37°C for 30 min. After incubation, these were analyzed by electrophoresis on a 1% agarose gel in TBE (45 mM Tris-borate and 1mM EDTA, pH 8.0) buffer at 100 voltages, followed by staining with ethidium bromide. Quantitative analysis of relative intensities of DNA bands was conducted by the UN-SCAN-IT software (Silk Scientific, Orem, UT, USA) [20].

Delivery of Peptide/plasmid Complex into Plant Cells

To examine the AID peptide-based plasmid DNA delivery, root cells of soybean (*Glycine max*) were used. Plant material preparation was described previously [14]. The AID peptide used here was a commercially synthesized nona-arginine (SR9) peptide [14, 20]. This SR9 peptide was dissolved in 30% acetic acid at a concentration of 7 mM (10 μg/μl) and frozen at -20°C until use. The molecular mass of SR9 peptide is 1423.7 Dalton. The pHBTsGFP(S65T)-NOS plasmid (kindly provided by Dr. Jen Sheen, Harvard University, MA, USA) with the GenBank Accession Number of EF090408 contains the GFP coding sequence under the control of the HBT promoter. HBT is a chimeric promoter with the 35S cauliflower mosaic virus enhancer fused to the basal promoter of the maize *C4PPDK* gene [21]. Three different preparations [10 μg of SR9 peptide, 10 μg of the pHBTsGFP(S65T)-NOS plasmid, and 10 μg of SR9 peptide plus 10 μg of the pHBTsGFP(S65T)-NOS plasmid] were individually added in phosphate buffered saline (PBS) to a final volume of 50 μl and incubated at 37°C for 30 min. These preparations were transferred into the eppendorf tubes containing 350 μl of double deionized water, respectively. Plant cells were treated with different preparations for 30 min at room temperature, washed with water, and followed by incubation for another 48 hours at room temperature.

Confocal Microscopy

Green fluorescent images were observed by the TCS SL confocal microscope system (Leica, Wetzlar, Germany). The confocal microscope was set up for the detection of GFP as follows: excitation was set at 488 nm and emission was collected using a 700 nm dichroic mirror and a 500–530 nm bandpass barrier filter.

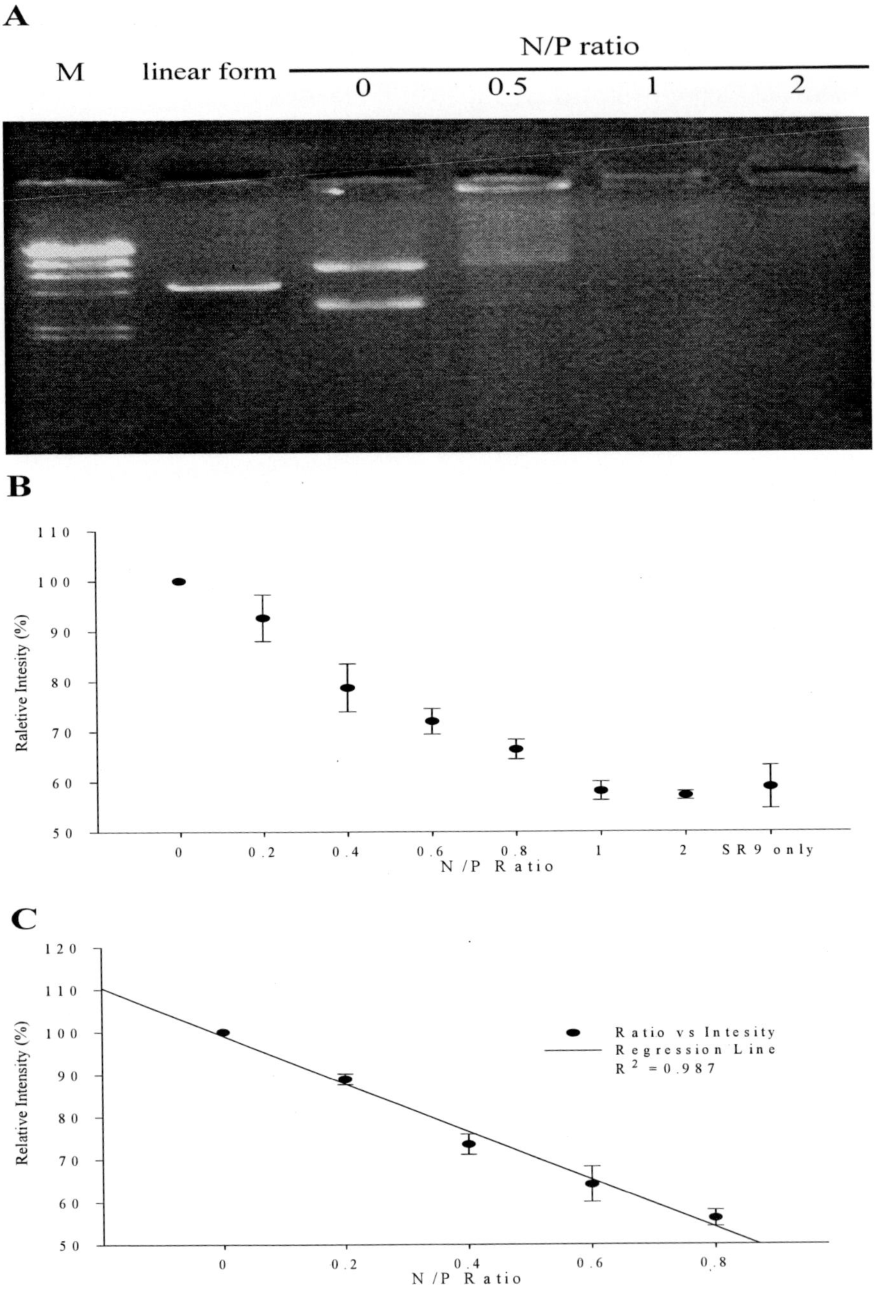

Figure 1. AID peptides interacted with plasmid DNA *in vitro*. (A) Gel retardation assay of AID peptide/plasmid DNA complexes. Lane 1 represents lambda/*Hind* III digested DNA markers (M). Linear (lane 2) or circular (lanes 3-6) form of pEGFP-C1 plasmid was pre-mixed with different amounts of SR9 peptide to form various N/P ratios (0, 0.5, 1, and 2), analyzed by electrophoresis, and stained by ethidium bromide. (B) Quantitative analysis of plasmid DNA intensity. Relative intensities of different N/P ratios (0, 0.2, 0.4, 0.6, 0.8, 1, 2, and ∞) were quantified. (C) Linear relationship between AID peptides and plasmid DNA. Amounts of SR9 peptide and intensities of plasmids presented an inverse ratio from 0 to 0.8 of N/P ratio.

RESULTS

Interaction of AID Peptides with Plasmid DNA

In our study of peptide/plasmid complexes, Figure 1A displays gel retardation assay that peptide/plasmid complexes were analyzed by electrophoresis on an agarose gel and stained by ethidium bromide. AID peptides bound to DNA *in vitro* with different N/P ratios and prevented plasmid DNA from the staining of the intercalating dye. Besides, DNA bands shifted when the amount of SR9 peptide increased (Figure 1A). The intensities were also getting lower as the N/P ratio got higher, and when the N/P ratio was 1, the intensity contained the same level as that of SR9 peptide only (Figure 1B). Figure 1C shows the relationship between the intensities of DNA bands and N/P ratios. The relative intensities decreased linearly and the R^2 value was about 0.987 in the regression line (Figure 1C). These results suggested that AID peptides could interact with DNA and form complexes. Relative intensities of DNA bands would also decrease linearly and present an inverse ratio with the amount of SR9 peptide.

Delivery of Peptide/plasmid Complex into Plant Cells

In order to investigate the delivery and expression of plasmid DNA mediated by AID peptides, a SR9 peptide was used in this study. Little green fluorescence was observed in root cells of soybean after 48 hours in which cells were treated with the pHBTsGFP(S65T)-NOS plasmid DNA alone (Figure 2A) or SR9 peptide alone (data not shown) for 30 min. In contrast, root cells of soybean treated with SR9 peptide/plasmid DNA complex at a molar N/P ratio of 3 clearly expressed green fluorescence with a high efficiency (Figure 2B). These results demonstrated that SR9 peptide could transfer plasmid DNA into plant cells and the target *gfp* gene of the transferred plasmid DNA was subsequently expressed.

CONCLUSION

In this chapter, we demonstrated that AID peptide was able to translocate across cell membranes and to carry exogenous DNA molecules into plant cells. The expression of target *gfp* gene of the transferred plasmid DNA mediated by SR9 peptide could be monitored by the emission of green fluorescence within cells. The fluorescent signal was detected evenly in almost all cells observed, and was predominantly located at higher levels in nuclei (data not shown). This result supported the potential role played by AID peptides as nuclear routing carriers.

The nontoxic SR9 peptide was able to interact with plasmid DNA electrostatically as revealed by gel retardation analysis [20]. These interactions were based on both parts between negative charges of plasmid DNA and positive charges of SR9 peptide. Recently, many reports further stated the ability of PTD peptides to transfer exogenous DNA molecules into cells [5, 22-29]. More detailed studies indicated that the DNA delivery mechanism mediated by AID peptides might involve macropinocytosis [20].

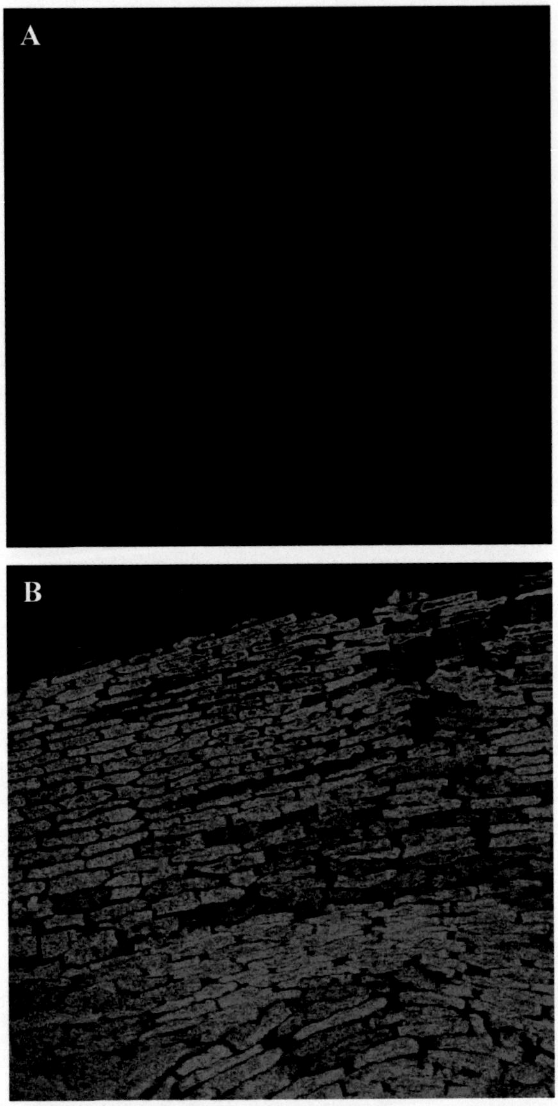

Figure 2. The delivery and expression of plasmid DNA mediated by AID peptide in plant cells. Root cells of soybean were treated with the pHBTsGFP(S65T)-NOS plasmid DNA only (A) or with SR9 peptide/plasmid DNA complex (B) for 30 min, and observed at a magnification of 400X by confocal microscopy after 48 hours of incubation.

Remarkably, this AID peptide-based DNA delivery strategy conferred several advantages, such as nuclear targeting, nontoxic effect, and easy protocol without protoplast preparation. However, there are still some problems needing to be overcome as we conducted this method in plants. First, the efficiency of DNA transfection is not high enough. The image taken from some local area after treatment showed promising effect. However, in general, approximately 20 percent of all plant cells could be transfected overall. Second, the plasmid DNA delivered by SR9 peptides can not stay in plant cells permanently. Gene expression of the delivered DNA plasmid in cells, GFP in this case, remains within one week. Third, the application of this method may be limited to certain plant cells we tested so far.

Acknowledgements

We thank Dr. Jen Sheen (Harvard University, MA, USA) for provision of the pHBTsGFP(S65T)-NOS plasmid, and Mr. Christopher Ciambarella for grammatical editing of this manuscript. This work was supported by the National Science Council, Taiwan.

References

[1] Lindsay, M. A. (2002) Peptide-mediated cell delivery: application in protein target validation. *Curr. Opin. Pharmacol.* 2, 587-594.

[2] Wadia, J. S. & Dowdy, S. F. (2002) Protein transduction technology. *Curr. Opin. Biotechnol.* 13, 52-56.

[3] Vives, E., Brodin, P. & Lebleu, B. (1997) A truncated HIV-1 Tat protein basic domain rapidly translocates through the plasma membrane and accumulates in the cell nucleus. *J. Biol. Chem.* 272, 16010-16017.

[4] Wadia, J. S. & Dowdy, S. F. (2003) Modulation of cellular function by TAT mediated transduction of full length proteins. *Curr. Protein Pept. Sci.* 4, 97-104.

[5] Dowdy, S. F. & Snyder, E. L. (2005) Recent advances in the use of protein transduction domains for the delivery of peptides, proteins and nucleic acids *in vivo*. *Expert Opin. Drug Deliv.* 2, 43–51.

[6] Wadia, J. S., Stan, R. V. & Dowdy, S. F. (2004) Transducible TAT-HA fusogenic peptide enhances escape of TAT-fusion proteins after lipid raft macropinocytosis. *Nat. Med.* 10, 310–315.

[7] Nakase, I., Niwa, M., Takeuchi, T., Sonomura, K., Kawabata, N., Koike, Y., Takehashi, M., Tanaka, S., Ueda, K., Simpson, J. C., Jones, A. T., Sugiura, Y. & Futaki, S. (2004) Cellular uptake of arginine-rich peptides: roles for macropinocytosis and actin rearrangement. *Mol. Therapy* 10, 1011-1022.

[8] Kaplan, I. M., Wadia, J. S. & Dowdy, S. F. (2005) Cationic TAT peptide transduction domain enters cells by macropinocytosis. *J. Control. Release* 102, 247-253.

[9] Nakase, I., Takeuchi, T., Tanaka, G. & Futaki, S. (2008) Methodological and cellular aspects that govern the internalization mechanisms of arginine-rich cell-penetrating peptides. *Adv. Drug Deliv. Rev.* 60, 598-607.

[10] Gump, J. M. & Dowdy S. F. (2007) TAT transduction: the molecular mechanism and therapeutic prospects. *Trends Mol. Med.* 13, 443-448.

[11] Chang, M., Chou, J. C. & Lee, H. J. (2005) Cellular internalization of fluorescent proteins via arginine-rich intracellular delivery peptide in plant cells. *Plant Cell Physiol.* 46, 482-488.

[12] Liu, K., Lee, H. J., Leong, S. S., Liu, C. L. & Chou, C. J. (2007) A bacterial indole-3-acetyl-L-aspartic acid hydrolase inhibits mungbean 2 (*Vigna radiata* L.) seed germination through arginine-rich intracellular delivery. *J. Plant Growth Regul.* 26, 278-284.

[13] Wang, Y. H., Chen, C. P., Chan, M. H., Chang, M., Hou, Y. W., Chen, H. H., Hsu, H. R., Liu, K. & Lee, H. J. (2006) Arginine-rich intracellular delivery peptides

noncovalently transport protein into living cells. *Biochem. Biophys. Res. Commun.* 346, 758-767.

[14] Chang, M., Chou, J. C., Chen, C. P., Liu, B. R. & Lee, H. J. (2007) Noncovalent protein transduction in plant cells by macropinocytosis. *New Phytol.* 174, 46–56.

[15] Hou, Y. W., Chan, M. H., Hsu, H. R., Liu, B. R., Chen, C. P., Chen, H. H. & Lee, H. J. (2007) Transdermal delivery of proteins mediated by noncovalently associated arginine-rich intracellular delivery peptides. *Exp. Dermatol.* 16, 999-1006.

[16] Liu, B. R., Chou, J. C. & Lee, H. J. (2008) Cell membrane diversity in noncovalent protein transduction. *J. Membr. Biol.* 222, 1-15.

[17] Rudolph, C., Plank, C., Lausiter, J., Schillinger, U., Muller, R. H. & Rosenecker, J. (2003) Oligomers of the Arginine-rich motif of the HIV-1 TAT protein are capable of transferring plasmid DNA into cells. *J. Biol. Chem.* 278, 11411-11418.

[18] Ignatovich, I. A., Dizhe, E. B., Pavlotskaya, A. V., Akifiev, B. N., Burov, S. V., Orlov, S. V. & Perevozchikov, A. P. (2003) Complexes of plasmid DNA with basic domain 47-57 of the HIV-1 Tat protein are transferred to mammalian cells by endocytosis-mediated pathways. *J. Biol. Chem.* 278, 42625-42636.

[19] Kim, H. H., Lee, W. S., Yang, J. M. & Shin, S. (2003) Basic peptide system for efficient delivery of foreign genes. *Biochim. Biophys. Acta* 1640, 129-136.

[20] Chen, C. P., Chou, J. C., Liu, B. R., Chang, M. & Lee, H. J. (2007) Transfection and expression of plasmid DNA in plant cells by an arginine-rich intracellular delivery peptide without protoplast preparation. *FEBS Lett.* 581, 1891-1897.

[21] Chiu, W., Niwa, Y., Zeng, W., Hirano, T., Kobayashi, H. & Sheen, J. (1996) Engineered GFP as a vital reporter in plants. *Curr. Biol.* 6, 325–330.

[22] Hellgren, I., Gorman, J. & Sylven, C. (2004) Factors controlling the efficiency of Tat-mediated plasmid DNA transfer. *J. Drug Target.* 12, 39-47.

[23] Jarver, P. & Langel, U. (2004) The use of cell-penetrating peptides as a tool for gene regulation. *Drug Discov. Today* 9, 395-402.

[24] Gupta, B., Levchenko, T. S. & Torchilin, V. P. (2005) Intracellular delivery of large molecules and small particles by cell-penetrating proteins and peptides. *Adv. Drug Deliv. Rev.* 57, 637-651.

[25] Choi, H. S., Kim, H. H., Yang, J. M. & Shin, S. (2006) An insight into the gene delivery mechanism of the arginine peptide system: role of the peptide/DNA complex size. Biochim. *Biophy. Acta* 1760, 1604-1612.

[26] Kim, H. H., Choi, H. S., Yang, J. M. & Shin, S. (2007) Characterization of gene delivery *in vitro* and *in vivo* by the arginine peptide system. *Int. J. Pharm.* 335, 70-78.

[27] Morris, M. C., Gros, E., Aldrian-Herrada, G., Choob, M., Archdeacon, J., Heitz, F. & Divita, G. (2007) A non-covalent peptide-based carrier for *in vitro* delivery of DNA mimics. *Nucl. Acids Res*. 35, e49.

[28] Ziegler, A. & Seeling, J. (2007) High affinity of the cell-penetrating peptide HIV-1 Tat-PTD for DNA. *Biochem.* 46, 8138-8145.

[29] Deshayes, S., Morris, M., Heitz, F. & Divita, G. (2008) Delivery of proteins and nucleic acids using a non-covalent peptide-based strategy. *Adv. Drug Deliv. Rev.* 60, 537-547.

In: Transgenic Plants: New Research
Editor: Oliver T. Chan
ISBN 978-1-60692-017-6
© 2009 Nova Science Publishers, Inc.

Chapter 1

WHEAT TRANSFORMATION OR HETEROLOGOUS TRANSFORMATION? FUNCTIONAL EVALUATION OF WHEAT ABIOTIC STRESS-RESPONSIVE GENES AND THEIR INTERACTION IN TRANSGENIC DICOT PLANT

Shigeo Takumi*[1]*, Akihiro Terashima*[1]*, Yumeto Kurahashi*[1]*,
***Chikako Egawa*[1] *and Fuminori Kobayashi*[2,3]**
[1]Laboratory of Plant Genetics, Graduate School of Agricultural Science, Kobe University, 1-1 Rokkodai-cho, Nada-ku, Kobe 657-8501, Japan
[2]Research Fellow of the Japan Society for the Promotion of Science, Japan
[3]Plant Genome Research Unit, National Institute of Agrobiological Sciences, 2-1-2 Kannondai, Tsukuba, Ibaraki 305-8602, Japan

ABSTRACT

Although wheat transformation has been succeeded by several research groups using particle bombardment and *Agrobacterium* infection methods, wheat cultivars suitable for transformation are limited and the transformation efficiency still remains to be low. Wheat transformation is effectively applied to molecular breeding of improved wheat varieties and functional analysis of newly identified wheat genes. Alternatively, heterologous plant species have been used for functional studies revealed by overexpression of transgenes from wheat. Our recent studies of abiotic-stress responsive genes showed that the heterologous tobacco transformation system was efficient for the evaluation of wheat gene function. GUS reporter assay using promoter regions of wheat abiotic stress-responsive genes reflected to their gene expression patterns in transgenic dicotyledonous plant species such as tobacco and *Arabidopsis*. Transgenic dicot plants overexpressing wheat transcription factor genes associated with abiotic stress responses significantly increased in tolerance levels against abiotic stress including freezing stress. Crossing among the transformants clearly proved the *cis-trans* relationship between the transcription factors and their direct target genes. Therefore, the heterologous transformation system gives us valuable and helpful information for molecular basis of the wheat signal networks in abiotic stress responses.

INTRODUCTION

Wheat is a plant group including some cultivated species such as bread wheat (*Triticum aestivum* ssp. *aestivum*) and durum wheat (*T. turgidum* ssp. *durum*). Common wheat is not only a major cereal but also an important material for plant science. Wheat has been a model plant for classical genetics and cytogenetics based on the production of inter- and intra-specific hybrids and the observation of chromosome pairing and banding. Compared with model plant species, *Arabidopsis thaliana* and rice, however, molecular biology studies in wheat are taking disadvantages due to the large genome size, the complex genome with polyploidy and difficulty of transformation. Common wheat is an allohexaploid species providing three homoeologous genomes (A, B and D genomes), and the genome size of common wheat is about 16,000-Mbp with the large content of repetitive sequences. Recent technological and experimental progresses in wheat science are going to overcome the disadvantages. Wheat consensus map was established based on simple sequence repeat (SSR) markers (Röder et al. 1998), several large DNA insert libraries were equipped (Akhunov et al. 2005), and a large number of expression-sequenced tags (ESTs) have been deposited in the DNA database (Zhang et al. 2004). The wheat genome information has been abundantly accumulated and shared in wheat researchers. The progresses make it possible to identify and isolate quantitative trait loci (QTLs) in common wheat. In the next decade, functional analyses of wheat genes will give us more and more significant, valuable and considerable information for wheat genetics and breeding.

Here, we describe current status of gene functional analysis in wheat. Wheat transformation requires a lot of efforts before and after the foreign gene transfer. Recently, some research groups successfully applied the wheat transformation systems to functional gene analysis. On the other hand, wheat genes have been transferred into the heterologous plant species such as *Arabidopsis*, rice and tobacco. The heterologous transformation systems are comparatively easy to use for the functional studies of wheat genes. We first summarize the wheat transformation systems and their application to functional gene analysis and determination and complementation of candidate gene in map-based cloning. Next, examples of gene functional analysis using the heterologous transformation system are explained. Finally, we mention our approaches to study wheat abiotic stress signal pathways using the heterologous transformation system, followed by the brief introduction of abiotic stress responses in *Arabidopsis* and wheat.

PROGRESS AND PROBLEM ON WHEAT TRANSFORMATION

Genetic transformation is a powerful tool for plant molecular genetics and breeding. There is a need for a reliable and efficient wheat transformation system for general application in a range of varieties. More than 10 years ago, a wheat transformation system was developed using scutellar tissues of cultured immature embryos via particle bombardment (Weeks et al. 1993; Vasil et al. 1993; Nehra et al. 1994; Becker et al. 1994; Takumi and Shimada 1996). The scutellar tissues of immature embryos are the most useful and efficient targets to deliver a foreign DNA through particle bombardment. In the previous wheat transformation studies, several common wheat cultivars such as Bobwhite, Fielder, Florida and Akadaruma have been used to generate transgenic plants. Out of them, Bobwhite is the most widely, commonly used cultivar in the wheat transformation experiments. Variation of wheat transformation frequency is generally caused by the difference in the *in vitro* culture response with the genotype, rather than the introduction efficiency of the transgene into wheat cells by particle bombardment (Takumi and Shimada 1997). Since the wheat transformation system using immature embryos is based on the somatic embryogenesis from the competent cells of the scutellar tissues, the transformation frequency is considered to be directly influenced by the peculiar ability of immature embryo culture. Regenerable cells from scutellar tissues of wheat immature embryos are generated on Murashige-Skoog (Murashige and Skoog 1962) or Linsmaier-Skoog (Linsmaier and Skoog 1965) medium in the presence of 2,4 dichlorophenoxy acetic acid (2,4-D) and sucrose. These media of the tissue culture of the wheat scutellar tissues are commonly used by most researchers of wheat transformation studies, but it might be necessary to improve the culture medium for spreading the wheat cultivar range available for the genetic transformation.

Now, low transformation frequency, limitation of varieties available for the transformation and unstable reproducibility of the transformation frequency are the main problems in wheat transformation. Large efforts on the optimization for wheat transformation have been still paid to overcome the problems. Optimization of transformation is essential in order to overcome the relatively low transformation frequencies encountered with wheat. Age of the donor wheat plants affects the reproducibility of transformation frequency of elite wheat varieties (Pastori et al. 2001). Subjecting the donor plants and/or explants to controlled abiotic stress markedly improves the transformation frequency (Bommineni et al. 1997; Pellegrineschi et al. 2002). Modification of the basal callus induction medium and optimization of the 2,4-D concentration are efficient for improvement of transformation of European elite wheat varieties (Rasco-Gaunt et al. 2001). These data make the wheat transformation more stable and reliable technique. Moreover, integration of low-copy-numbered transgenes into the wheat genome is required for stable inheritance and expression of the transgene (Takumi et al. 1999a; Yao et al. 2006; Figure 1A). Multiple copy integration of the transgene frequently occurs in particle bombardment-mediated transformation. The frequency of single copy, un-rearranged insertions in particle bombardment-generated wheat plants is rare (Pawlowski and Somers 1996). The multiple transgenes produce genomic rearrangement in the progeny and inhibit stable expression through DNA methylation and gene silencing (reviewed in Depicker et al. 2005). Towards stabilizing expression of transgenes, the common approach is to select transgenic pants containing a single intact copy of the transgene (Finnegan and McElroy 1994). Confirmation of the stable expression of the transgene should be needed in the progeny of regenerated transformants.

Figure 1. Wheat transformation and its application. (A) Mendelian segregation of the bialaphos-resistance in transgenic wheat expressing the *bar* gene under the control of the maize ubiquitin 1 gene promoter. The transgene was integrated into a single locus of the wheat genome. (B) Bialaphos-resistant callus from transgenic wheat expressing the *Ac* transposase gene. A plasmid construct containing the *Ds* element located between the rice actin1 gene (*Act1*) promoter and the *GUS* reporter gene was introduced into the cultured embryos of the transgenic wheat plants expressing the *Ac* transposase gene co-transformed with the *bar* gene. The blue tissues indicate integration and expression of *GUS* under the control of the *Act1* promoter after excision of the *Ds* element from the introduced plasmid. (C) Bialaphos-resistant plants regenerated from the callus described in (B). All tissues of the left plant were stained as blue, indicating the *GUS* expression. However, the right plant was not partly stained, meaning the chimeric expression of the introduced *GUS*.

Agrobacterium-mediated wheat transformation has been developed in the last decade (Cheng et al. 1997). It is perceived to have several advantages over other transformation techniques such as particle bombardment, including the ability to transfer large segments of DNA with minimal rearrangement and with fewer copies of transgenes at higher efficiencies (Hiei et al. 1997). Cultured scutella of immature embryos are target tissues of the *Agrobacterium*-mediated transformation as well as the biolistic transformation in wheat. Because the *Agrobacterium* transformation relies on the tissue-culture phase, the successful wheat transformation remains genotype-dependent (Wu et al. 2003). Herbicide phosphinothricin/bialaphos is the most successful selectable marker in wheat transformation. Alternatively, glyphosate selection is also successful in the *Agrobacterium*-mediated wheat transformation (Hu et al. 2003). Therefore, selectable markers should be considered in the efficient selection of wheat transformants again.

In spite of the previous large efforts, wheat transformation still remains to be a difficult technique, and the transformation frequency is relatively low and unstable. Chimeras plants containing cells with and without transgenes are frequently regenerated on the selection media in wheat transformation, especially in the particle bombardment-mediated system. Because of the chimerism, transgene inheritance should be confirmed in the next selfed-seed generation of the regenerated wheat plants. Further trials using various wheat varieties might make the wheat transformation systems more and more reliable and efficient.

Application of Wheat Transformation System to Functional Analysis

Bread-making quality is one of the most important traits for wheat breeding. Therefore, the wheat transformation technique was first applied to manipulate both the quality and quantity of high-molecular-weight glutenin subunits, a class of seed-storage proteins determining bread-making quality of wheat flour (Altpeter et al. 1996; Blechl and Anderson 1996). The introduced glutenin subunit genes are expressed and translated to functional subunit proteins in grains of transgenic plants, and the bread-making characters can be altered by the transgene expression (Barro et al. 1997; Rooke et al. 1999; Darlington et al. 2003). Besides the high-molecular-weight glutenin subunits, it was recently reported that co-suppression of the introduced and endogenous puroindoline genes altered grain hardiness in transgenic wheat (Xia et al. 2007). Over the next decade, the wheat transformation technique helps our understanding of gene function associated with the bread-making quality (Jones 2005).

The mutagenic potential of transposable elements has been widely exploited in many organisms to clone genes based on transposon tagging (Walbot 1992; Sundaresan 1996). Especially in *Arabidopsis* and rice, maize *Ac*/*Ds* transposons are vigorously used as efficient molecular tools for functional gene analysis (Kolesnik et al. 2004; Kuromori et al. 2006). Introduction of maize *Ac*/*Ds* transposable elements has been tried in transgenic wheat callus and plants (Takumi 1996; Takumi et al. 1999a, 1999b). Stably *Ac* transposase-expressing wheat plants were established, and *Ds* excision events were observed in the cells of the *Ac* transgenic lines (Takumi et al. 1999a, 2000a; Figure 1B, 1C). Reintegration of the excised *Ds* elements has never been found in the crossed progeny between the *Ds*-containing lines and the transposase-expressing lines in wheat, implying that transposition of the *Ac*/*Ds* elements has been unsuccessful in wheat yet. It was recently reported that the *Ac*/*Ds* elements were mobile in the barley genome, and the gene tagging system is improving towards the efficient functional analysis of the barley genes (Singh et al. 2006). This progress suggests the possibility to develop the *Ac*/*Ds* transposon tagging system in wheat.

Now, RNA interference (RNAi) is the most effective tool for the functional gene analysis (reviewed in Fu et al. 2007). This RNA-based post-transcriptional gene silencing technology is now useful for the functional confirmation of wheat genes isolated by map-based cloning

(Yan et al. 2004; Uauy et al. 2006a). Two examples of the stable RNAi transformation experiments are introduced as followed. First, vernalization requirement is a component character of major genetic determinants in wheat flowering control. One of the vernalization requirement loci, *Vrn-2*, was a repressor of another vernalization requirement loci, *Vrn-1*. *Vrn-1*, encoding a homolog of the *Arabidopsis* APETALA1 MADS-box protein, is an activator of phase transition from vegetative to reproductive growth (Yan et al. 2003; Murai et al. 2003). Physical mapping of the *Vrn-2* locus exhibited presence of two candidate genes, *ZCCT1* and *ZCCT2* (Yan et al. 2004). Transgenic wheat plants with the RNAi construct for *ZCCT1* clearly showed early flowering phenotype caused by repression of the *Vrn-2* function, indicating that *Vrn-2* is the *ZCCT1* gene but not *ZCCT2* (Yan et al. 2004). Another example of application of the RNAi technique is about gene functional analysis for *Gpc-B1*. *Gpc-B1* is a major QTL for grain protein content in the mapping population between wild tetraploid wheat, *T. dicoccoides*, and an emmer wheat cultivar, 'Langdon' (Joppa and Cantrell 1990; Joppa et al. 1997; Olmos et al. 2003). The high protein content allele of *Gpc-B1* showed pleiotropic effects such as rapid senescence of flag leaf and peduncle (Uauy et al. 2006b). The positional cloning of the *Gpc-B1* locus showed that *Gpc-B1* encodes a NAC transcription factor, and reduction of the transcript levels of multiple *Gpc-B1* homologs by RNAi delayed senescence and reduced the grain protein content in transgenic wheat (Uauy et al. 2006a). Thus, wheat RNAi has been used successfully to target a wide range of genes (Fu et al. 2007).

Virus-induced gene silencing (VIGS) also reduces the transcript accumulation levels of target genes (reviewed in Dalmay 2005). The silencing effect of VIGS is transient, but VIGS is an important tool for the analysis of gene function in plants. In wheat, VIGS based on the viral genomes of barley stripe mosaic virus is successfully applied to clarify function of target genes such as *Lr21* disease resistant gene (Scofield et al. 2005). Cloutier et al. (2007) applied VIGS to confirm compatible interaction between an *Lr1* allele from a wheat cultivar 'Thatcher' and leaf rust *Avr1*. Therefore, VIGS is a powerful tool for dissecting the genetic pathways of disease resistance in hexaploid wheat (Scofield et al. 2005). On the other hand, when the effect of the target gene has to be modified over a long developmental process, the use of stable RNAi transformation would be a better strategy as mentioned by Fu et al. (2007).

RNAi and VIGS basically result in loss-of-function mutation-like phenotype, although imperfect repression of the transcript accumulation levels of target genes causes intermediate phenotype. On the other hands, overexpression and ectopic expression of the target genes generate gain-of-function mutation phenotype, giving us valuable information about the gene function.

The third vernalization requirement gene, *Vrn-3*, was isolated by map-based cloning approach, and the final complementation analysis was conducted by introduction of the gain-of-function mutation allele into the wheat genome (Yan et al. 2006). *Vrn-3* is orthologous to *Arabidopsis FLOWERING LOCUS T* (*FT*). Expression of the wheat *FT* ortholog, *TaFT*, under the control of the *TaFT* promoter of the 'Hope' allele results in early flowering phenotype in the transgenic wheat plants (Yan et al. 2006).

Wheat *Q* gene governs the free-threshing character and square-head spike phenotype. In addition, *Q* pleiotropically affects a repertoire of other characters important for domestication such as rachis fragility, glume shape and tenacity. The candidate gene of *Q* was identified as a wheat homolog of *Arabidopsis APETALA2*, *WAP2*, by positional cloning (Faris et al. 2003). Overexpression of *WAP2* under the control of the native promoter showed subcompactoid and compactoid spikes and reduced the height in transgenic plants of cultivar 'Bobwhite' which has two endogenous copies of *Q* (Simons et al. 2006). These phenotypic alterations resemble severe phenotype of the multi-copied *Q* genes, strongly indicating that *WAP2* is the *Q* gene.

Out of more than 50 wheat leaf rust resistance (*Lr*) genes, *Lr1*, *Lr10* and *Lr21* were recently isolated by map-based cloning (Huang et al. 2003; Feuillet et al. 2003; Cloutier et al. 2007). For the complementation analysis of *Lr21*, a cosmid clone with the *Lr21* functional allele was introduced into the genome of a susceptible cultivar 'Fielder', and improvement of the resistance was confirmed in the transgenic wheat (Huang et al. 2003). Overexpression of *Lr10* also enhanced the leaf rust disease resistance in the susceptible wheat cultivar 'Bobwhite' (Feuillet et al. 2003). *Lr1* was finally confirmed by recovering the resistant phenotype in transgenic Fielder with the candidate gene (Cloutier et al. 2007).

FUNCTIONAL ANALYSIS OF WHEAT GENES IN THE HETEROLOGOUS PLANT

An alternative way to study the wheat gene functions is use of transformation systems in heterologous species. Biological and biochemical information in model plant species such as *Arabidopsis* and rice are quite useful and valuable in wheat gene functional analyses. Especially in the fields of development, flowering and abiotic stress response, a lot of wheat studies have been referred to the data from the model plants. Here, we introduce successful, functional studies of wheat genes associated with determination of development, flowering time and abiotic stress tolerance using the heterologous plants.

At the plant shoot apical meristem (SAM), *Knotted1*-type homeobox (*KNOX*) genes play important roles to initiate, establish and maintain the meristematic activities (Vollbrecht et al. 1991; Lincoln et al. 1994). Ectopic overexpression of the *KNOX* genes results in production of dissected and lobed leaves in dicots and in knot-formation on leaves of monocots often by ectopic shoot meristem formation (Matsuoka et al. 1993; Lincoln et al. 1994; Figure 2A, 2B). Significant importance of the *KNOX* genes in SAM implies evolutionary conservation of the KNOX functions. For confirmation of the functional conservation of the *KNOX* genes, the *Arabidopsis*, maize, rice and tomato studies give us a lot of information about alterations of transgenic plants overexpressing the *KNOX* genes. One of the wheat *KNOX* genes, *Wknox1*, is an ortholog of maize *Knotted1* and rice *OSH1* (Takumi et al. 2000b). Expression of *Wknox1* under the control of the CaMV35S promoter induces the morphological alteration with dissected and lobed leaves and the ectopic SAMs in transgenic tobacco, and the altered morphology resembles the previously reported phenotypes of the transgenic dicot plants (Morimoto et al. 2005; Figure 2C). Overexpression of another wheat member of the *KNOX* genes, *WRS1*, also dramatically alters the leaf morphology of transgenic tobacco as well as the case of maize *rough sheath1 KNOX* gene (Schneeberger et al. 1995; Morimoto et al. 2009; Figure 2D). Moreover, transgenic tobacco plants with a *GUS* reporter gene under the control of the *Wknox1* promoter region confer the meristematic expression patterns of the *KNOX* gene (Morimoto et al. 2005; Figure 2E). Thus, the heterologous transformation system is efficient to confirm the orthologous function and expression pattern of the evolutionary conserved genes associated with wheat development.

Figure 2. Heterologous transformation of wheat development-related genes. (A, B) Overexpression of a rice *OSH1* homeobox gene under the control of the CaMV35S promoter resulting in ectopic formation of a ligule/auricle-like tissue on leaf blade. Arrows indicate the ectopic formed tissues. (C, D) Transgenic tobacco plants with *CaMV35S promoter::Wknox1* (C) and *CaMV35S promoter::WRS1* (D) chimeric genes. (E) Histochemical localization of the GUS activity near the shoot apices of the transgenic tobacco plant with the chimeric GUS gene under the control of the 2-kb *Wknox1d* promoter region.

Arabidopsis CONSTANS (*CO*) and its rice ortholog, *Heading date1* (*Hd1*), play central roles in controlling flowering time (Mouradov et al. 2002). The *CO*/*Hd1*-containg flowering pathway is associated with day length and photoperiodic response. *Arabidopsis CO* shows rhythmic expression with a 24-h period circadian rhythm, and the CO protein acts in the phloem to positively regulate a downstream *FT* expression under long-day condition (Suárez-López et al. 2001; An et al. 2004). On the other hand, *Hd3a*, a rice *FT* ortholog, is negatively regulated by *Hd1* in rice which is a short-day plant (Hayama et al. 2003). Expression of *TaHd1*, a wheat homolog of *CO*/*Hd1*, under control of the *TaHd1* promoter region promotes heading under shot-day condition and delays under long-day condition in transgenic rice plants (Nemoto et al. 2003). These transgenic plants with *TaHd1* prove that the *TaHd1* gene can complement the rice *Hd1* gene and function normally in the rice background. *Arabidopsis GIGANTEA* (*GI*) gene acts upstream of *CO* and functions in mediating photoperiodic flowering (Araki and Komeda 1993). Overexpression of a rice *GI* ortholog, *OsGI*, also increases the *Hd1* expression under both short- and long-day conditions in transgenic rice (Hayama et al. 2003). *TaGI1*, a wheat *GI* homolog, can complement the *gi* mutant phenotype of *Arabidopsis*, and overexpression of *TaGI1* alters flowering time in transgenic *Arabidopsis* (Zhao et al. 2005). Moreover, *SUPPRESSOR OF OVEREXPRESSION OF CO 1* (*SOC1*) functions as an integrator of the flowering pathways and plays an important role in floral transition of *Arabidopsis* (Moon et al. 2003). Transgenic *Arabidopsis* plants with a wheat *SOC1* homolog, *WSOC1*, under the control of CaMV35S promoter exhibit early-flowering phenotype under the short-day condition but not under the long-day condition (Shitsukawa et al. 2007). This function of *WSOC1* under the *Arabidopsis* background is corresponding to that of the *Arabidopsis SOC1*. These results indicate that the wheat photoperiod-responsive pathway determining flowering time includes the same members consisting of the central pathway of flowering in *Arabidopsis* and rice.

Except for the flowering-associated genes, wheat abiotic stress-responsive genes have been introduced into the heterologous genomes of *Arabidopsis*, tobacco and rice. Overexpression of *WhUCP*, a wheat gene of mitochondrial uncoupling protein, under the control of CaMV35S promoter results in a significant increase in tolerance against oxidative stress and cold stress in transgenic rice plants (Ozawa et al. 2006). This experiment indicates that the wheat UCP located in the mitochondrial inner membrane is positively involved in the pathway for abiotic stress tolerance, probably through a decrease in cellular oxidative damage. Similarly, overexpression of wheat alternative oxidase, WAOX1a (Takumi et al. 2002), improves mitochondrial respiration capacity and decreases levels of reactive oxygen species under low temperature stress in the transgenic *Arabidopsis* plant (Sugie et al. 2006). The result in the transgenic *Arabidopsis* with *Waox1a* supports the hypothesis that mitochondrial alternative oxidase alleviates oxidative stress when the cytochrome pathway of respiration is inhibited under abiotic stress conditions. These heterologous transgenic studies clearly demonstrated that the decrease of oxygen species in mitochondria plays an important role in development of abiotic stress tolerance.

Reversible protein phosphorylation is central to the perception and response to abiotic stress such as water deficit (Hirt 1997). Some of genes encoding the catalytic subunit of Ser/Thr protein phosphatase 2A (PP2Ac) are up-regulated by drought and high salinity in plants (Yu et al. 2003). *TaPP2Ac-1* is one of the catalytic subunit genes of PP2A, and its transcript accumulation level is enhanced in wheat seedlings (Xu et al. 2007). Transgenic tobacco plants overexpressing *TaPP2Ac-1* show less wilting under water-deficit conditions than non-tranformed controls, indicating that the *TaPP2Ac-1* gene is associated with increasing plant resistance to drought stress (Xu et al. 2007).

Thus, functional gene analyses using the heterologous transformation systems are widely conducted in wheat studies about abiotic-stress signal pathways and development of tolerance. Signal transduction pathways in response to abiotic stress such as low temperature, dehydration and salt are well understood and established in a model organism, *Arabidopsis thaliana* (Thomashow 1999; Shinozaki and Yamaguchi-Shinozaki 2000; Shinozaki et al. 2003). We summarize below the knowledge of abiotic-stress signal pathways in *Arabidopsis* and wheat, and then report in detail the results obtained using heterologously transgenic plants with wheat abiotc stress-responsive genes.

ABIOTIC STRESS RESPONSES IN *ARABIDOPSIS* AND WHEAT

Abiotic stress including cold/freezing, drought and salt has large influences on plant growth and yield. Cold/freezing tolerance is acquired through the cold acclimation process that is triggered in response to low temperature (Thomashow 1999). Low temperature induces and/or enhances expression of numerous cold-responsive (*Cor*)/late embryogenesis-abundant (*Lea*) genes, and accumulated COR/LEA proteins promote and sustain the development of freezing tolerance (Thomashow 1999). In *Arabidopsis*, a functional *cis*-acting element, i.e. the CCGAC core motif known as a CRT (C-repeat)/DRE (dehydration-responsive element) sequence, was proven to play a pivotal role in the promoter function of *COR15A/RD29A* genes (Yamaguchi-Shinozaki and Shinozaki 1994; Baker et al. 1994). It is well known that the CBF/DREB1 protein family dominantly regulates the low temperature-responsive induction of the *Cor/Lea* transcripts (Shinozaki and Yamaguchi-Shinozaki 2000; Thomashow 2001). These transcription factors contain a DNA binding domain found in the ethylene-responsive element binding protein/APETALA2 (EREBP/AP2) family (Stockinger et al. 1997; Liu et al. 1998). In *Arabidopsis*, it has been considered that the CBF/DREB1 regulon is involved in abscisic acid (ABA)-independent regulatory systems (Shinozaki and Yamaguchi-Shinozaki 2000; Shinozaki et al. 2003; Zhang et al. 2004). In *Arabidopsis*, three tandemly duplicated *CBF/DREB1* genes, *CBF1/DREB1B*, *CBF2/DREB1C* and *CBF3/DREB1A*, are induced by low temperature. The *CBF/DREB1* transcript levels reach a maximum at about 3 h LT treatment and then decline significantly, and the transcripts remain at a low level over the course of the 3-week LT treatment (Zarka et al. 2003). The CBF-mediated cold-response system in *Arabidopsis* appears to be conserved in both dicotyledonous and monocotyledonous plants including rape, tomato, wheat and rye (Jaglo et al. 2001). Another transcription factor recognizing CRT/DRE is DREB2 which is mainly activated by drought stress (Liu et al. 1998). Thus, cross-talk between low temperature and drought stress response occurs at the transcription level through binding of the CRT/DRE *cis*-element (Shinozaki and Yamaguchi-Shinozaki 2000).

ABA is one of key plant hormones responding to environmental stresses. The ABA level is transiently increased by low temperature (Chen et al. 1983; Ryu and Li 1994; Lång et al. 1994). Several *Cor/Lea* genes, whose expression levels correlate with increased freezing tolerance, are also responsive to exogenous ABA (Palva 1994), and ABRE sequences exist in the promoter regions of these *Cor/Lea* genes in *Arabidopsis* (Lång and Palva 1992; Baker et al. 1994; Yamaguchi-Shinozaki and Shinozaki 1994). The *Arabidopsis* bZIP-type AREB/ABF proteins as dimers bind to ABRE and activate ABA-dependent gene expression (Choi et al. 2000; Uno et al. 2000; Kim 2006). Their activities are reduced in both ABA-deficient *aba2* and ABA-insensitive *abi1* mutants, but enhanced in the ABA-hypersensitive *era1* mutant, indicating that the activation of the AREB/ABF proteins requires an ABA-mediated signal (Uno et al. 2000). Overexpression of *ABF3*, *AREB1/ABF2* or *AREB2/ ABF4* induces higher expression levels of *RD29B* and *RAB18*, and then improves freezing tolerance levels in transgenic plants (Kim 2006). These findings suggest that the ABA-dependent

AREB/ABF regulon may be partly involved in cold acclimation and freezing tolerance. Several ABA-inducible C_2H_2-type zinc-finger proteins have also been reported to affect cold response (Zhang et al., 2004; Yamaguchi-Shinozaki and Shinozaki, 2005). Expression of an *Arabidopsis CBF4* gene, as another member of the *CBF/DREB* family, is also up-regulated by drought stress through the ABA-dependent pathway (Haake et al. 2002). Because many low temperature- and drought-inducible genes contain both CRT/DRE and ABA-responsible element (ABRE) motifs in their promoters, these *cis* elements are considered to function independently. A study of the *Arabidopsis RD29A* promoter showed that the CRT/DRE functions cooperatively with ABRE as a coupling element in the ABA-responsive gene expression in response to drought stress (Narusaka et al. 2003). Such regulatory pathways in *Arabidopsis* provide considerably informative models to study the low temperature/drought signal response in monocotyledonous crop species.

In Triticeae species such as common wheat (*T. aestivum* L.), barley (*Hordeum vulgare* L.) and rye (*Secale cereale* L.), a number of low temperature-inducible *Cor/Lea* and *CBF/DREB*-related genes have been isolated and characterized (Thomashow 1999; Kobayashi et al. 2004a), and genetic maps of major loci and QTLs affecting cold/freezing tolerance have been constructed and compared (Cattivelli et al. 2002). Transcript accumulation levels of these wheat *Cor/Lea* and *CBF/DREB*-related genes show significant differences between cultivars with distinct levels of freezing tolerance (Ohno et al. 2001, 2003; Takumi et a. 2003; Kobayashi et al. 2004a). In common wheat, *Vrn-1* and *Fr-1* are well known major quantitative trait loci (QTLs) respectively determining vernalization requirement for flowering and winter hardiness. The two loci are linked on the long-arms of homoeologous group 5 chromosomes (Galiba et al. 1995; Sutka et al. 1999). Another locus, *Fr-A2*, was identified and mapped on a distal region of the long arm of chromosome $5A^m$ in diploid wheat *T. monococcum* (Vágújfalvi et al. 2003). A cluster of 11 *CBF* genes in einkorn wheat was located to the $Fr\text{-}A^m2$ locus (Miller et al. 2005). The $Fr\text{-}A^m2$ locus controls *Cor/Lea* gene expression and freezing tolerance (Vágújfalvi et al. 2003).

The CRT/DRE, ABRE and other cold-responsive motifs have been identified in the 5' upstream regions of many *Cor/Lea* genes in common wheat (Takumi et al. 2003; Kobayashi et al. 2008a). Wheat *CBF* homologs such as *TaCBF*, *TaDREB1* and *Wcbf2* encode a transcription factor recognizing the CRT motif (Jaglo et al. 2001; Shen et al. 2003; Kume et al. 2005). The WCBF2 cascade is tightly associated with cold acclimation and the *Fr-1* alleles, whereas expression of the cold-responsive *Vrn-1* gene is apparently not related to expression of the *Wcbf2* gene (Kobayashi et al. 2005). Many wheat *Cor/Lea* genes such as *Wdhn13*, *Wrab17*, *Wrab18* and *Wrab19* are responsive to low temperature, drought and ABA (Ohno et al. 2003; Kobayashi et al. 2004a, 2006), and their promoter regions contain ACGT core motifs (Kobayashi et al. 2008a). These observations suggest that bZIP-type transcription factors play significant roles in regulation of *Cor/Lea* gene expression. Two genes encoding a bZIP-type transcription factor, *Wlip19* and *Wabi5*, have been isolated from common wheat (Kobayashi et al. 2008b, 2008c). Expression of other cold-responsive transcription factor genes, *Wdreb2* and *Wlip19*, are weakly controlled by *Fr-1* (Kobayashi et al. 2004b), while *Wabi5* involved in the ABA-dependent pathway appears to act independently of *Fr-1* (Kobayashi et al. 2008c). Mutant analyses suggest that ABA sensitivity is not associated with cold acclimation, while ABA is involved in determination of the basal level of freezing tolerance in wheat (Kobayashi et al. 2006, 2008d). Thus, the *Cor/Lea* gene expression mediated by CBF/DREB and bZIP-type transcription factors might play central roles in development of abiotic stress tolerance in wheat. However, no direct evidence has shown an association of the CBF/DREB and bZIP-type transcription factors with stress tolerance or activation of stress-responsive *Cor/Lea* genes. We describe below our approach using the heterologous transformation systems to clarify the contribution of the CBF/DREB and bZIP-type transcription factors on abiotic stress tolerance and their direct activation of the wheat *Cor/Lea* genes.

APPLICATION OF HETEROLOGOUS TRANSFORMATION TO ANALYSIS OF WHEAT ABIOTIC STRESS RESPONSES

Here we show our recent progresses using the heterologous transformation systems on abiotic stress responses in common wheat. As mentioned above, some transcription factor genes assumed to control *Cor/Lea* gene family under low temperature and drought stress conditions have been isolated based on high homology of their putative amino acid sequences to transcription factors identified in *Arabidopsis*. However, information about roles of the transcription factors is still limited in wheat. In the following section, we briefly introduce (1) chloroplast targeting of wheat COR proteins, (2) abiotic stress-responsiveness of wheat gene promoters of *Cor/Lea* and their transcription factors, (3) overexpression of the *Cor/Lea* and their transcription factors in transgenic plants, and (4) *cis-trans* relationship between the *Cor/Lea* promoters and their transcription factors.

Analysis of Intracellular Localization of Wheat COR Proteins

Wcor14 and *Wcor15* belong to the wheat *Cor/Lea* gene family, and their encoded proteins are targeting to chloroplast (Tsvetanov et al. 2000; Takumi et al. 2003). Their expression is specifically induced by low temperature in wheat leaves, and light illumination markedly increases the steady-state level of its transcripts. The WCOR14 and WCOR15 proteins share an identical chloroplast-targeting signal (Ohno et al. 2006), suggesting that *Wcor14* and *Wcor15* are paralogs possibly generated by duplication in the ancestral diploid genomes. Green fluorescent protein (GFP) is widely used as a reporter for analysis of intracellular protein localization. Chloroplast-targeting of WCOR14 and WCOR15 was proved by the bombardment-mediated transient expression of chimeric *CaMV35S promoter::Wcor14-GFP* and *CaMV35S promoter::Wcor15-GFP* constructs in the guard cell chloroplasts of excised spiderwort leaves (Takumi et al. 2003; Takumi and Nakamura 2005). Transgenic tobacco lines stably expressing the *Wcor15-GFP* fusion gene showed transport and abundant accumulation of the WCOR15 protein in the stromal compartment of the chloroplasts (Shimamura et al. 2006; Figure 3A). Immunoblot analysis also demonstrated that both WCOR14 and WCOR15 were accumulated in the stromal compartment of chloroplasts isolated from cold-acclimated wheat leaves (Shimamura et al. 2006). Barley COR14b and wheat WCS19 are similarly targeted into chloroplasts revealed by the immunoblot analysis of the chloroplast protein fraction (Crosatti et al. 1999) or immuno-localization by electron microscopy of the purified chloroplasts (NDong et al. 2002).

The GFP fusion protein can be used as a reporter for *in vivo* monitoring in transgenic plants. Shimamura et al. (2006) reported the *in vivo* monitoring of the WCOR15 accumulation kinetics during cold acclimation in two types of transgenic tobacco plants. Two different plasmid constructs, *CaMV35S promoter::Wcor15-GFP* and *Wcor15 promoter::Wcor15-GFP*, were introduced into the tobacco genome. Increase of the GFP activity controlled by the *Wcor15* promoter showed a good agreement with the corresponding increase revealed by the immunoblot analysis of the WCOR15 protein. The GFP-detection system, therefore, serves as a useful means in directly monitoring the kinetics of *in vivo* expression patterns of the *Cor/Lea* gene in transgenic plants.

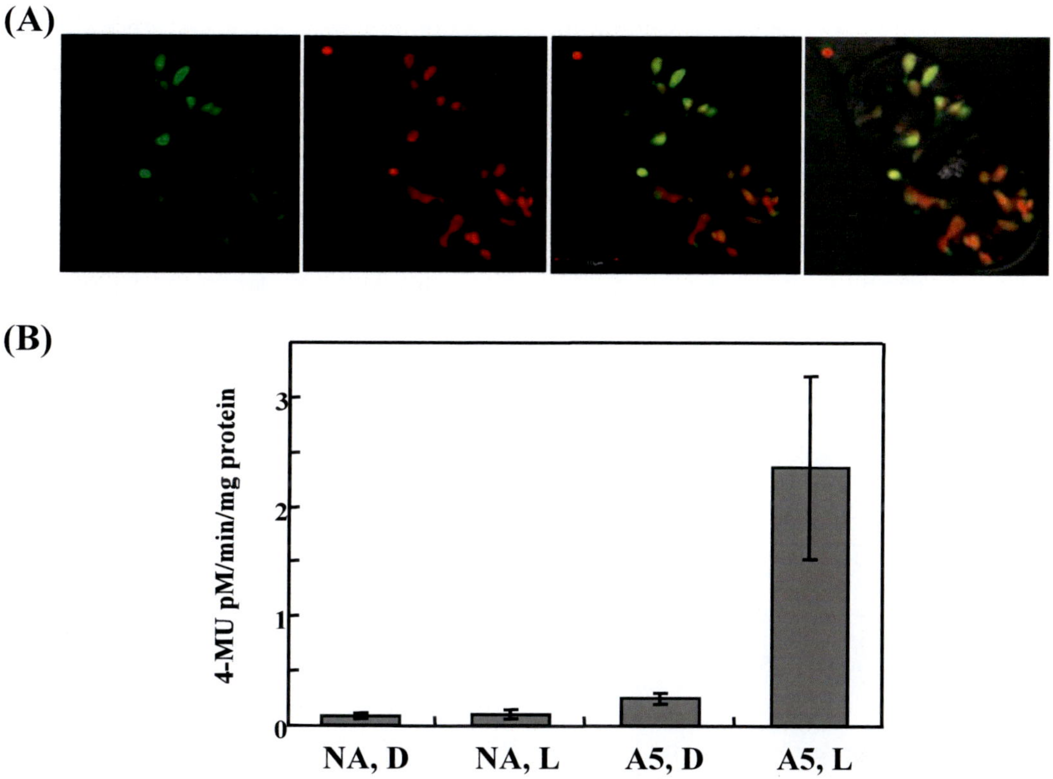

Figure 3. Analyses of the wheat *Cor/Lea* genes using transgenic tobacco plants. (A) Chloroplast targeting of the WCOR15-GFP fusion protein and its effect on freezing tolerance in transgenic tobacco plants. Highly magnified images of the GFP fluorescence in a trichome at shoot apex of transgenic tobacco plants with the *CaMV35S promoter::Wcor15-GFP* construct. Right two photos are the merged images. (B) Cold induction of the reporter genes under control of the 5' upstream region of *Wcor15* in transgenic tobacco plants. Cold-specific and light-stimulated induction of the *GUS* reporter gene in a transgenic tobacco plant with *Wcor15 promoter::GUS*. The histogram illustrates the quantified GUS activity. NA, Non-acclimated; A5, 5-d acclimated; D, continuous dark; L, 16 hlight/8-h dark.

Monitoring of Promoter Responses to Abiotic Stress

Expression of *Wcor14* and *Wcor15* are specifically induced by low temperature similar to that of *Wlt10* and *Wcs120* (Houde et al. 1992; Tsvetanov et al. 2000; Ohno et al. 2001; Takumi et al. 2003). In the 5' upstream region of *Wcor15*, three CRT/DRE core motifs are located, similar to *Wcs120* (Vazquez-Tello et al. 1998; Takumi et al. 2003). To examine if the cold-responsiveness of the 5' upstream sequence, transgenic plants with a chimeric *GUS* reporter gene under the control of the 1.7-kb upstream sequence of *Wcor15* were produced. The T_2 homozygous population for the introduced *GUS* gene was established. The seedlings were grown under the condition of 16 h day-length and 27°C for 7 d, and then treated with 4°C under 24 h light or dark for 5 d. Remarkable GUS expression in the leaves was observed only in the low temperature-treated seedlings under the light condition (Figure 3B). The *Wcor15* promoter region functions for cold-responsive and light-stimulated expression as previously reported by Takumi et al. (2003). It is noted that the promoter of *Wcs120* contains two CRT/DRE motifs in the regions showing cold responsiveness and it is functional in both monocotyledonous and dicotyledonous transgenic plants (Quellet et al. 1998). The cold-responsiveness of the wheat *Cor*/*Lea* gene expression, therefore, was successfully demonstrated in the heterologous tobacco transformation system.

Several putative stress- and ABA-responsive *cis*-elements such as CRT/DRE and ABRE are recognized in the 5' upstream sequences of other wheat *Cor*/*Lea* genes, *Wdhn13*, *Wrab17*, *Wrab18* and *Wrab19* (Kobayashi et al. 2008a). Expression of the four *Cor*/*Lea* genes is responsive to low temperature, drought salt and exogenous ABA in seedling leaves of common wheat (Kobayashi et al. 2004a; Egawa et al. 2006). The expression levels of a *GUS* reporter gene under control of the *Cor*/*Lea* promoter sequences are enhanced by cold, drought and ABA treatment in transgenic tobacco plants, which are consistent with the results obtained from RNA gel blot analyses. The 5' upstream regions of the four *Cor*/*Lea* genes function as promoters of their transcription under low temperature, drought and ABA treatment. At least some of the putative *cis*-elements found in their promoters are essential for the *Cor*/*Lea* gene expression. The heterologous tobacco system is useful to analyze abiotic stress responsiveness of wheat *Cor*/*Lea* promoters.

Arabidopsis CBF/DREB transcription factors transcriptionally activate the downstream *Cor*/*Lea* genes under the abiotic stress conditions as mentioned above, and expression of the transcription factor genes are also abiotic stress-responsive. *Wcbf2* is one of wheat CBF-type transcription factor genes. Expression of the *Wcbf2* gene is induced rapidly by low temperature, and the first up-regulation occurs within 1-4 h in seedling leaves (Kume et al. 2005). The *GUS* reporter gene under the control of 1.3-kb 5' upstream sequence of *Wcbf2* was introduced into *Arabidopsis* genome, and cold-responsive GUS expression is observed in the transgenic *Arabidopsis* plants with the *Wcbf2 promoter::GUS* construct (Takumi et al. unpublished data). WDREB2 is a wheat DREB homolog, and three transcript forms are processed through alternative splicing. Only two of the three spliced forms can encode a functional WDREB2 transcription factor, and the post-transcriptional regulation of alternative splicing is stress-responsible (Egawa et al. 2006; Terashima and Takumi 2009). The *Wdreb2* expression is also transcriptionally activated by cold, drought, salt and ABA treatment (Egawa et al. 2006). The 1.7-kb upstream region of *Wdreb2* containing the CRT/DRE and ABRE core motifs was connected to the *GUS* reporter gene, and the *Wdreb2 promoter::GUS* chimeric plasmid was introduced into *Arabidopsis* genome. Seedlings of the transgenic

Arabidopsis plants were treated with abiotic stress such as cold, drought, salt and exogenous ABA, and then the GUS assay was conducted. Histochemical GUS staining showed increase in GUS accumulation levels of the stress-treated seedlings compared with non-stressful plants (Figure 4A). The GUS quantification demonstrated that low temperature (4˚C) enhanced the GUS levels after 1-d treatment, whereas drought, NaCl and ABA rapidly induced the GUS expression within 6 h in the transgenic *Arabidopsis* plants (Figure 4B). The different responses of the *Wdreb2* promoter to cold and other stress agreed with our previous results that the *Wdreb2* gene expression was differentially controlled by low temperature and drought/salt stress (Egawa et al. 2006). Thus, it is suggested that two independent pathways are related with the transcriptional activation of *Wdreb2*, and that one pathway for drought and salt stress is ABA-independent and another for cold stress seems to be ABA-independent. Interestingly, the *Arabidopsis* DREB2 transcription factors are activated only by drought and salt stress through ABA-independent pathway (Liu et al. 1998). The abiotic stress responses of the *Wdreb2* promoter, therefore, were effectively confirmed in transgenic *Arabidopsis* plants.

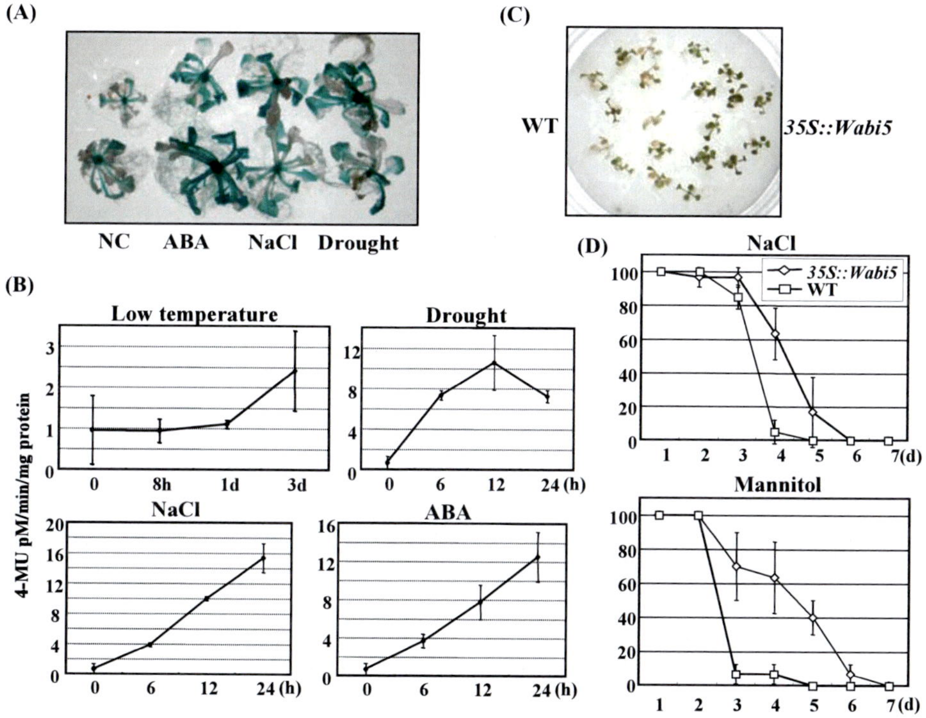

Figure 4. Analyses of the stress-responsive transcription factor genes using transgenic *Arabidopsis* plants. (A, B) Expression patterns of a *GUS* reporter gene under control of the *Wdreb2* 5' upstream region in transgenic *Arabidopsis* plants. Histochemical GUS staining (A) and GUS activity (B) were performed after abiotic stress treatments such as cold (4°C), drought (27°C), 0.4 M NaCl and exogenous 20 µM ABA. (C) Increase in salt stress tolerance in the *CaMV35S promoter::Wabi5* transgenic *Arabidopsis* plants (*35S::Wabi5*). Growth of the transgenic and wild-type (WT) plants 3 days after treatment of 0.2 M NaCl was shown. (D) Salt (upper) and osmotic (lower) stress tolerance. Percentages of plants with green cotyledons were compared after supplementation with 0.2 M NaCl and 0.5 M mannitol solution, respectively.

Overexpression of Abiotic Stress-Responsive Genes

Overall expression profiles of the *Cor/Lea* genes are well correlated with the time-dependent development and the level of freezing tolerance under low temperature in two common wheat cultivars, 'Mironovskaya 808 (M808)' and 'Chinese Spring (CS)', showing distinct levels of freezing tolerance (Ohno et al. 2001, 2003; Takumi et al. 2003; Kobayashi et al. 2004a). Overexpression of one wheat COR/LEA protein WCS19 leads to increased freezing tolerance in *Arabidopsis* (Ndong et al. 2002). *Wcor15*-overexpression also improves the freezing tolerance in transgenic tobacco plants (Shimamura et al. 2006). The improvement by WCOR15, however, occurs only under limited freezing conditions. Influence of each COR/LEA protein on cold acclimation and freezing tolerance thus might not only be small but also spatially restricted, and a set of COR/LEA proteins should cooperatively contribute to freezing tolerance. M808 develops much higher levels of freezing tolerance than CS throughout cold acclimation periods (Ohno et al. 2001; Kobayashi et al 2004a). During the acclimation period, M808 accumulates more COR/LEA proteins such as WDHN13, WCOR14 and WCOR15 than CS (Ohno et al. 2003; Kobayashi et al. 2004a). The protein accumulation agrees very well with that of the corresponding transcripts and the developmental time-course of freezing tolerance. Cultivar differences in the level of freezing tolerance in wheat thus are probably due to the different levels of transcript accumulation of *Cor/Lea* genes that are likely regulated through the same or overlapping signal pathways.

TaDREB1 is a member of the DREB transcription factor genes in wheat, and the transcript accumulation level is increased under the drought stress condition (Shen et al. 2003). The *TaDREB1* expression activates expression of the downstream *rd29A* gene under non-stress condition in transgenic *Arabidopsis*, indicating that TaDREB1 functions as a DRE-binding transcription factor in wheat (Shen et al. 2003). This result indicates that the wheat abiotic stress-responsive transcription factors induce a battery of downstream genes such as *Cor/Lea* and efficiently improve the levels of abiotic stress tolerance in transgenic plants. For example, overexpression of the wheat CBF transcription factor might lead to higher levels of increased freezing tolerance than that of individual COR/LEA proteins. In fact, expression of an *Arabidopsis CBF* gene greatly increased transcript accumulation levels of many *Cor/Lea* genes and the freezing tolerance level in transgenic plants (Jaglo-Ottosen et al. 1998; Liu et al. 1998; Kasuga et al. 1999; Pellegrineschi et al. 2004).

Wcbf2 also encodes a CRT/DRE-binding transcription factor in wheat seedlings (Kume et al. 2005), and is assigned to the long arm of chromosome 5D (Takumi et al. unpublished result). Introduction of the *Wcbf2* gene under control of the CaMV35S promoter significantly increases the freezing tolerance levels in seedlings of transgenic tobacco plants without any phenotypic alteration (Takumi et al. 2008). The transgenic tobacco expressing *Wcbf2* shows ectopic expression of some cold-responsive genes under the normal temperature condition. Thus, the high levels of freezing tolerance might be due to ectopic expression of endogenous tobacco genes via WCBF2. *Wcbf2* can alter endogenous gene expression patterns in heterologous plant species, resulting in improved tolerance to freezing stress as previously reported for other *CBF* genes (Jaglo-Ottosen et al. 1998; Liu et al. 1998; Shen et al. 2003; Pellegrineschi et al. 2004). *Wdreb2* expression under the control of the CaMV35S promoter also improves significantly freezing, salt and osmotic stress tolerance in transgenic tobacco plants (Kobayashi et al. 2008a). Drought and salt stress-responsive expression of *Wdreb2* is assumed to be ABA-dependent (Egawa et al. 2006). The magnitude of inhibition of root growth by exogenous ABA treatment is significantly greater in the *35S promoter::Wdreb2* tobacco than in wild-type tobacco (Kobayashi et al. 2008a). Primary root elongation of the *35S promoter::Wdreb2* tobacco is hypersensitive to exogenous ABA during post-germination growth. This observation suggests that *Wdreb2* expression affects ABA sensitivity as well as the ABA-dependent stress response in heterologous plants, although the molecular mechanisms increasing ABA sensitivity remain unknown. *Wdreb2* takes part directly in ABA-mediated stress-responsive pathways. In addition, *Wdreb2* might contribute indirectly to development of abiotic stress tolerance through an increase in ABA sensitivity.

Besides of the CBF/DREB transcription factors, bZIP-type transcription factors can activate *Cor*/*Lea* genes through binding to an ABRE motif (Choi et al. 2000; Uno et al. 2000). *Wlip19* and *Wabi5* encode the bZIP-type transcription factor, and the gene expression of *Wlip19* and *Wabi5* respond to low temperature, drought, salt and exogenous ABA in seedling leaves of common wheat (Kobayashi et al. 2008b, 2008c). Rice *lip19* (low-temperature-induced protein 19) and maize m*lip15* are strongly induced by low temperature in the seedlings (Aguan et al. 1993; Kusano et al. 1995). *Wlip19* is a counterpart of the rice *lip19* and maize m*lip15*. Barley HvABI5, classified into *Arabidopsis* AREB/ABF family, contains a bZIP domain and directly regulates the ABA-mediated *HVA1*, one of barley *Cor*/*Lea* genes, gene expression in leaves under drought condition (Casaretto and Ho 2003; Xue and Loveridge 2004). *Wabi5* is an ortholog of *HvABI5*. *Wlip19* and *Wabi5* are transcriptionally activated by low temperature, drought and exogenous ABA treatment in wheat seedlings (Kobayashi et al. 2008b, 2008c). Information about direct roles of the stress-responsive bZIP-type transcription factors on development of abiotic stress tolerance is still limited in cereals. Transgenic tobacco plants with *Wlip19* under the control of the CaMV35S promoter show a significant increase in abiotic stress tolerance, especially freezing tolerance without cold acclimation (Kobayashi et al. 2008b). *Wabi5*-expressing transgenic tobacco also exhibits significantly increased tolerance to abiotic stresses such as freezing, osmotic and salt stresses (Kobayashi et al. 2008c). These two types of transgenic tobacco plants are hypersensitive to exogenous ABA in the seedling stage compared with wild-type plants. The ABA-hypersensitivity of the *CaMV35S promoter::Wlip19* and *CaMV35S promoter::Wabi5* tobacco is similar to that observed in the *CaMV35S promoter::Wdreb2* transgenic plants.

The *CaMV35S promoter::Wabi5* chimeric construct was also introduced into *Arabidopsis* genome through *Agrobacterium*-mediated transformation. The *CaMV35S promoter::Wabi5* transgenic *Arabidopsis* showed a significant increase in salt and osmotic stress tolerance (Figure 5). To assay stress tolerance, 7-day-old seedlings of wild-type and *35S::Wabi5 Arabidopsis* plants were placed on two sheets of filter paper (55 mm in diameter) wetted with 3 ml of 0.2 M NaCl solution or 0.5 M mannitol solution in a glass □ypoc dish (60 mm in diameter and 15 mm in depth) under normal temperature conditions. Until 7 days after treatment with NaCl and mannitol, the number of plants with green cotyledons was scored. The experiment was performed in triplicate three times. Under mannitol and NaCl stress, cotyledons of wild-type seedlings of *Arabidopsis* yellowed (Figure 4C). The tolerance level of the *35S::Wabi5 Arabidopsis* plants were significantly higher than that of wild-type plants under the NaCl and osmotic stress (Figure 4D). The performance of abiotic stress tolerance in the transgenic *Arabidopsis* was quite similar to that in transgenic tobacco. Therefore, abiotic stress tolerance of dicotyledonous plants could be dramatically improved by expression of the wheat bZIP-type transcription factors. The increased stress tolerance in the heterologous transgenic plants strongly indicated that the wheat stress-responsive transcription factors are associated with development of abiotic stress tolerance in wheat.

Trans-activation of *Cor*/*Lea* via Abiotic Stress-responsive Transcription Factors

As mentioned above, it is strongly suggested that *Fr-1* controls development of freezing tolerance and *Cor*/*Lea* gene expression through *CBF* transcriptional activation in wheat (Kobayashi et al. 2005). *Fr-1* is epistatic to *Fr-2* and affects the *CBF* gene expression (Vágújfalvi et al. 2005). Wheat *CBF* genes form a gene cluster on the *Fr-2* loci (Vágújfalvi et al. 2003). To clarify the signal transduction through *Fr-1* and *Fr-2* during cold acclimation, it should be needed to prove direct relationships between CBF and *Cor*/*Lea* gene expression. For confirmation of the relationships, we conducted two different approaches. One was transient expression study using wheat cultured cell line, and another crosses among transgenic tobacco lines. The 5' upstream regions of *Wcor15* and *Wdhn13* were combined with the *GUS* reporter gene. The *Wcor15 promoter::GUS* and *Wdhn13 promoter::GUS* chimeric constructs were introduced via particle bombardment into wheat cultured cell line 'HY-1'. After co-transformed with the *CaMV35S promoter::Wcbf2* construct, the GUS activities were estimated. In both *Wcor15 promoter::GUS* and *Wdhn13 promoter::GUS* constructs, bombardment with the *CaMV35S promoter::Wcbf2* construct gave higher GUS activities than that without the *CaMV35S promoter::Wcbf2*, although the differences were not statistically significant (Kume et al. 2005). On the other hand, the *CaMV35S promoter::Wcbf2* transgenic tobacco was crossed with the *Cor*/*Lea promoter::GUS* transgenic plants, and the F_1 plants were assessed to clarify the interaction between the wheat CBF transcription factor and *Cor*/*Lea* promoter regions. The 5' upstream sequences of *Wdhn13* and *Wrab17* significantly increased GUS activity in F_1 seedlings compared with those in the parental transgenic plants under the normal temperature condition (Takumi et al. 2008). This heterologous tobacco system clearly indicated that the WCBF2 protein directly and positively regulates *Wdhn13* and *Wrab17* gene expression.

DRF1E is identified as the HvDRF1 recognition sequence, and HvDRF1 activates the promoter activity of a stress- and ABA-responsive *HVA1* gene in barley leaves (Xue and Loveridge 2004). WDREB2 is a wheat orthologue of HvDRF1 based on their high sequence homology and transcriptional characteristics (Egawa et al. 2006). One or two DRF1E motifs are found in the 5' upstream regions of *Wdhn13*, *Wrab18* and *Wrab19* (Kobayashi et al. 2008a). Similar to the case of *Wcbf2*, we first conducted the transient expression study using the HY-1 callus line to estimate the interaction between the WDREB2 transcription factor and *Cor*/*Lea* promoter regions. In the *Wrab19 promoter::GUS* construct, introduction together with *CaMV35S promoter::Wdreb2* significantly increased the GUS activities (Egawa et al. 2006). No significant differences were observed in the *Wrab17 promoter::GUS* and *Wrab18 promoter::GUS* constructs introduced with *CaMV35S promoter::Wdreb2*. Next, the GUS activities were compared in the F_1 progeny and their parental transgenic tobacco lines. GUS activities in four types of the *Cor*/*Lea promoter::GUS* plants, *Wdhn13*, *Wrab17*, *Wrab18* and *Wrab19*, were significantly increased by constitutive expression of *Wdreb2* under the normal temperature condition (Kobayashi et al. 2008a). These results indicate that *Wdhn13*, *Wrab17*, *Wrab18* and *Wrab19*, are direct targets of the WDREB2 transcription factor and that expression of the four *Cor/Lea* genes is regulated at least by WDREB2. The *Wrab17* promoter region contains a CRT/DRE motif but not DRF1E. Therefore, cereal DREB2-type transcription factors are able to recognize both CRT/DRE and DRF1E *cis*-elements in the four *Cor/Lea* promoters.

The bZIP-type transcription factors LIP19 and mLIP15 dimerize respectively with rice OsOBF1 and maize OBF1, which then bind to ABRE *cis*-elements containing ACGT core motif (Shimizu et al. 2005). However, downstream target genes of LIP19 and mLIP15 remain unknown. Promoter regions of *Wdhn13*, *Wrab17*, *Wrab18* and *Wrab19* contain the ACGT core motifs (Kobayashi et al. 2008a), suggesting that *Wlip19* plays a significant role in regulation of *Cor/Lea* gene expression. The transient expression assay showed that *GUS* expression under control of the *Wdhn13* and *Wrab17* promoters was significantly increased by *Wlip19* (Kobayashi et al. 2008b). The interaction between WLIP19 and wheat *Cor/Lea* promoters was also confirmed in the F_1 progeny derived from crossing *Cor/Lea promoter::GUS* tobacco plants with the *CaMV35S prmoter::Wlip19* tobacco lines. GUS quantification showed that GUS activity significantly increased in the F_1 seedlings compared with parental transgenic plants under non-stressful condition (Kobayashi et al. 2008b). Therefore, WLIP19 functions as a transcriptional activator and positively regulates *Wdhn13*, *Wrab17*, *Wrab18* and *Wrab19* gene expression. Recently we demonstrated that expression of a *GUS* reporter gene under the control of promoters of three wheat *Cor/Lea* genes, *Wdhn13*, *Wrab18* and *Wrab19*, was enhanced by ectopic *Wabi5* expression in wheat callus and tobacco plants (Kobayashi et al. 2008c). At least two types of bZIP proteins, WLIP19 and WABI5, act as important transcription factors in abiotic stress responses and tolerance of common wheat.

The transient expression studies and heterologous transformation systems led us to clarify positive contribution of the stress-responsive transcription factors on development of abiotic stress tolerance and direct relationship between the transcription factors and downstream *Cor/Lea* genes in common wheat. Together with a lot of previous expression studies using various wheat experimental lines and cultivars, these results allow us to represent putative signal networks for low temperature and drought stress responses (Figures 5 and 6). WABI5 and WDREB2 act cooperatively in the activation of *Wrab18/Wrab19* expression, and this cooperation in the stress signal transduction pathway is conserved in wheat and barley. Expression of other *Cor/Lea* genes such as *Wdhn13* and *Wrab17* is regulated by WCBF2, WDREB2 and WLIP19 (Takumi et al. 2008; Kobayashi et al. 2008a, 2008b). These transcription factors cooperatively activate *Cor/Lea* gene expression under abiotic stress conditions. Therefore, various stress-responsive transcription factor genes participate in the wheat abiotic stress signal network, in which *Cor/Lea* expression is finally regulated through the *Fr-1*-dependent and -independent signal transduction pathways. The cooperative interaction between transcription factors possibly has a synergistic effect on the activation of gene expression, resulting in development of abiotic stress tolerance.

FOR FURTHER FUNCTIONAL STUDIES OF WHEAT GENES

Wheat lacks the whole genome sequence information, but provides a nearly comprehensive EST database. Functional analysis of newly identified genes is going to be more and more significant in wheat bioscience. Despite abundant genetic resources and experimental lines, wheat remains difficult to transform compared with the model plants, *Arabidopsis* and rice. Continuous efforts should be still required for improvement of wheat transformation system. Therefore, the heterologous transformation systems will continue to be valuable for wheat gene functional analyses in the next decade. On the other hand, recently reported technologies such as RNAi and VIGS will have more significant impacts on functional analyses of wheat genes.

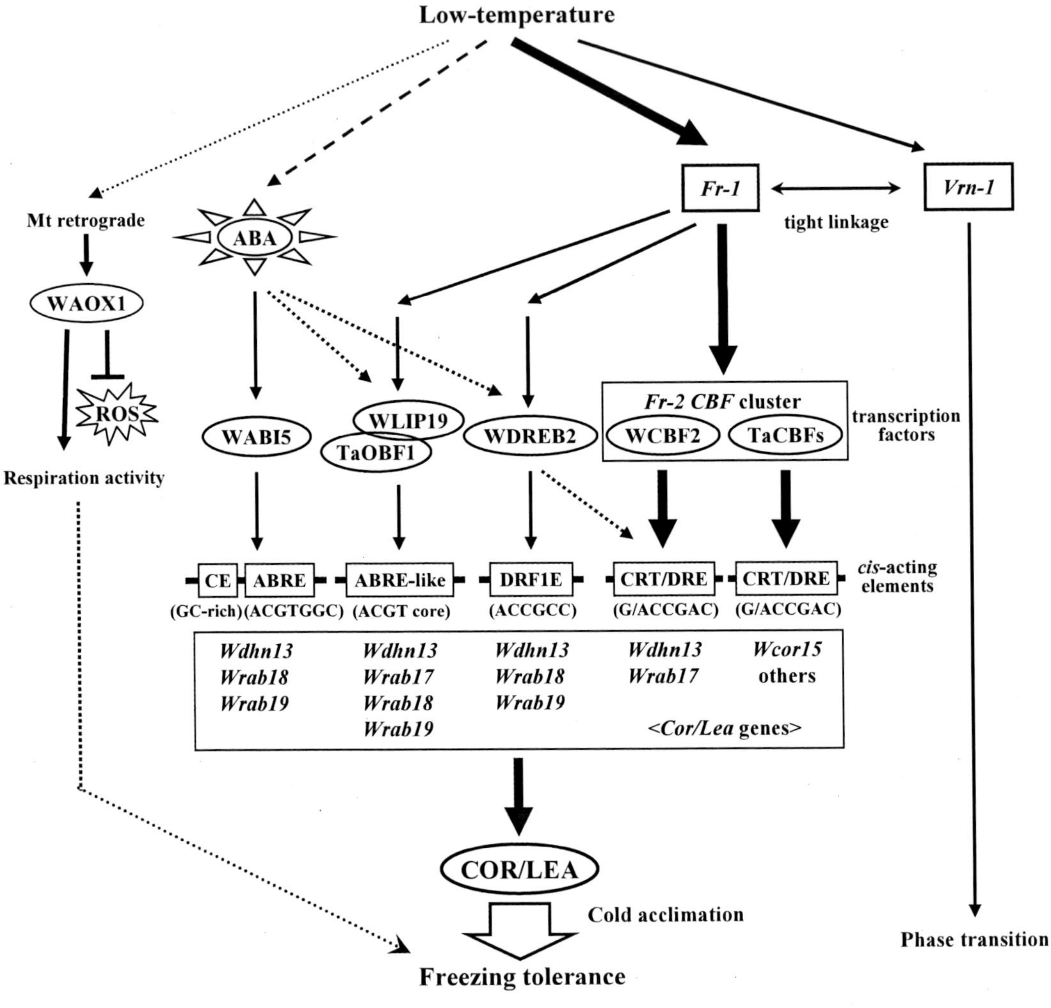

Figure 5. Schematic representation of the low temperature signal pathways in wheat. The *Fr-1* loci are major QTLs for freezing tolerance by resulting various *Cor/Lea* genes mainly through the ABA-independent pathway. Several transcription factors directly bind *cis*-elements of the *Cor/Lea* promoter and coordinately activate the *Cor/Lea* expression. Mitochondrial alternative pathway is also associated with the cultivar difference of freezing tolerance (Mizuno et al. 2007). The *Vrn-1* loci are separated from the *Fr-1* loci, while these two loci are closely linked on the long arms of homoeologous group 5 chromosomes.

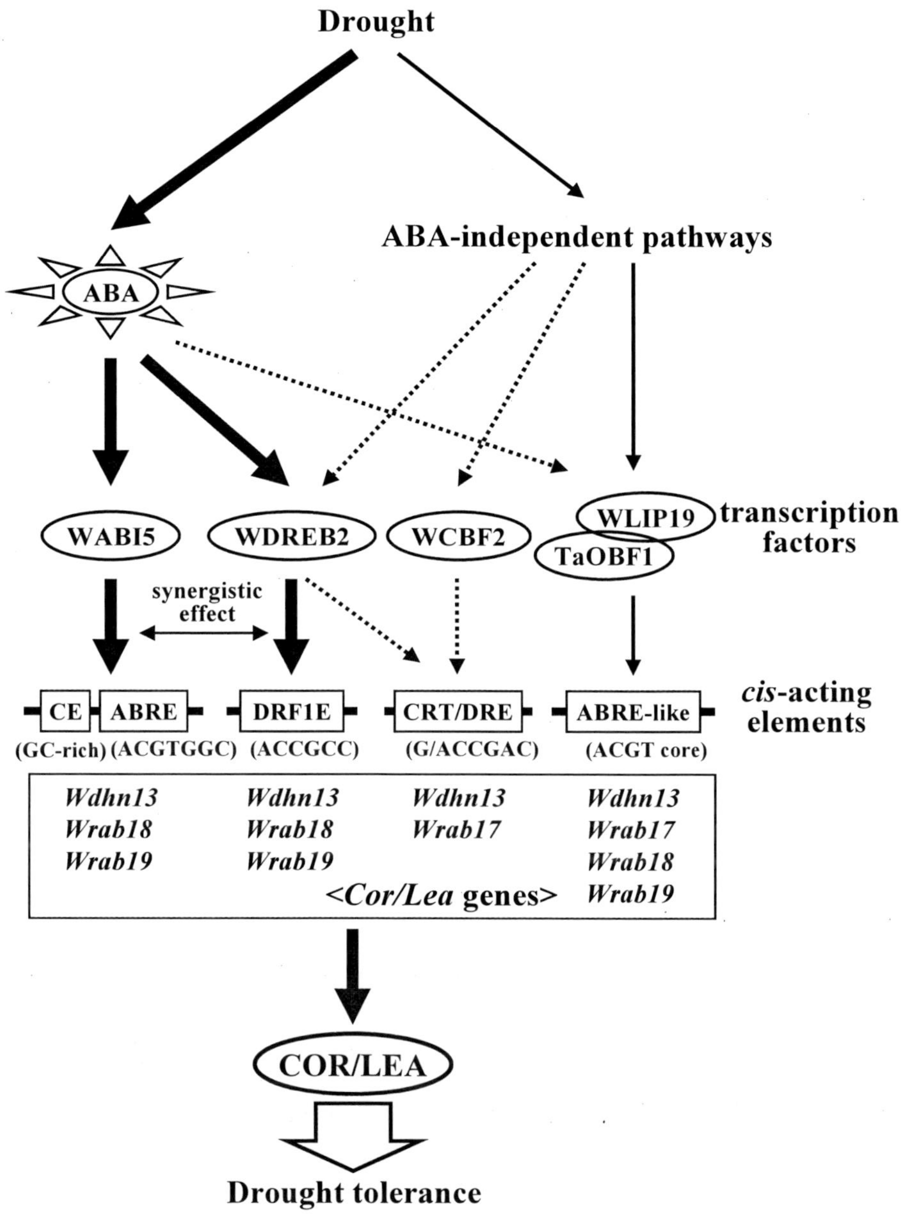

Figure 6. Schematic representation of the drought signal pathways in wheat. Drought tolerance is mainly developed through the ABA-dependent pathway.

Similar to the model plants, a powerful reverse genetics approach was recently established through the combination of EMS-mediated mutagenesis and TILLING (targeting induced local lesions in genomes) technology in wheat (Slade et al. 2005). The TILLING strategy makes it suitable for large-scale screening of chemically induced mutations in *Arabidopsis* and other plants (Colbert et al. 2001). TILLING has several advantages over other reverse genetic strategies. Unlike methods providing only knockout mutations, TILLING yields a traditional allelic series of point mutations. For identification of novel alleles, natural variation should be alternatively useful. In *Arabidopsis*, natural variation has been enthusiastically studied for isolation of novel alleles useful for understanding gene function (Koonneef et al. 2004). A large number of accessions of wheat and its relatives are stored as genetic resources for wheat breeding. As mentioned by Fu et al. (2007), however, the effect of single-knockouts can be masked by the functional redundancy of homoeologous genes present in the other wheat genomes. For phenotypic effect of the mutations, double and triple mutants should be generated and then the mutant phenotype should be evaluated. For time consuming, alternative approaches are surely required for gene functional analysis. In fact, a single RNAi-inducing silencing can dominantly repress multiple orthologs (Lawrence and Pikaard 2003). The RNAi strategy for overcoming gene redundancy in polyploids becomes more valuable for further functional analyses of wheat genes.

A lot of genetic phenomena have been observed and studied in a long history of wheat research. The wheat genes are still waiting for identification, isolation and functional characterization. Studies towards wheat gene cloning and functional analyses of the cloned genes will be surely accelerated in the next decade. Both wheat transformation and heterologous transformation approaches become more and more significant in further wheat bioscience.

REFERENCES

Aguan, K., Sugawara, K., Suzuki, N. & Kusano, T. (1993) Low-temperature-dependent expression of a rice gene encoding a protein with a leucine-zipper motif. *Mol. Gen. Genet.*, 240, 1-8.

Akhunov, E. D., Akhunova, A. R. & Dvorák, J. (2005) BAC libraries of *Triticum urartu*, *Aegilops speltoides* and *Ae. tauschii*, the diploid ancestors of □ypocotyls wheat. *Theor. Appl. Genet.*, 111, 1617-1622.

An, H., Roussot, C., Suárez-López, P., Corbesier, L., Vincent, C., Pineiro, M., Hepworth, S., Mouradov, A., Justin, S., Turnbull, C. & Coupland, G., (2004) CONSTANS acts in the phloem to regulate a systemic signal that induce photoperiodic flowering of *Arabidopsis*. *Development*, 131, 3615-3626.

Altpeter, F., Vasil, V., Srivastava, V. & Vasil, I. K. (1996) Integration and expression of the high-molecular-weight glutenin subunit 1Ax1 gene into wheat. *Nature Biotechnol.*, 14, 1155-1159.

Araki, T. & Komeda, Y. (1993) Analysis of the role of the late-flowering locus, *GI*, in the flowering of *Arabidopsis thaliana*. *Plant J.*, 3, 231-239.

Baker, S. S., Wilhelm, K. S. & Thomashow, M. F. (1994) The 5'-region of *Arabidopsis thaliana cor15a* has cis-acting elements that confer cold-, drought- and ABA-regulated gene expression. *Plant Mol. Biol.*, 24, 701-713.

Barro, F., Rooke, L., Bekes, F., Gras, P., Tathyam, A. S., Fido, R., Lazzeri, P. A., Shewry, P. R. & Barcelo, P. (1997) Transformation of wheat with high molecular weight subunit genes results in improved functional properties. *Nature Biotechnol.*, 15, 1295-1299.

Becker, D., Brettschneider, R. & Lörz, H. (1994) Fertile transgenic wheat from microprojectile bombardment of scutellar tissue. *Plant J.*, 2, 299-307.

Blechl, A. E. & Anderson, O. D. (1996) Expression of a novel high-molecular-weight glutenin subunit gene in transgenic wheat. *Nature Biotechnol.*, 14, 875-879.

Bommineni, V. R., Jauhar, P. P. & Peterson, T. S. (1997) Transgenic durum wheat by microprojectile bomcardment of isolated scutella. *J. Hered.*, 88, 474-481.

Casaretto, J. & Ho, T. H. D. (2003) The transcription factors HvABI5 and HvVP1 are required for the abscisic acid induction of gene expression in barley aleurone cells. *Plant Cell*, 15, 271-284.

Cattivelli, L., Baldi, P., Crosatti, C., Di Fonzo, N., Faccioli, P., Grossi, M., Mastrangelo, A. M., Pecchioni, N. & Stanca, A. M. (2002) Chromosome regions and stress-related sequences involved in resistance to abiotic stress in *Triticeae*. *Plant Mol. Biol.*, 48, 649-665.

Chen, H. H., Li, P. H. & Brenner, M. L., (1983) □ypocotyls□ of abscisec acid in potato cold acclimation. *Plant Physiol.*, 71, 362-365.

Cheng, M., Fry, J. E., Pang, S. Z., Zhou, H. P., Hironaka, C. M., Duncan, D. R., Conner, T. W. & Wan, Y. C. (1997) Genetic transformation of wheat mediated by *Agrobacterium tumefaciens*. *Plant Physiol.*, 115, 971-980.

Choi, H., Hong, J., Ha, J., Kang, J. & Kim, S. Y. (2000) ABFs, a family of ABA-responsive element binding factors. *J. Biol. Chem.*, 275, 1723-1730.

Cloutier, S., McCallum, B. D., Loutre, C., Banks, T. W., Wicker, T., Feuillet, C., Keller, B. & Jordan, M. C. (2007) Leaf rust resistance gene *Lr1*, isolated from bread wheat (*Triticum aestivum* L.) is a member of the large psr567 gene family. *Plant Mol. Biol.* 65, 93-106.

Colbert, T., Till, B. J., Tompa, R., Reynolds, S., Steine, M. N., Yeung, A. T., McCallum, C. M., Comai, L. & Henikoff, S. (2001) High-throughput screening for induced point mutations. *Plant Physiol.*, 126, 480-484.

Crosatti, C., de Laureto, P. P., Bassi, R. & Cattivelli, L. (1999) The interaction between cold and light controls the expression of the cold-regulated barley gene *cor14b* and the accumulation of the corresponding protein. *Plant Physiol.*, 119, 671-680.

Dalmay, T. (2005) Virus-induced gene silencing. In P. Meyer (Ed.), *Plant Epigenetics* (pp. 223-243). Oxford: Blackwell Publishing Ltd.

Darlington, H., Fido, R., Tatham, A. S., Jones, H., Salmon, S. E. & Shewry, P. R. (2003) Milling and baking properties of field grown wheat expressing HMW subunit transgenes. *J. Cereal. Sci.*, 38, 301-306.

Depicker, A., Sanders, M. & Meyer, P. (2005) Transgene silencing. In P. Meyer (Ed.), *Plant Epigenetics* (pp. 1-32). Oxford: Blackwell Publishing Ltd.

Egawa, C., Kobayashi, F., Ishibashi, M., Nakamura, T., Nakamura, C. & Takumi, S. (2006) Differential regulation of transcript accumulation and alternative splicing of a *DREB2* homolog under abiotic stress conditions in common wheat. *Genes Genet. Syst.*, 81, 77-91.

Faris, J. D., Fellers, J. P., Brooks, S. A. & Gill, B. S. (2003) A bacterial artificial chromosome contig spanning the major domestication locus *Q* in wheat and identification of a candidate gene. *Genetics*, 164, 311-321.

Feuillet, C., Travella, S., Stein, N., Albar, L., Nublat, A. & Keller, B. (2003) Map-based isolation of the leaf rust disease resistance gene *Lr10* from the hexaploid wheat (*Triticum aestivum* L.) genome. *Proc. Natl. Acad. Sci. USA*, 100, 15253-15258.

Finnegan, J. & McElroy, D. (1994) Transgene inactivation: plants flight back! *Bio/Technol.*, 12, 883-888.

Fu, D., Uauy, C., Blechl, A. & Dubcovsky, J. (2007) RNA interference for wheat functional gene analysis. *Transgenic Res*., 16, 689-701.

Galiba, G., Quarrie, S. A., Sutka, J., Morgounov, A. & Snape, J. W. (1995) RFLP mapping of the vernalization (*Vrn1*) and frost resistance (*Fr1*) genes on chromosome 5A of wheat. *Theor. Appl. Genet*., 90, 1174-1179.

Haake, V., Cook, D., Riechmann, J. L., Omaira, P., Thomashow, M. F. & Zang, J. Z. (2002) Transcription factor CBF4 is a regulator of drought adaptation in *Arabidopsis*. *Plant Physiol.*, 130, 639-648.

Hayama, R., Yokoi, S., Tamaki, S., Yano, M. & Shimamoto, K. (2003) Adaptation of photoperiodic control pathways produces short-day flowering in rice. *Nature*, 422, 719-722.

Hiei, Y., Komari, T. & Kubo, T. (1997) Transformation of rice mediated by *Agrobacterium tumefaciens*. *Plant Mol. Biol*., 35, 205-218.

Hirt, H. (1997) Multiple roles of MAP kinases in plant signal transduction. *Trend. Plant Sci*., 2, 11-15.

Houde, M., Danyluk, J., Laliberté, J.F., Rassart, E., Dhindsa, R.S. & Sarhan, F. (1992) Cloning, characterization and expression of a cDNA encoding a 50 kilodalton protein specifically induced by cold acclimation in wheat. *Plant Physiol*., 99, 1381-1387.

Hu, T., Metz, S., Chay, C., Zhou, H. P., Biest, N., Chen, G., Cheng, M., Feng, X., Radionenko, M., Lu, F. & Fry, J. (2003) *Agrobacterium*-mediated large-scale transformation of wheat (*Triticum aestivum* L.) using glyphosate selection. *Plant Cell Rep*., 21, 1010-1019.

Huang, L., Brooks, S. A., Li, W., Fellers, J. P., Trick, H. N. & Gill, B.S. (2003) Map-based cloning of leaf rust resistance gene *Lr21* from the large and □ypocotyls genome of bread wheat. *Genetics*, 164, 655-664.

Jaglo-Ottosen, K. R., Gilmour, S. J., Zarka, D. G., Schabenberger, O. & Thomashow, M. F. (1998) *Arabidopsis CBF1* overexpression induces *COR* genes and enhances freezing tolerance. *Science*, 280, 104-106.

Jaglo, K. R., Kleff, S., Amundsen, K. L., Zhang, X., Haake, V., Zhang, J. Z., Deits, T. & Thomashow, M. F. (2001) Components of the *Arabidopsis* C-repeat/dehydration responsive element binding factor cold-responsive pathway are conserved in *Brassica napus* and other plant species. *Plant Physiol*., 127, 910-917.

Jones, H. D. (2005) Wheat transformation: current technology and applications to grain development and composition. *J. Cereal Sci*., 41, 137-147.

Joppa, L. R. & Cantrell, R. G. (1990) Chromosomal location of genes for grain protein content in wild tetraploid wheat. *Crop Sci*., 30, 1059-1064.

Joppa, L. R., Du, C., Hart, G. E., & Hareland, G. A. (1997) Mapping gene(s) for grain protein in tetraploid wheat (*Triticum turgidum* L.) using a population of recombinant inbred chromosome lines. *Crop Sci*., 37, 1586-1589.

Kasuga, M., Liu, Q., Miura, S., Yamaguchi-Shinozaki, K. & Shinozaki, K. (1999) Improving plant drought, salt, and freezing tolerance by gene transfer of a single stress-inducible transcription factor. *Nature Biotechnol*., 17, 287-291.

Kim, S. Y. (2006) The role of ABF family bZIP class transcription factors in stress response. *Physiol. Plant*., 126, 519-527.

Kobayashi, F., Takumi, S., Nakata, M., Ohno, R., Nakamura, T., & Nakamura, C. (2004a) Comparative study on the expression profiles of the *Cor*/*Lea* gene family during cold acclimation in two wheat cultivars with contrasting levels of freezing tolerance. *Physiol. Plant.*, 120, 585-594.

Kobayashi, K., Takumi, S. & Nakamura, C. (2004b) Regulation of cold-responsive *Cor/Lea* genes and their transcription factors by the major freezing tolerance locus *Fr-1* in wheat. *Rec. Res. Dev. Plant Sci.*, 2, 249-266.

Kobayashi, F., Takumi, S., Kume, S., Ishibashi, M., Ohno, R., Murai, K. & Nakamura, C. (2005) Regulation by *Vrn-1/Fr-1* chromosomal intervals of CBF-mediated *Cor/Lea* gene expression and freezing tolerance in common wheat. *J. Exp. Bot.*, 56, 887-895.

Kobayashi, F., Takumi, S., Egawa, C., Ishibashi, M. & Nakamura, C. (2006) Expression patterns of low temperature responsive genes in a dominant ABA-less-sensitive mutant of common wheat. *Physiol., Plant.*, 127, 612-623.

Kobayashi, F., Ishibashi, M. & Takumi, S. (2008a) Transcriptional activation of *Cor/Lea* genes and increase in abiotic stress tolerance through expression of a wheat *DREB2* homolog in transgenic tobacco. *Transgenic Res.*, 17, 755-767.

Kobayashi, F., Maeta, E., Terashima, A., Kawaura, K., Ogihara, Y. & Takumi, S. (2008b) Development of abiotic stress tolerance via a bZIP-type transcription factor LIP19 in common wheat. *J. Exp. Bot.*, 59, 891-905.

Kobayashi, F., Maeta, E., Terashima, A. & Takumi, S. (2008c) Positive role of a wheat *HvABI5* ortholog in abiotic stress response of seedlings. *Physiol., Plant.*, 134, 74-86.

Kobayashi, F., Takumi, S. & Nakamura, C. (2008d) Increased freezing tolerance in an ABA-hypersensitive mutant of common wheat. *J. Plant Physiol.*, 165, 224-232.

Kolesnik, T., Szeverenyi, I., Bachmann, D., Kumar, C. S., Jiang, S., Ramamoorthy, R., Cai, M., Ma, Z. G., Sundaresan, V. & Ramachandran, S. (2004) Establishing an efficient *Ac/Ds* tagging system in rice: large-scale analysis of *Ds* flanking sequences. *Plant J.*, 37, 301-314.

Koonneef, M., Alonso-Blanco, C. & Vreugdenhil, D. (2004) Naturally occurring genetic variation in *Arabidopsis thaliana. Annu. Rev. Plant Biol.*, 55, 141-172.

Kume, S., Kobayashi, F., Ishibashi, M., Ohno, R., Nakamura, C. & Takumi, S. (2005) Differential and coordinated expression of *Cbf* and *Cor/Lea* genes during long-term cold acclimation in two wheat cultivars showing distinct levels of freezing tolerance. *Genes Genet. Syst.*, 80, 185-197.

Kuromori, T., Wada, T., Kamiya, A., Yuguchi, M., Yokouchi, T., Imura, Y., Takabe, H., Sakurai, T., Akiyama, K., Hirayama, T., Okada, K. & Shinozaki, K. (2006) A trial of phenome analysis using 4000 *Ds*-insertional mutants in gene-conding regions of Arabidopsis. *Plant J.*, 47, 640-651.

Kusano, T., Berberich, T., Harada, M., Suzuki, N. & Sugawara, K. (1995) A maize DNA-binding factor with a bZIP motif is induced by low temperature. *Mol. Gen. Genet.*, 248, 507-517.

Lång, V. & Palva, E. T. (1992) The expression of a *rab*-related gene, *rab18*, is induced by abscisic acid during the cold acclimation process of *Arabidopsis thaliana* (L.) Heynh. *Plant Mol. Biol.*, 20, 951-962.

Lång, V., Mäntylä, E., Welin, B., Sundberg, B. & Palva, E. T. (1994) Alteration in water status, endogenous abscisic acid content and expression of *rab18* gene during the development of freezing tolerance in *Arabidopsis thaliana. Plant Physiol.*, 104, 1341-1349.

Lawrence, R. J. & Pikaard, C. S. (2003) Transgene-induced RNA interference: a strategy for overcoming gene redundancy in polyploids to generate loss-of-function mutations. *Plant J.*, 36, 114-121.

Lincoln, C., Long, J., Yamaguchi, J., Serikawa, K. & Hake, S. (1994) A *knotted1*-like homeobox gene in Arabidopsis is expressed in the vegetative meristem and dramatically alters leaf morphology when overexpressed in transgenic plants. *Plant Cell*, 6, 1859-1876.

Linsmaier, E. M. & Skoog, F. (1965) Organic growth factor requirement of tobacco tissue cultures. *Physiol. Plant.*, 18, 100-126.

Liu, Q., Sakuma, Y., Abe, H., Kasuga, M., Miura, S., Yamaguchi-Shinozaki, K. & Shinozaki, K. (1998) Two transcription factors, DREB1 and DREB2, with an EREBP/AP2 DNA binding domain, separate two cellular signal transduction pathways in drought- and low temperature-responsive gene expression, respectively, in *Arabidopsis*, *Plant Cell*, 12, 165-178.

Matsuoka, M., Ichikawa, H., Saito, A., Tada, Y., Fujimura, T. & Kano-Murakami, Y. (1993) Expression of a rice homeobox gene causes altered morphology of transgenic plants. *Plant Cell*, 5, 1039-1048.

Miller, A. K., Galiba, G. & Dubcovsky, J. (2005) A cluster of 11 *CBF* transcription factors is located at the frost tolerance locus *Fr-A^{m}2* in *Triticum monococcum*. *Mol. Genet. Genomics*, 275, 193-203.

Mizuno, N., Sugie, A., Kobayashi, F. & Takumi, S. (2008) Mitochondrial alternative pathway is associated with development of freezing tolerance in common wheat. *J. Plant Physiol.*, 165, 462-467.

Moon, J., Suh, S. S., Lee, H., Choi, K. R., Hong, C. B., Paek, N. C., Lim, S. G. & Lee, I. (2003) The *SOC1* MADS-box gene integrates vernalization and gibberellin signals for flowering in *Arabidopsis*. *Plant J.*, 35, 613-623.

Morimoto, R., Kosugi, T., Nakamura, C. & Takumi, S. (2005) Intragenic diversity and functional conservation of the three homoeologous loci of the *KN1*-type homeobox gene *Wknox1* in common wheat. *Plant Mol. Biol.*, 57, 907-924.

Morimoto, R., Nishioka, E., Murai, K. & Takumi, S. (2009) Functional conservation of wheat orthologs of maize *rough sheath1* and *rough sheath2*. *Plant Mol. Biol.*, 69, (doi:10.1007/s11103-008-9422-5)

Mouradov, A., Cremer, F. & Coupland, J. (2002) Control of flowering time: Interacting pathways as a basis for diversity. *Plant Cell*, 14 (Suppl), S111-S130.

Murai, K., Miyamae, M., Kato, H., Takumi, S. & Ogihara, Y. (2003) *WAP1*, a wheat *APETALA1* homolog, plays a central role in the phase transition from vegetative to reproductive growth. *Plant Cell Physiol.*, 44, 1255-1265.

Murashige, T. & Skoog, F. (1962) A revised medium for rapid growth and bioassays with tobacco tissue cultures. *Physiol. Plant.*, 15: 473-497.

Narusaka, Y., Nakashima, K., Shinwari, Z.K., Sakuma, Y., Furihata, T., Abe, H., Narusaka, M., Shinozaki, K. & Yamaguchi-Shinozaki, K. (2003) Interaction between two *cis*-acting elements, ABRE and DRE, in ABA-dependent expression of *Arabidopsis rd29A* gene in response to dehydration and high-salinity stresses. *Plant J.*, 34, 137-148.

Ndong, C., Danyluk, J., Wilson, K. E., Pocock, T., Huner, N. P. A. & Sarhan, F. (2002) Cold-regulated cereal chloroplast late embryogenesis abundant-like proteins. Molecular characterization and functional analyses. *Plant Physiol.*, 129, 1368-1381.

Nehra, N. S., Chibbar, R. N., Leung, N., Caswell, K., Mallard, C., Steinhauer, L., Baga, M. & Kartha, K. K. (1994) Self-fertile transgenic wheat plants regenerated from isolated scutellar tissues following microprojectile bombardment with two distinct gene constructs. *Plant J.*, 5, 285-297.

Nemoto, Y., Kisaka, M., Fuse, T., Yano, M. & Ogihara, Y. (2003) Characterization and functional analysis of three wheat genes with homology to the *CONSTANS* flowering gene in transgenic rice. *Plant J.*, 36, 82-93.

Ohno, R., Takumi, S. & Nakamura, C. (2001) Expression of a cold-responsive *Lt-Cor* gene and development of freezing tolerance during cold acclimation in wheat (*Triticum aestivum* L.). *J. Exp. Bot.*, 52, 2367-2374.

Ohno, R., Takumi, S. & Nakamura, C. (2003) Kinetics of transcript and protein accumulation of a low-molecular-weight wheat LEA D-11 dehydrin in response to low temperature. *J. Plant Physiol.*, 160, 193-200.

Ohno, R., Takumi, S. & Nakamura, C. (2006) Phosphorylation of wheat chloroplast-targeting COR/LEA proteins via 50-kDa protein kinase. *Wheat Inform. Serv.*, 101, 1-3.

Olmos, S., Distelfeld, A., Chicaiza, O., Schlatter, A. R., Fahima, T., Echenique, V. & Dubcovsky, J. (2003) Precise mapping of a locus affecting grain protein content in durum wheat. *Theor. Appl. Genet.*, 107, 1243-1251.

Ozawa, K., Murayama, S., Kobayashi-Uehara, A. & Handa, H. (2006) Overexpression of wheat mitochondrial uncoupling protein in rice plants confers tolerances to oxidative stresses promoted by exogenous hydrogen peroxide and low □ypocotyls□. *Mol. Breed.*, 18, 51-56.

Palva, E. T. (1994) Gene expression under low temperature stress. In: Basra AS (Ed.), *Stress Induced Gene Expression in Plants* (pp103-130) New York: Harwood Academic Publishers.

Pastori, G. M., Wilkinson, M. D., Steele, S. H., Sparks, C. A., Jones, H. D. & Parry, M. A. J., (2001) Age-dependent transformation frequency in elite wheat varieties. *J. Exp. Bot.*, 52, 857-863.

Pawlowski, W. P., & Somers, D. A. (1996) Transgene inheritance in plants genetically engineered by microprojectile bombardment. *Mol. Biotechnol.*, 6, 17-30.

Pellegrineschi, A., Brito, R.M., Velazquez, L., Noguera, L.M., Pfeiffer, W., McLean, S. & Hoisington, D. (2002) The effect of pretreatment with mild heat and drought stresses on the explant and biolistic transformation frequency of three durum wheat cultivars. *Plant Cell Rep.*, 20, 955-960.

Pellegrineschi, A., Reynolds, M., Pacheco, M., Brito, R.M., Almeraya, R., Yamaguchi-Shinozaki, K. & Hoisington, D. (2004) Stress-induced expression in wheat of the *Arabidopsis thaliana* DREB1A gene delays water stress symptoms under greenhouse conditions. *Genome*, 47, 493-500.

Quellet, F., Vazquez-Tello, A., & Sarhan, F. (1998) The wheat *wcs120* promoter is cold-inducible in both monocotyledonous and dicotyledonous species. *FEBS Lett.*, 423, 324-328.

Rasco-Gaunt, S., Riley, A., Cannell, M., Barcelo, P. & Lazzeri, P. A. (2001) Procedures allowing the transformation of a range of European elite wheat (*Triticum aestivum* L.) varieties via particle bombardment. *J. Exp. Bot.*, 52, 865-874.

Röder, M. S., Korzun, V., Wendehake, K., Plaschke, J., Tixier, M. H., Leroy, P. & Ganal, M. W. (1998) A microsatellite map of wheat. *Genetics*, 149, 2007-2023.

Rooke, L., Bekes, F., Fido, R., Barro, F., Gras, P., Tatham, A. S., Barcelo, P., Lazzeri, P. & Shewry, P. R. (1999) Overexpression of a gluten protein in transgenic wheat results in greatly increased dough strength. *J. Cereal Sci.*, 30, 115-120.

Ryu, S. B. & Li, P. H. (1994) Potato cold hardiness development and abscisic acid. II. *De novo* protein synthesis is required for the increase in free abscisic acid during potato (*Solanum commersonii*) cold acclimation. *Plant Physiol.*, 90, 21-26.

Schneeberger, R. G., Becraft, P. W., Hake, S. & Freeling, M. (1995) Ectopic expression of the *knox* homeo box gene *rough sheath1* alters cell fate in the maize leaf. *Genes Dev.*, 9, 2292-2304.

Scofield, S. R., Huang, L., Brandt, A. S. & Gill, B. S. (2005) Development of a virus-induced gene-silencing system for hexaploid wheat and its use in functional analysis if the *Lr21*-mediated leaf rust resistance pathway. *Plant Physiol.*, 138, 2165-2173.

Shen, Y. G., Zhang, W. K., He, S. J., Zhang, J. S., Liu, Q. & Chen, S. Y. (2003) An EREBP/AP2-type protein in *Triticum aestivum* was a DRE-binding transcription factor induced by cold, dehydration and ABA stress. *Theor. Appl. Genet.*, 106, 923-930.

Shimamura, C., Ohno, R., Nakamura, C. & Takumi, S. (2006) Improvement of freezing tolerance in transgenic tobacco plants expressing a cold-responsive and chloroplast-targeting protein WCOR15 of common wheat. *J. Plant Physiol.*, 163, 213-219.

Shimizu, H., Sato, K., Berberich, T., Miyazaki, A., Ozaki, R., Imai, R. & Kusano, T. (2005) LIP19, a basic leucine zipper protein, is a Fos-like molecular switch in the cold signaling in rice Plant. *Plant Cell Physiol.*, 46, 1623-1634,

Shinozaki, K. & Yamaguchi-Shinozaki, K. (2000) Molecular responses to dehydration and low temperature: differences and cross-talk between two signaling pathways. *Curr. Opin. Plant Biol.*, 3, 217-223.

Shinozaki, K., Yamaguchi-Shinozaki, K. & Seki, M. (2003) Regulatory network of gene expression in the drought and cold responses. *Curr. Opin. Plant Biol.*, 6, 410-417.

Shitsukawa, N., Ikari, C., Mitsuya, T., Sakiyama, T., Ishikawa, A., Takumi, S. & Murai, K. (2007) Wheat *SOC1* functions independently of *WAP1*/*VRN1*, an integrator of vernalization and photoperiod flowering promotion pathways. *Physiol. Plant.*, 130, 627-636.

Simons, K. J., Fellers, J. P., Trick, H. N., Zhang, Z., Tai, Y. S., Gill, B. S. & Faris, J. D. (2006) Molecular characterization of the major wheat domestication gene *Q*. *Genetics*, 172, 547-555.

Singh, J., Zhang, S., Chen, C., Cooper, L., Bregitzer, P., Sturbaum, A., Hayes, P. H. & Lemaux, P. G. (2006) High-frequency *Ds* remobilization over multiple generations in barley facilities gene tagging in large genome cereals. *Plant Mol. Biol.*, 62, 937-950.

Slade, A. J., Fuerstenberg, S. I., Loeffler, D., Steine, M. N. & Facciotti, D. (2005) A reverse genetic, nontransgenic approach to wheat crop improvement by TILLING. *Nature Biotechnol.*, 23, 75-81.

Stockinger, E. J., Gilmour, S. J. & Thomashow, M. F. (1997) *Arabidopsis thaliana CBF1* encodes an AP2 doomain-containing transcription activator that binds to the C-repeat/DRE, a *cis*-acting DNA regulatory element that stimulates transcription in response to low temperature and water deficit. *Proc. Natl. Acad. Sci. USA*, 94, 1035-1040.

Suárez-López, P., Wheatley, K., Robson, F., Onouchi, H., Valverde, F. & Coupland, G. (2001) *CONSTANS* mediates between the circadian clock and the control of flowering in *Arabidopsis*. *Nature*, 410, 1116-1120.

Sugie, A., Naydenov, N., Mizuno, N., Nakamura, C. & Takumi, S. (2006) Overexpression of wheat alternative oxidase gene *Waox1a* alters respiration capacity and response to reactive oxygen species under low temperature in transgenic *Arabidopsis*. *Genes Genet. Syst.*, 81, 349-354.

Sundaresan, V. (1996) Horizontal spread of transposon mutagenesis: new use for old elements. *Trends Plant Sci.*, 1, 184-190.

Sutka, J., Galiba, G., Vágújfalvi, A., Gill, B. S. & Snape, J. W. (1999) Physical mapping of the *Vrn-A1* and *Fr1* genes on chromosome 5A of wheat using deletion lines. *Theor. Appl. Genet.*, 99, 199-202.

Takumi, S. & Shimada, T. (1996) Production of transgenic wheat through particle bombardment of scutellar tissues: frequency is influenced by culture duration. *J. Plant Physiol.*, 149, 418-423.

Takumi, S. (1996) Hygromycin-resistant calli generated by activation and excision of maize *Ac/Ds* transposable elements in diploid and hexaploid wheat cultured cells. *Genome*, 39, 1169-1175.

Takumi, S. & Shimada, T. (1997) Variation in transformation frequencies among six common wheat cultivars through particle bombardment of scutellar tissues. *Genes Genet. Syst.*, 72, 63-69.

Takumi, S., Murai, K., Mori, N. & Nakamura, C. (1999a) *Trans*-activation of a maize *Ds* transposable element in transgenic wheat plants expressing the *Ac* transposase gene. *Theor. Appl. Genet.*, 98, 947-953.

Takumi, S., Murai, K., Mori, N. & Nakamura, C. (1999b) Variations on the maize *Ac* transposase transcript level and the *Ds* excision frequency in transgenic wheat callus lines. *Genome*, 42, 1234-1241.

Takumi, S., Ikawa, Y., Solis, R., Mori, N. & Nakamura, C. (2000a) Excision of maize *Ac/Ds* transposable elements and its frequency in transgenic wheat revealed by expression of marker genes. *Czech J. Genet. Plant Breed.*, 36, 101-105.

Takumi, S., Kosugi, T., Murai, K., Mori, N. & Nakamura, C. (2000b) Molecular cloning of three homoeologous cDNAs encoding orthologs of the maize KNOTTED1 homeobox protein from young spikes of hexaploid wheat. *Gene*, 249, 171-181.

Takumi, S., Tomioka, M., Eto, K., Naydenov, N. & Nakamura, C. (2002) Characterization of two non-homoeologous nuclear genes encoding mitochondrial alternative oxidase in common wheat. *Genes Genet. Syst.*, 77, 81-88.

Takumi, S., Koike, A., Nakata, M., Kume, S., Ohno, R. & Nakamura, C. (2003) Cold-specific and light-stimulated expression of a wheat (*Triticum aestivum* L.) *Cor* gene *Wcor15* encoding a chloroplast-targeted protein. *J. Exp. Bot.*, 54, 2265-2274.

Takumi, S. & Nakamura, C. (2005) Abiotic stress signal pathways associated with development of freezing tolerance after cold acclimation in common wheat. In: Frontiers of Wheat Bioscience (eds. Tsunewaki, K., and Nishikawa, K.), *Wheat Inform. Serv.*, 100, 89-107.

Takumi, S., Shimamura, C. & Kobayashi, F. (2008) Increased freezing tolerance through up-regulation of downstream genes via the wheat *CBF* gene in transgenic tobacco. *Plant Physiol., Biochem.*, 46, 205-211.

Terashima, A. & Takumi, S. (2009) Allopolyploidization reduces alternative splicing efficiency for transcripts of the wheat *DREB2* homolog, *WDREB2*. *Genome*, 52, (in press)

Thomashow, M. F. (1999) Plant cold acclimation: freezing tolerance genes and regulatory mechanisms. *Ann. Rev. Plant Physiol. Plant Mol. Biol.*, 50, 571-599.

Thomashow, M. F. (2001) So what's new in the field of plant cold acclimation? Lots! *Plant Physiol.*, 125, 89-93.

Tsvetanov, S., Ohno, R., Tsuda, K., Takumi, S., Mori, N., Atanassov, A. & Nakamura, C. (2000) A cold-responsive wheat (*Triticum aestivum* L.) gene *wcor14* identified in a winter-hardy cultivar 'Mironovska 808'. *Genes Genet. Syst.*, 75, 49-57.

Uauy, C., Distelfeld, A., Fahima, T., Blechl, A. & Dubcovsky, J. (2006a) A NAC gene regulating senescence improves grain protein, zinc, and iron content in wheat. *Science*, 314, 1298-1301.

Uauy, C., Brevis, J. C. & Dubcovsky, J. (2006b) The high grain protein content gene *Gpc-B1* accelerates senescence and has pleiotropic effects on protein content in wheat. *J. Exp. Bot.*, 57, 2785-2794.

Uno, Y., Furihata, T., Abe, H., Yoshida, R., Shinozaki, K. & Yamaguchi-Shinozaki, K. (2000) *Proc. Natl. Acad. Sci. USA*, 97, 11632-11637.

Vágújfalvi, A., Galiba, G., Cattivelli, L. & Dubcovsky, J. (2003) The cold-regulated transcriptional activator *Cbf3* is linked to the frost-tolerance locus *Fr-A2* on wheat chromosome 5A. *Mol. Gen. Genet.*, 269, 60-67.

Vágújfalvi, A., Aprile, A., Miller, A., Dubcovsky, J., Delugu, G., Galiba, G. & Cattivelli, L. (2005) The expression of several *Cbf* genes at the *Fr-A2* locus is linked to frost resistance in wheat. *Mol. Gen. Genomics*, 275, 506-514.

Vasil, V., Srivastava, V., Castillo, A. M., Fromm, M. E. & Vasil, I. K. (1993) Rapid production of transgenic wheat plants by direct bombardment of cultured immature embryos. *BioTechnol.*, 11, 1553-1558.

Vazquez-Tello, A., Quellet, F. & Sarhan, F. (1998) Low temperature-stimulated phosphorylation regulates the binding of nuclear factors to the promoter of *Wcs120*, a cold-specific gene in wheat. *Mol. Gen. Genet.*, 257, 157-166.

Vollbrecht, E., Veit, B., Sinha, N. & Hake, S. (1991) The developmental gene *Knotted-1* is a member of a maize homeobox gene family. *Nature*, 350, 241-243.

Walbot, V. (1992) Strategies for mutagenesis and gene cloning using transposon tagging and T-DNA insertional mutagenesis. *Annu. Rev. Plant Physiol. Plant Mol. Biol.*, 43, 49-82.

Weeks, J. T., Anderson, O. D. & Blechl, A. E. (1993) Rapid production of multiple independent lines of fertile transgenic wheat (*Triticum aestivum*). *Plant Physiol.*, 102, 1077-1084.

Wu, H., Sparks, C., Amoah, B. & Jones, H. D. (2003) Factors influencing successful *Agrobacterium*-mediated genetic transformation of wheat. *Plant Cell Rep.*, 21, 659-668.

Xia, L., Geng, H., Chen, X., He, Z., Lillemo, M. & Morris, C. F. (2007) Silencing of puroindoline a alters the kernel texture in transgenic bread wheat. *J. Cereal Sci.*, 47, 331-338.

Xu, C., Jing, R., Mao, X., Jia, X. & Chang, X. (2007) A wheat (*Triticum aestivum* L.) protein phosphatase 2A catalytic subunit gene provides enhanced drought tolerance in tobacco. *Ann. Bot.*, 99, 439-450.

Xue, G. P. & Loveridge, C. W. (2004) *HvDRF1* is involved in abscisic acid-mediated gene regulation in barley and produces two forms of AP2 transcriptional activators, interacting preferably with a CT-rich element. *Plant J.*, 37, 326-339.

Yamaguchi-Shinozaki, K. & Shinozaki, K. (1994) A novel *cis*-acting element in an *Arabidopsis* gene is involved in responsiveness to drought, low temperature, or high-salt stress. *Plant Cell*, 6, 251-264.

Yamaguchi-Shinozaki, K. & Shinozaki, K. (2005) Organization of *cis*-acting regulatory elements in osmotic- and cold-stress-responsive promoters. *Trends Plant Sci.*, 10, 88-94.

Yan, L., Loukoianov, A., Tranquilli, G.,Helguera, M., Fahima, T. & Dubcovsky, J. (2003) Positional cloning of what vernalization gene *VRN1*. *Proc. Natl. Acad. Sci. USA*, 100, 6263-6268.

Yan, L., Loukoianov, A., Blechl, A., Tranquilli, G., Ramakrishna, W., SanMiguel, P., Bennetzen, J.L., Echenique, V. & Dubcovsky, J. (2004) The wheat *VRN2* gene is a flowering repressor down-regulated by vernalization. *Science*, 303, 1640-1644.

Yan, L., Fu, D., Li, C., Blechl, A., Tranquilli, G., Bonafede, M., Sanchez, A., Valarik, M., Ysuda, S. & Dubcovsky, J. (2006) The wheat and barley vernalization gene *VRN3* is an orthologue of *FT*. *Proc. Natl. Acad. Sci. USA*, 103, 19581-19586.

Yao, Q., Cong, L., Chang, J. L., Li, K. X., Yang, G. X. & He, G. Y. (2006) Low copy number gene transfer and stable expression in a commercial wheat cultivar via particle bombardment. *J. Exp. Bot.*, 57, 3737-3746.

Yu, R. M., Zhou, Y., Xu, Z. F., Chye, M. L., & Kong, R. Y. (2003) Two genes encoding protein phosphatase 2A catalytic subunits are differentially expressed in rice. *Plant Mol. Biol.*, 51, 295-311.

Zarka, D. G., Vogel, J. T., Cook, D. & Thomashow, M. F. (2003) Cold induction of Arabidopsis CBF genes involves multiple ICE (inducer of CBF expression) promoter elements and a cold-regulatory circuit that is desensitized by low temperature. *Plant Physiol.*, 133, 910-918.

Zhang, D., Choi, D. W., Wanamaker, S., Fenton, R. D., Chin, A., Malatrasi, M., Turuspekov, Y., Walia, H., Akhunov, E. D., Kianian, P., Otto, C., Simons, K., Deal, K. R., Echenique, V., Stamova, B., Ross, K., Butler, G. E., Strader, L., Verhey, S. D., Johnson, R., Altenbach, S., Kothari, K., Tanaka, C., Shah, M. M., Laudencia-Chingcuanco, D., Han, P., Miller, R. E., Crossman, C. C., Chao, S., Lazo, G. R., Klueva, N., Gustafson, J.P., Kianian, S. F., Dubcovsky, J., Walker-Simmons, M. K., Gill, K. S., Dvorák, J., Anderson, O.D., Sorrells, M. E., McGuire, P. E., Qualset, C. O., Nguyen, H. T. & Close, T. J. (2004) Construction and evaluation of cDNA libraries for large-scale expressed sequenced tag sequencing in wheat (*Triticum aestivum* L.). *Genetics*, 168, 595-608.

Zhang, J. Z., Creelman, R. A. & Zhu, J. K. (2004) From laboratory to field. Using information from Arabidopsis to engineer salt, cold and drought tolerance in crops. *Plant Physiol.*, 135, 615-621.

Zhao, X. Y., Liu, M. S., Li, J. R., Guan, C. M. & Zhang, X. S. (2005) The wheat *TaGI*, involved in photoperiodic flowering, encodes an Arabidopsis *GI* ortholog. *Plant Mol. Biol.*, 58, 53-64.

In: Transgenic Plants: New Research
Editor: Oliver T. Chan
ISBN 978-1-60692-017-6
© 2009 Nova Science Publishers, Inc.

Dear author, there is the mention of only table 2 in several places the text, while the submitted table is Table 1. Should the mentioning of table 2 be changed in table 1 in all cases? Please check and correct.

Chapter 2

TRANSGENIC ORNAMENTAL PLANTS: MAJOR ACHIEVEMENTS, PERSPECTIVES AND POSSIBILITIES

Jaime A. Teixeira da Silva**, *Michio Tanaka
Faculty of Agriculture, Kagawa University, Miki-cho, Ikenobe 2393, Kagawa-ken, 761-0795, Japan

ABSTRACT

Ornamental plants (Ops) play important roles in society, religion and science. A wealth of species are considered to be Ops, and this makes this group of plants an increasingly important group as model plants for a range of physiological, biochemical, genetic and biotechnological studies. The introduction of horticulturally and agronomically useful traits into Ops through well-developed and fairly well-established transgenic systems would be one way in which to overcome some of the barriers of conventional breeding programmes and to shorten what might easily take years or decades to achieve, especially in some Ops whose vegetative cycles or vegetative-transition stage takes months or even years. The success of a transformation protocol invariably depends on the inherent success of the tissue culture protocol. Genetic engineering to incorporate selector, marker or useful genes into Ops has been primarily achieved by gene transfer mediated by *Agrobacterium* and particle bombardment, although other methods have been employed. In this chapter, I provide a brief overview of how the main technologies have been used to develop genetically modified Ops. The desire for new traits (color, improved scent, cut-flower longevity and post-harvest robustness, modified architecture, acclimatization efficiency, and tolerance to biotic and abiotic stresses) and a greater overall acceptance of GM Ops, more than GM fruits or vegetables, make this group of plants important for every fundamental type of plant science.

* Correspondence: jaimetex@yahoo.com

Keywords: genetic engineering, molecular breeding, ornamental, recombinant DNA, tissue culture, transformation, transgenic plants

1. INTRODUCTION

Ornamental plants (Ops) form the bulwark of natural decoration, whether in a traditional setting such as a wedding or in less conventional applications such as *in vitro* bouquets (see Volumes II and IV of Teixeira da Silva 2006). The introduction of novel traits can be achieved through mutation and conventional breeding which would allow the introduction of single or multiple traits. However, at the pace at which the world is currently evolving, and the rate at which the consumer market is literally demanding change and innovation, much pressure is placed on breeders to develop methods to rapidly insert genes of interest into horticulturally important cultivars or varieties, providing more valuable genetic resources with value-added traits to satisfy the thirst of the OP industry. At present there are a few genetic engineering strategies available to the OP molecular breeder, and these will be the topic of discussion of this chapter. To date, the only genetically modified Ops that have already reached the market are the Moon series of transgenic carnations (presenting mauve, purple, or violet flowers), developed as a joint venture between Florigene in Australia and Suntory, a Japanese company, have seen limited market implementation. Despite the perception of transgenic Ops not being in the same light as that for transgenic consumables, there exist, nonetheless, reticent attitudes towards transgenic plants (in general) that may hamper that practical application and integration of transgenics in a widely-accepted market. These factors are discussed by Dobres (2008) in much more detail.

The exact value of the ornamental horticulture industry across the world has very unique trends (for a review see Volume IV of Teixeira da Silva 2006) and the global production of cut flowers, pot plants and ornamental foliage is continually expanding. What is certain, as shown b the explosive interest in OP research and biotechnology in the not too distant past (Teixeira da Silva 2006), is that Ops have been are and will continue to take center stage in several fields of plant biology and biotechnology.

The most desirable characteristics to consumers are the modification in plant or flower shape, architecture and size, and the form and color of the flowers and foliage. This hunger for change drives the transgenic OP market. Others important traits have been manipulated, including the modification of scent, introduction of insect and disease (caused by virus, bacteria, and fungus) resistance, improved post-harvest longevity, and drought and salt tolerance.

The success of genetic engineering depends on the totipotency of plant cells, i.e. the inherent ability of plant tissue to produce cells that can be regenerated into a completely viable plant. The use of genetic engineering to incorporate useful genes into elite breeding lines depends on the development of suitable tissue culture techniques to generate organogenic/embryogenic cells, and of protocols for the delivery of transgenes into competent cells, to make possible the regeneration of stably transformed transgenic plants. This chapter will provide an overview of the regeneration and genetic modification of Ops. Readers are referred to recent publications by Chandler (2003), Misra and Srivastava (2004), Teixeira da

Silva (2004), Chandler and Lu (2005), Tanaka *et al.* (2005) and Teixeira da Silva (2006), in which a much broader range of aspects of OP biotechnology are discussed.

2. Genetic Engineering

Plant tissue culture has been the base for the economic viability of a massive global OP industry. The base of a successful genetic transformation protocol is the tissue culture of Ops, which has been discussed in detail in Volume II of Teixeira da Silva (2006). Moreover, the role of thin cell layers or TCLs, a simple but effective means for establishing regeneration systems, not only for ornamental plants but for all plants, will undoubtedly increase (Teixeira da Silva *et al.* 2007).

There is considerable interest in the introduction of agronomically useful traits into Ops and tissue culture procedures are important for both "classical" and molecular genetic manipulations. Molecular genetic manipulation for applied purposes is only possible with the development of an efficient protocol for transformation. Transformation procedures will be efficient only if an efficient protocol of regeneration has previously been established. Regeneration systems have been developed for several ornamental species, outlined in Volume II of Teixeira da Silva (2006) allowing genetic manipulation by *in vitro* mutation and selection and by the use of the recombinant DNA technology. Nevertheless, for many species regeneration systems still need to be developed or improved, and the field of genetic engineering of Ops is bright and rewarding.

Desirable genes can be cloned from any organism, introduced, and expressed in plants across sexual boundaries. Thus, the source of variability available for plant improvement has been expanded. This technology, associated with other "classical" breeding methods, will accelerate the production of new varieties with useful traits. Although there is potential use of genetic engineering for flower breeding programs, few ornamentals species have been genetically modified, lagging behind main food crops (Vainstein *et al.* 2001; Misra and Srivastava 2004; Tanaka *et al.* 2005). Nevertheless, transformation systems are available for all the major flower species. The majority of transformation protocols used in ornamental species continues to be based on gene transfer mediated by *Agrobacterium tumefaciens* (Table 2), although this profile is changing with *in vitro* floral dip and *in planta* transformation (Kojima *et al.* 2006; Chumakov 2007) and sonication-assisted *Agrobacterium*-mediated transformation or SAAT becoming more popular as their use was shown to be successful, simple and reproducible (Teixeira da Silva and Fukai 2003; several cases in Teixeira da Silva 2006, 2008). The main limiting factors were the lack of efficient transformation systems for floricultural species and the rather small markets for each species in comparison with food crops (Tanaka *et al.* 2005), although this is now changing as cut-flower markets bring large and consistent income to developing nations (Teixeira da Silva 2006, 2008). An efficient, reproducible, cultivar-independent transformation system is critical to generate the requested number of elite transgenic lines.

Table 1. Summary of select ornamental/floricultural species genetic transformation systems

Species	Genes	Explant	Method	References
Antirrhinum majus	*nptII, gus, Plena*	□ypocotyls	At	Heidmann *et al.* 1998
A. majus	*gus, rolA, B, C* and *D* of Ri T-DNA, recombinase ® (MAT vector)	leaf	At	Cui *et al.* 2001
A. majus	*nptII, hpt*	□ypocotyls	At	Cui *et al.* 2004
A. majus	*nptII, gfp*	□ypocotyls	At	Cui *et al.* 2003
Curcuma alismatifolia	*gus, nptII, ACS1*	retarded shoot	Ar	Mahadtanapuk *et al.* 2006
Curcuma longa	*gus, bar*	leaf	B	Shirgurkar *et al.* 2006
Cymbidium	*gus, nptII*	PLB	B	Yang *et al.* 1999
Cymbidium hybrids (x8)	*gus, nptII, bar*	PLB	B	Teixeira da Silva and Tanaka 2008a, 2008b
Cymbidium niveo-marginum	ORSV *cp*	PLB	At	Chen *et al.* 2006
Dendranthema grandiflora	*gus, nptII*	stem, leaf, TCL	At, B, AtB, SAAT	Teixeira da Silva and Fukai 2003
Dendrobium	*nptII*	PLB	B	Kuehnle and Sugii 1992
Dendrobium	*luc*	PLB	B	Chia *et al.* 1994
Dendrobium phalaenopsis, D. nobile	*hpt, gus*	callus, PLB	B	Men *et al.* 2003a, 2003b
Dianthus caryophyllus	*nptII, gus*	stem	AtB	Zuker *et al.* 1999
D. caryophyllus	*nptII, gus*	stem node	At	Nontaswatsri *et al.* 2004
D. caryophyllus	*DC-ACS1, s/aACS*	leaflet	At	Kinouchi *et al.* 2006
Euonymus alatus	*gus*	cotyledon, □ypocotyls	At	Chen *et al.* 2006
Eustoma grandiflorum (syn. *Lisianthus russelianus*)	*nptII*	leaf disc	At	Deroles *et al.* 1993
Forsythia intermedia	*hpt, dfr, ans, nptII*	callus	At	Rosati *et al.* 1996, 2003
Gentiana triflora × *G.scabra*	*gus, hpt*	leaf	B	Hosokawa *et al.* 2000
Gentiana dwarf type	*agropine*	leaf	Ar	Mishiba *et al.* 2006
Gladiolus	*GUBQ1* (promoter tests)	leaf	At	Joung and Kamo 2006
Hyacinthus orientalis	*nptII, pepII*	leaf	At	Popowich *et al.* 2007
Hypericum perforatum	*gus*	seedling	Ar	Vinterhalter *et al.* 2006

Table 1. Continued

Species	*Genes*	Explant	Method	References
Ipomoea purpurea	*rolB*	cotyledon, hypocotyl	Ar	Tan *et al.* 2007
Lilium	*hpt, nptII, gus*	filament-derived callus	At	Hoshi *et al.* 2004
Lobelia erinus	*nptII, gus, hpt*	leaf, cotyledon	At	Tsugawa *et al.* 2004
Nemesia strumosa	*nptII, gfp*	stem segments	At	Cui and Ezura 200
Nemesia strumosa	*IbMADS4*	leaf, shoots from internodes	At	Aswath *et al.* 2004
Nierembergia caerulea	MAT vector	leaf	At	Khan *et al.* 2006
Oncidium	*gus, hpt*	PLB	At	Liau *et al.* 2003a
Pelargonium graveolens	*Ri*	leaf, petiole, internode segment	Ar	Saxena *et al.* 2007
Phalaenopsis	*hpt nptII, gus*	germinated protocorm	At	Mishiba *et al.* 2005
Phalaenopsis	meropenem-degrading gene	germinated protocorm	At	Sjahril and Mii 2006; Sjahril *et al.* 2006
Phalaenopsis (*Doritaenopsis* × *Phalaenopsis*)	*nptII, gus, hpt*	flower	At	Belarmino and Mii 2000
Rhipsalidopsis gaertneri	*nptII, gus*	aeriole	At	Al-Ramamneh *et al.* 2006
Rhododendron	*gus, smGFP, nptII*	leaf	B	Knapp *et al.* 2001
Rosa hybrida	*nptii, gus, luc*	filament	At, Ar	Firoozabady *et al.* 1994
Torenia fournieri	*chs, dfr* (sense and antisense orientation)	leaf	At	Aida *et al.* 2000a
T. fournieri	*manA*	leaf	At	Li *et al.* 2007
Verbena hybrida	*sGFP, nptII*	shoot internode	At	Tamura *et al.* 2003
Verticordia grandis	*gus, nptII*	leaf disc	Ar, At	Stummer *et al.* 1995

At = *Agrobacterium tumefaciens*; Ar = *Agrobacterium rhizogenes*; AB = Agrobalistics (combinations of biolistics- and *Agrobacterium*-mediated transformation system); B = biolistic; PLB = protocorm-like body; SAAT = sonication-assisted *Agrobacterium*-mediated transformation; TCL = thin cell layer.

The biolistic (or particle bombardment) process has been utilized to transformed some floricultural species, especially orchids, such as *Dendrobium* (Kuehnle and Sugii 1992; Chia *et al.* 1994) and *Cymbidium* (Yang *et al.* 1999; Teixeira da Silva and Tanaka 2008a, 2008b) (Table 2), although a deeper discussion of orchid transformation, which will not be covered in detail here, since it has been covered by Chai and Yu (2007). The basis of the process is the acceleration of DNA-coated microparticles (mainly particles of gold or tungsten) at high speed (above 1,500 km/h) towards living cells. After penetration in the cell, the DNA dissociates from the particles and integrates into the chromosomes (Aragão 2003). The biolistic process allows the bombardment of intact plant cells turning possible the transformation of cells that are recalcitrant to tissue cultural manipulations. This minimizes the necessity for treatment with phytohormones and thus the opportunity for somaclonal variation. An efficient protocol based on the biolistic process depends on optimizations of several biological and physics parameters. For *Cymbidium*, among three different sizes of gold particles tested (0.6, 1.0, and 1.6 μm in diameter), the smallest were most efficient in delivering the genes (Yang *et al.* 1999), although this trend was not true or reproducible for several (eight) *Cymbidium* cultivars (Teixeira da Silva and Tanaka 2008a, 2008b). Different levels of helium gas pressures (1,100, 1,350, and 1,550 psi) did not affect transformation efficiency (Yang *et al.* 1999). For *Rhododendron*, 1 μm diameter gold particles and helium pressure of 1,100 psi was used (Knapp *et al.* 2001).

The combination of the biolistic- and *Agrobacterium*-mediated transformation methods (called Agrobombardment, Agrobalistic, or Agrolistics) has been used to transform carnation and chrysanthemums (Zuker *et al.* 1999; Teixeira da Silva and Fukai 2003; Teixeira da Silva 2004) and model plant tobacco (Teixeira da Silva 2005). In this method, the biolistic process is used to accelerate uncoated microprojectiles, causing micro-wounds in the explants prior to its co-cultivation with *Agrobacterium.* By using this methodology, it was possible to achieve a high efficient transformation of carnation (Zuker *et al.* 1999). For chrysanthemums, the Agro-bombardment method was compared with transformation mediated by *Agrobacterium*, biolistics and *Agrobaterium*/SAAT (Teixeira da Silva and Fukai 2003). The results have shown that the regeneration capacity was affected by the transformation system. Sonication stimulated regeneration while the treatment with *Agrobacterium* had a negative impact. The Agrobombardment increased shoot regeneration for *in vitro* transverse thin cell layer explants, but completely inhibited normal shoot formation in explants from the greenhouse, despite a high explant survival. In general, shoot regeneration decrease with an increase in tissue damage (Agro-bombardment > biolistics > *Agro*infection > sonication) (Teixeira da Silva and Fukai 2003).

Selective marker genes are introduced into plant genome to express a protein with, generally, an enzymatic activity, allowing to distinguish transformed from non-transformed cells (profound discussion in several chapters in Teixeira da Silva 2006, Vol II). The purpose of the use of a selective marker gene is to give to the transformed cells a selective advantage, allowing them to grow faster and better, and to kill the non-transformed cells. The transformation systems available for floricultural plants use several marker genes. These genes confer tolerance to antibiotics, mainly *nptII* (tolerance to kanamycin) and *hpt* (tolerance to hygromycin) genes, and herbicide, such as the *bar* gene (confers tolerance to glufosinate ammonium). Paromicin was used to select transgenic shoots of *Antirrhinum majus* transformed with the *nptII* gene (Heidmann et al. 1998). Reporter genes were also utilized, such as the *gus* (coding for β-glucuronidase) and *gfp* (coding for a green fluorescent protein) genes (Heidmann *et al.* 1998; Mercuri *et al.* 2001; Cui *et al.* 2001; Mercuri *et al.* 2002; Cui and Ezura 2003, Liau *et al.* 2003a, Tsugawa *et al.* 2004; Dirks and Tanke 2006; Pédelacq *et al.* 2006; Stewart Jr 2006; Table 2) although positive-selection systems, such as the native tryptophan synthase beta 1 gene (Hsiao *et al.* 2007) will continue to play an increasingly important role, especially in 2008 in the EU where negative selection systems will no longer be allowed to be used.

The green fluorescent protein (GFP) from the jellyfish *Aequorea victoria* is a spontaneously fluorescent protein. *mGFP5* gene was modified to remove an aberrant cryptic mRNA splicing site, in order to display an optimal codon usage for plants and to permit a better protein folding (Haseloff and Amos 1995; Siemering *et al.* 1996; Haseloff *et al.* 1997; Sacchetti and Alberti 1999; Sacchetti *et al.* 2000). Moreover, the addition of an N-terminal signal peptide and of a C-terminal HDEL sequence to mGFP5 directs GFP to the endoplasmic reticulum, and allows a non-toxic accumulation of GFP to high levels (Haseloff *et al.* 1997). In addition, mGFP5 contains an I167T (substitution of a isoleucine by a threonine) mutation that induces dual excitation peaks of similar intensity at 400 nm and 475 nm, allowing visualization with both long UV and blue light (Heim *et al.* 1994; Mercuri *et al.* 2001, 2002). mGFP5 gene was expressed in flower tissues from *Eustoma grandiflorum* (syn. *Lisianthus russelianus*) and *Osteospermum ecklonis*, that displayed intense fluorescence when illuminated with UV light (Mercuri *et al.* 2001 2002). In addition, it was found that petals of transgenic *Limonium* were expressing the *gfp* gene in spite of the lack of macroscopic green fluorescence, probably to the opacity to the UV and blue light used to excite GFP (Mercuri *et al.* 2001 2002). The sGFP, with the mutation S65T (substitution of a serine by a threonine) was expressed in *A. majus*, which presented shoot fluorescence (Cui *et al.* 2003). A deeper discussion of the GFP protein in plant transformation is provided by Malabadi *et al.* (2008).

The selection method using the expression luciferase gene from firefly has also been developed (Chia *et al.* 1994; Yang *et al.* 1999). In this system, light-emitting cells are screened and isolated using a photon-counting video camera-photomultiplier system and high-power dissecting microscope. The screening and isolation process is repeated until a pure transgenic cell line is isolated. The isolated cell lines are then regenerated into a whole plant (Chia *et al.* 1994; Yang *et al.* 1999). A good review that places the use of GFP into a wider perspective was provided by Jach and Winter (2006).

Regarding biosafety concerns, complex evaluations have been accomplished, aiming to study the potential impacts of antibiotic resistance genes present in transgenic plants on human health and the environment. Despite no evidence of deleterious effects on the use of transgenic plants carrying these genes, their removal has already been stated as a "good practice of laboratory," by several regulatory committees (Aragão and Brasileiro 2002). In addition, selective agents generally have a negative effect on proliferation and differentiation of plant cells and may retard the regeneration of adventitious shoots during the transformation process. Some plant species, however, are insensitive to the selective agents, so it is difficult to separate the transformed and non-transformed cells or tissues (Cui *et al.* 2000). I have found this to be true in several *Cymbidium* cultivars which, even though not genetically transformed, are capable of surviving on low to moderate levels (5-25 μg/ml) of kanamycin (Teixeira da Silva, unpublished findings).

Although OPs are not generally used for animal feeding, several strategies to remove the selection marker genes have been developed. One of these strategies applied to ornamental species is called "multi-auto-transformation system" (MAT). The MAT system is firstly based on the visual selection of transgenic plants containing the *ipt* or *rol* genes (Ebinuma *et al.* 1997; Cui *et al.* 2000). In the presence of *ipt* gene, the transformed plant loses the apical dominance and ability to rooting, while in presence of *rol* genes, from Ri plasmids of *Agrobacterium rhizogenes*, hairy roots are induced. The abnormal phenotype was called "extreme shooty phenotype". In a second step, the unsuitable *ipt* gene is removed of the transgenic plant through the transposition of an *Ac* transposable element, which was transferred along the selection marker gene. In this way, marker-free transgenic plants could be generated with the normal phenotype restored, containing only the gene of interest. MAT system has been successfully applied to *A. majus* (Cui *et al.* 2001, 2000).

Flower symmetry or asymmetry is an important trait in floricultural species. *A. majus* is used in studies of the asymmetric development of flowers. This species, which is a well-known model specie that produce asymmetric flowers that consist of five sepals, two upper (dorsal) petals, three lower petals (two lateral petals and one ventral petal), four stamens and a fused carpel, has been transformed with *Agrobacterium* (Heidmann *et al.* 1998; Cui *et al.* 2001, 2004). The ornamental plant *Nemesia strumosa* was proposed as a model plant for the study of asymmetric floral development. Stable transformants of *N. strumosa* were obtained with *A. tumefaciens* harboring vectors containing the *nptII* and *gfp* genes, reaching a transformation efficiency of 9% (Cui and Ezura 2003).

In many gene-transfer experiments, the constitutive cauliflower mosaic virus (CaMV) 35S RNA promoter is mostly used. However, other promoters have been used in order to lead to higher or tissue specific expression. To achieve high expression in petal tissue of ray florets of chrysanthemum an endogenous *ubiquitin extension protein* (*UEP1*) promoter was cloned and tested with the *gus* reporter gene. The *UEP1* promoter expression levels were 1.5 to 4-fold higher when compared with the promoters from *chalcone synthase* (*chs-A*) and *zinc finger transcription factor* (*EPF*2-5) from petunia, *eceriferum* (*CER*6) from *Arabidopsis* and *multicystatin* (*PMC*) from potato. Moreover, the expression of *gus* gene under control of the *UEP*1 promoter was 50-fold of the expression observed when the gene was under control of the CaMV 35S promoter (Annadana *et al.* 2002a). Recently, the light harvesting cab promoter *Lhca3.St.*1 was identified as a strong promoter for expression in chrysanthemum leaves and stems (Annadana *et al.* 2002b), but the expression in ray florets was 6-fold lower than in leaf tissue.

3. Introduction of Useful Traits

Flower color is probably the most important trait in OPs. The generation of new colors is one of the most critical targets for the floricultural plant breeding. Flower color is primarily determined by anthocyanins and carotenoids. There are six main anthocyanidins: pelargonidin; cyanidin; peonidin; delphinidin; petunidin; and malvidin. As the number of hydroxyl groups in the B-ring increases, the visible absorption maximum shifts to become longer, so that the flower looks bluer (Aida *et al.* 2000a). Yellow and orange flowers normally contain carotenoids. The anthocyanins pelargonidin, cyanidin, and delphinidin 3-glucosides are colored pigments, responsible for pink, mauve, red, and blue shades of flowers. Flowers that produce delphinidin-based pigments generally have a violet-blue shade (reviewed by Mol *et al.* 1998; Forkmann and Heller 1999; Springob *et al.* 2003; Forkmann and Martens 2001; Martens *et al.* 2003a, 2003b; Ono and Nakayama 2007). By genetically manipulating genes related to the biosynthetic pathway of flower color pigments, it is possible to breed novel varieties.

Flower color has been modified in several OPs mainly by the overexpression or silencing of genes encoding enzymes from the flavonoid biosynthesis pathway. The flavonoid pathway leads to the first colored anthocyanins. Chalcones are intermediates in the biosynthesis of all flavonoids and, in some species, constitute the major yellow pigments present in flowers.

Various strategies have been used to modify the intensity of color flower, mainly by introducing the chalcone synthase (*chs*) and dihydroflavonol-4-redutase (*dfr*) genes in antisense or sense orientations, to obtain pure white alternated colored flower lines of several species, such as petunia, *Chrysanthemum*, gerbera, rose and torentia (Davies *et al.* 1998; Mol *et al.* 1995; Aida *et al.* 2000b; Tsuda *et al.* 2004; molecular mechanisms reviewed in Ono and Nakayama 2007).

The first reports of success in the genetic manipulation of flower color through plant transformation demonstrate the feasibility in creating an orange pelargonidin-producing petunia by expression of the gihydroflavonol-4-redutase gene (*dfr*) from maize (Meyer *et al.* 1987). However, flowers of many transgenic lines presented a pale brick red and showed silencing of the transgenes. Further, transgenic petunia lines containing the *dfr* gene were crossed with elite breeding lines, creating bright-orange varieties (Oud *et al.* 1995). The expression of a *chr* gene (chalcone redutase) from *Medicago sativa* into acyanic or cyanic-flowered lines of *Petunia* changed the flower color from either white to pale yellow or deep purple to pale purple (Davies *et al.* 1998). Transgenic lines presented accumulation up to 60% of their petals flavonoids as 6′-deoxychalcones.

Inactivation of the *dfr* in *Torenia fournieri* produced bluer flowers, showing that a co-pigmentation effect caused by the accumulation of flavones and flovonols likely to be the reason for the bluing phenomenon (Aida *et al.* 2000a). Indeed, when anthocyanins make complexes with co-pigments such as flavones or flavonols (by co-pigmentation), the visible absorption maximum shifts to become longer (phenomenon called bathochromic shift) (Tanaka *et al.* 1998). Flower color was modified in transgenic forsythia (*Forsythia* × *intermedia*) transformed with the *dfr* gene from *Antirrhinum majus* and *ans* (anthocyanidin synthase) gene from *Matthiola incana* displayed a novel bronze-orange petal color, caused by the *de novo* accumulation of cyanidin-derived anthocyanins over the carotenoid yellow background of wild type, and intense pigmentation of vegetative organs (Rosati *et al.* 2003). Transgenic lisianthus (*Eustoma grandiflorum*) containing a homologous *chs* gene in antisense presented an altered flower color pattern ranging from small streaks of white on the wild type purple background through to completely white flowers showing reduction of ca. 98% of total flavonoids. Northern analysis revealed that these phenotypes were associated with a loss of *chs* gene transcript. In addition, some lines also exhibited alterations in flower form (formation of frilled petal tips and reduced flower opening) (Deroles *et al.* 1998). Transgenic lisianthus plants line expressing antisense endogenous *fls* gene (flavonol synthase) produced flowers more red in color (magenta) than the untransformed plant (Nielsen *et al.* 2002).

The *F3'5'H* gene (coding a flavonoid 3,5-hydroxylase, a key enzyme in the synthesis of the blue flower-specific anthocyanidins, e.g., delphinidin) from *Vinca major*, was expressed in transgenic petunia hybrida. Transgenic plants revealed flower color alteration from red to deep red with deep purple sectors (Mori *et al.* 2004).

Carnation plants were transformed with the *f3h* gene, encoding the flavanone 3-hydroxylase (a key enzyme in the anthocyanin pathway), in antisense orientation. Plants presenting endogenous *f3h* gene suppression exhibited flower color modifications, ranging from attenuation to complete loss of their original orange-reddish color. In addition, transgenic plants with severe color modification were more fragrant than control plants, showing a possible interrelation between pathways leading to anthocyanin and fragrance production (Zuker *et al.* 2002).

Another important goal for flower breeders is the creation of longer-life flowers. The post-harvest life of many flowers is determined by the onset of petal senescence leading to the rapid flower deterioration. It is one of the major obstacles for the market (Mol *et al.* 1995). Plant senescence is regulated by a coordinated genetic program mediated in part by changes in ethylene, abscisic acid, and cytokinin content. In order to generate long-life flowers, one target is the conversion of S-adenosyl methionine (SAM) to 1-aminocyclopropane-1-carboxylic acid (ACC) and the conversion of ACC to ethylene. These steps of catalyzed by ACC synthase and ACC oxidase, respectively. Genes from ACC synthase (*acs*) and ACC oxidase (*aco*) have been isolated and characterized from several species. Transgenic carnation, transformed with an *aco* gene in antisense orientation, in order to silencing the endogenous gene, presented low levels of ethylene and a delay in petal senescence (Savin *et al.* 1995). Transgenic petunia expressing the *ipt*, a cytokinin biosynthetic gene from *A. tumefaciens*, under the control of the promoter from a senescence-associated gene (SAG12) revealed delayed senescence in 6 to 10 d relative to wild-type flowers (Chang *et al.* 2003; Satoh *et al.* 2006; Tanase *et al.* 2008).

Flowers of transformed petunia plants carrying the *boers*, a mutated allele of *BOERS*, an ethylene receptor-sensor gene of *Brassica oleracea*, retained turgidity, and pigmentation considerably longer. However, higher mortality was observed, suggesting that the ethylene insensitive petunia plants were also lower in disease resistance (Shaw *et al.* 2002).

During the process of production of some commercial OPs, such as chrysanthemum, it is often necessary to manipulate their growth and development to produce plants with uniform morphology. Therefore, production relies on the use of growth regulation agrochemicals in order to produce uniform plants. Several genes that are involved with flower and plant morphology have been cloned and characterized. Most genes are transcriptional factors that control vegetative growth and regulate organ morphogenesis or are regulatory genes involved in plant phytohormones pathway. These genes can be expressed in OPs in order to achieve changes in plant architecture. The MADS-box *IbMADS4* gene, isolated from *Ipomoea batatas* was introduced in transgenic chrysanthemum plants and resulted in a disrupted growth pattern (Aswath *et al.* 2004). Plants conditioned a highly branched phenotype, exhibited a three–four-fold reduction in the plant height, and increased biomass of plant body, compared with the wild type. However, plants presented suppression of flowering, even after six months following transplanting (Aswath *et al.* 2004). These characteristics could be useful foliage and potted plants. Transgenic *Chrysanthemum morifolium* were produced expressing the *gai* gene, which encodes a nuclear localized repressor of gibberellic acid (GA) responses isolated from GA insensitive *Arabidopsis thaliana* (Petty *et al.* 2003). The transgenic chrysanthemum lines exhibit a range of dwarf phenotypes, related to the reduction in the response to GA, demonstrating the feasibility of producing a dwarf plants without use of growth retardant compounds (Petty *et al.* 2003). The expression of the *MASAKO C1* and *D1* MADS-box genes isolated from rose in transgenic *Torenia fournieri* produced morphological changes of flowers, with sepals completely changed into carpels (Kitahara *et al.* 2004).

The manipulation of flower scent would have a great economic impact. Although, the number of different floral volatiles molecules is very large, these compounds are biosynthesized by relatively small number metabolic pathways (Mol *et al.* 1995; Vainstein *et al.* 2001; Hsiao *et al.* 2008). Research into floral scent genes using high throughput tools has been initiated in several floricultural species and some genes have been isolated and characterized, such as genes from monoterpene synthases, acetyltransferases, and methyltransferases. Whilst several groups have reported the genetic manipulation of plant volatile compounds (Vainstein *et al.* 2001), the first attempt to manipulate plant volatile molecules and modify fragrance production in flowers was reported in petunia transformed with the linalool synthase (*lis*) gene from *Clarkia breweri* (Lewinsohn *et al.* 2001). The product of this gene catalyzes the production of the monoterpene linalool from geranyl diphosphate. The aim was to produce the monoterpene linalool in transgenic plants. However, petunia did not result in linalool emission; instead, the non-volatile linalool glycoside accumulated (Lücker *et al.* 2001). On the other hand, further introduction of a *lis* gene construct into carnation led to the emission of linalool from petals as well as from leaves but it did not lead to olfactorally detectable changes in flower fragrance (Lavy *et al.* 2002).

The efficiency of acclimatization and tolerance to environmental stress under both greenhouse and external conditions are essentials for production of ornamental species. Recently, the gene coding an auxin-regulated tobacco glutathione S-transferase (GST), an enzyme that conjugates reduced glutathione (GSH) (*NT107*) to produce a wide variety of compounds that are lipophilic, was introduced in *Dianthus superbus*. Transgenic plants over expressing the *NT107* gene showed an enhanced acclimatization in the soil, against environmental stresses such as increased temperature, light intensity, and decreased humidity for the hardening of *in vitro* cultured transgenic plants. After four weeks, the survival rate was of 81.4% and 78.3% in two transgenic lines, much higher then observed in the wild type plants (29.9%) (Lim *et al.* 2005). In addition, assays carried out with leaf disks of transgenic plants suggested tolerance to environmental stress. High light intensity and water deficits can increase electron flow to O_2, causing photoinhibition, with a greater production of O^{-2} and H_2O_2. GST is in limiting the oxidative damage associated with photoinhibition. Indeed, GST over expressing plants was found to synthesize phytochelatin, which functions by sequestering and detoxifying excess copper ions (Lim *et al.* 2005).

There are still few examples of genetic modified floricultural species to introduce genes for disease resistance (Chandler 2003; Teixeira da Silva 2006 Vols II and III). In one of these examples *Oncidium* lines transformed with the *pflp* gene, coding an antimicrobial ferredoxin-like protein, showed enhanced resistance to *Erwinia carotovora*, the causal agent for the soft rot disease (Liau *et al.* 2003b).

4. Conclusions and Perspectives

Most developed transformation systems are based on cell transfection by *Agrobaterium*. However, it limits the transformation of some species when an efficient regeneration protocol is not yet established. In this regards, the particle bombardment allows the bombardment of intact plant cells turning possible the transformation of plants for which a regeneration system was not available. In some cases, this selection could not be efficient by using antibiotics. A system was developed to select transgenic meristematic cells based on the use of imazapyr, an herbicidal molecule capable of systemically translocating and concentrating in the apical meristematic region of the plant. This selectable marker system, combined with an improved multiple shooting induction protocol may result in a significant increase in the recovery of fertile transgenic lines (Aragão *et al.* 2000). Other herbicide molecules, such as glyphosate and imazetapyr, could also be used. These are however, negative selection systems that will not hold well in a "green" future and the use of positive selection systems will take center stage (Teixeira da Silva 2006; Hsiao *et al.* 2007; Teixeira da Silva 2008).

Readers are directed to recent, excellent reviews written recently on specific ornamentals, whose coverage here would be redundant: *Alstroemeria* (Hoshino 2008), *Narcissus* (Jowkar and Sticklen 2008), gentian (Nishihara *et al.* 2008), marigold (Godoy-Hernández and Miranda-Ham 2007) and turfgrasses (Tzfira and Citovsky 2006; Wang and Ge 2006). More general aspects of ornamentals and yearly updates can be found in *Global Plant Letters* (see http://www17.plala.or.jp/gsbjapan/).

During the last two to three decades, efforts to achieve efficient methodologies for regeneration and transformation of some OPs advanced, while the last 5-10 have seen a remarkable explosion and attention to OPs. In parallel, the availability of characterized genes (coding and regulatory sequences) has increased as a direct result of several plant structural and functional genome projects. These facts generated an excellent scenario for introduction of useful traits as well as study gene function in OPs. Besides manipulation of flowers characteristics, such as color, structure, and fragrance, there is a considerable interest in the introduction of genes for several useful traits in floricultural plants, such as virus, insect, bacteria and fungi resistance, and environmental stress tolerance. In addition, manipulation of plant architecture and phenological characteristics might facilitate management, increasing yield, quality, and diversity. Several traits might be manipulated by genetic engineering, such as plant life cycle and the transition from vegetative to reproductive growth, controlling flowering time. Indeed, the genes have already been characterized. Biotechnological tools complement those from classical breeding and have the potential to accelerate the generation of new varieties containing genes from agronomic traits, which are difficult to be found in the primary or secondary genepool.

REFERENCES

Aida R, Kishimoto S, Tanaka Y, Shibata M (2000a) Modification of flower color in torenia (*Torenia fournieri* Lind.) by genetic transformation. *Plant Science* 153, 33-42.

Aida R, Yoshida K, Kondo T, Kishimoto S, Shibata M (2000b) Copigmentation gives bluer flowers on transgenic torenia plants with the antisense dihydroflavonol-4-reductase gene. *Plant Science* 160, 49-56.

Alderson PG, Taeb AG (1990) Influence of culture environment on shoot growth and bulbing of tulip *in vitro*. *Acta Horticulturae* 266, 91-94.

Al-Ramamneh EA, Sriskandarajah S, Serek M (2006) *Agrobacterium tumefaciens*-mediated transformation of *Rhipsalidopsis gaertneri*. *Plant Cell Reports* 25, 1219-1225.

Annadana S, Beekwilder MJ, Kuipers G, Visser PB, Outchkourov N, Pereira A, Udayakumar M, De Jong J, Jongsma MA (2002a) Cloning of the chrysanthemum *UEP*1 promoter and comparative expression in florets and leaves of *Dendranthema grandiflora*. *Transgenic Research* 11, 437-445.

Annadana S, Mlynárová L, Udayakumar M, de Jong J, Nap J-P (2002b) The potato *Lhca3.St.1* promoter confers high and stable transgene expression in chrysanthemum, in contrast to CaMV-based promoters. *Molecular Breeding* 8, 335-344.

Aragão FJL (2003) Genetic transformation of common bean via particle bombardment. In: Jaiwal PK, Singh RP (eds) *Applied Genetics of Leguminosae Biotechnlogy*, Kluwer Academic Publishers, Dordrecht, The Netherlands, pp 35-45.

Aragão FJL, Brasileiro ACM (2002) Positive, negative and marker-free strategies for transgenic plant selection. *Brazilian Journal of Plant Physiology* 14, 1-10.

Aragão FJL, Sarokin L, Vianna GR, Rech EL (2000) Selection of transgenic meristematic cells utilizing a herbicidal molecule in the recovery of fertile transgenic soybean (*Glycine max* (L) Cerril) plants at a high frequency. *Theoretical and Applied Genetics* 101, 1-6.

Aswath CR, Mo SY, Kim S-H, Kim DH (2004) *IbMADS4* regulates the vegetative shoot development in transgenic chrysanthemum (*Dendrathema grandiflora* (Ramat.) Kitamura) *Plant Science* 166, 847-854.

Barrueto Cid LP, Cruz ARR, Castro LHR (2004) Somatic embryogenesis from three coffee cultivars: 'Rubi', 'Catuai Vermelho 81', and 'IAPAR 59'. *HortScience* 39, 130-131.

Belarmino MM, Mii M (2000) *Agrobacterium*-mediated genetic transformation of a phalaenopsis orchid. *Plant Cell Reports* 19, 435-442.

Beruto M, Lanteri L, Portogallo C (2004) Micropropagation of tree peony (*Paeonia suffruticosa*). *Plant Cell, Tissue and Organ Culture* 79, 249-255.

Chae WB, Chai GW, Chung IS (2004) Plant regeneration depending on explant type in *Chysanthemum coronarium* L. *Journal of Plant Biotechnology* 6, 253-258.

Chai D, Yu H (2007) Recent advances in transgenic orchid production. *Orchid Science and Biotechnology* 1, 34-39.

Chandler SF (2003) Commercialization of genetically modified ornamental plants. *Journal of Plant Biotechnology* 5, 69-77.

Chandler SF, Lu C-Y (2005) Biotechnology in ornamental horticulture. *In Vitro Cellular and Developmental Biology - Plant* 41, 591-601.

Chang H, Jones ML, Banowetz GM, Clark DG (2003) Overproduction of cytokinins in petunia flowers transformed with P_{SAG12}-IPT delays corolla senescence and decreases sensitivity to ethylene. *Plant Physiology* 132, 2174-2183.

Chen FC, Kuchnle AR, Sugü N (1997) *Anthurium* roots for micropropagation and *Agrobacterium tumefasciens* mediated gene transfer. *Plant Cell, Tissue and Organ Culture* 49, 71-74.

Chen JT, Chang C, Chang WC (1999) Direct somatic embryogenesis on leaf explants of *Oncidium gower* Ransey and subsequent plant regeneration. *Plant Cell Reports* 19, 143-149.

Chen JT, Chang WC (2003) Effects of GA_3, ancymidol, cycocel and paclobutrazol on direct somatic embryogenesis of *Oncidium in vitro*. *Plant Cell, Tissue and Organ Culture* 72, 105-108.

Chen L, Kawai H, Oku T, Takahashi C, Niimi Y (2006) Introduction of *Odontoglossum ringspot virus* coat protein gene from ORSV-infected *Epidendrum* into *Cymbidium niveo-marginum* using *A. tumefaciens*. *Journal of the Japanese Society for Horticultural Science* 75, 249-255.

Chen Y, Liu X, Liu Y (2005) *In vitro* plant regenetion from the immature seeds of *Cymbidium faberi*. *Plant Cell, Tissue and Organ Culture* 81, 247-251.

Chen Y, Lu L, Deng W, Yang X, McAvoy R, Zhao D, Pei Y, Luo K, Duan H, Smith W, Thammina C, Zheng X, Ellis D, Li Y (2006) *In vitro* regeneration and

Agrobacterium-mediated genetic transformation of *Euonymus alatus*. *Plant Cell Reports* 25, 1043-1051.

Chia TF, Chan YS, Chua NH (1994) The firefly luciferase gene as a non-invasive reporter for *Dendrobium* transformation. *Plant Journal* 6, 441-446.

Chumakov MI (2007) *Agrobacterium*-mediated plant transformation under *in planta* conditions. *Transgenic Plant Journal* 1, 60-65.

Cui M, Ezura H (2003) *Agrobacterium*-mediated transformation of *Nemesia strumosa* Benth, a model plant for asymmetric floral development. *Plant Science* 165, 863-870.

Cui M, Ezura H, Nishimura S, Kamada H, Handa T (2004) A rapid *Agrobacterium*-mediated transformation of *Antirrhinum majus* L. by using direct shoot regeneration from hypocotyl explants. *Plant Science* 166, 873-879.

Cui M, Handa T, Ezura H (2003) An improved protocol for *Agrobacterium*-mediated transformation of *Antirrhinum majus* L. *Molecular and General Genomics* 270, 296-302.

Cui M, Takayanagi K, Kamada H, Nishimura S, Handa T (2000) Transformation of *Antirrhinum majus* L. by a *rol*-type multi-auto-transformation (MAT) vector system. *Plant Science* 159, 273-280.

Cui M, Takayanagi K, Kamada H, Nishimura S, Handa T (2001) Efficient shoot regeneration from hairy roots of *Antirrhinum majus* L. transformed by the rol type MAT vector system. *Plant Cell Reports* 20, 55-59.

Cui ML, Takayanagi K, Handa T (2004) High frequency of shoot regeneration from hypocotyls and stem segments of *Antirrhinum majus* (Snapdragon). *Plant Cell, Tissue and Organ Culture* 78, 51-53.

Davies KM, Bloor SJ, Spiller GB, Deroles SC (1998) Production of yellow colour in flower: redirection of flavonoid biosynthesis in petunia. *Plant Journal* 13, 259-266.

Deroles SC, Bradley JM, Schwinn KE, Markham KR, Bloor S, Manson DG, Davies KM (1998) An antisense chalcone synthase cDNA leads to novel colour patterns in lisianthus (*Eustoma grandiflorum*) flowers. *Molecular Breeding* 4, 59-66.

Deroles SC, Ledger SE, Miller RM, Davies KM, Given NK (1993) Transformation in *Eustoma grandiflorum* (Lisianthus). In: Bajaj YPS (Ed), *Biotechnology in Agriculture and Forestry, Plant Protoplasts and Genetic Engineering III* (Vol 22) Springer-Verlag, Berlin, pp 202-212.

Dias MC, Almeida R, Romano A (2002) Rapid clonal multiplication of *Lavandula viridis* L'Hér through *in vitro* axillary shoot proliferation. *Plant Cell, Tissue and Organ Culture* 68, 99-109.

Dirks RW, Tanke HJ (2006) Advances in fluorescent tracking of nucleic acids in living cells. *BioTechniques* 40, 489-496.

Dobres MS (2008) Barriers to genetically engineered ornamentals: an industry perspective. In: Teixeira da Silva JA (Ed) Floriculture, Ornamental and Plant Biotechnology: Advances and Topical Issues, Global Science Books, Isleworth, UK, Volume V, in press.

Ebinuma H, Sugita K, Matsunaga E, Yamakado M (1997) Selection of marker-free transgenic plants using the isopentenyl transferase gene. *Proceedings of the National Academic of Sciences USA* 99, 2117-2121.

Firoozabady E, Moy Y, Courtney-Gutterson N, Robinson K (1994) Regeneration of transgenic rose (*Rosa hybrida*) plants from embryogenic tissue. *Bio/Technology* 12, 609-613.

Forkmann G, Heller W (1999) Biosynthesis of flavonoids. In: Sankawa U (ed) *Polyketides and Other Secondary Metabolites Including Fatty Acid and Their Derivatives*, Elsevier, Amsterdam, pp 713-748.

Forkmann G, Martens S (2001) Metabolic engineering and applications of flavonoids. *Current Opinion in Biotechnology* 12, 155-160.

Godoy-Hernández G, Miranda-Ham ML (2007) Marigold biotechnology: tissue culture and genetic transformation. *Transgenic Plant Journal* 1, 169-174.

Haensch KT (2004) Thidiazuron-induced morphogenetic response in petiole cultures of *Pelargonium* x *hortorum* and *Pelargonium* x *domesticum* and its histological analysis. *Plant Cell Reports* 23, 211-217.

Hahn EJ, Paek KY (2005) Multiplication of chysenthemum shoots in bioreactor as affected by culture method and inoculation density of single node stems. *Plant Cell, Tissue and Organ Culture* 81, 301-306.

Haseloff J, Amos B (1995) GFP in plants. *Trends in Genetics* 11, 328-329.

Haseloff J, Siemering KR, Prasher DC, Hodge S (1997) Removal of a cryptic intron and subcellular localization of green fluorescent protein are required to mark transgenic *Arabidopsis* plants brightly. *Proceedings of the National Academy of Sciences USA* 94, 2122-2127.

Heidmann I, Efremova N, Saedler H, Schwarz-Sommer Z (1998) A protocol for transformation and refeneration of *Antirrhinum majus*. *Plant Journal* 13, 723-728.

Heim R, Prasher DC, Tsien RY (1994) Wavelength mutations and posttranslational autoxidation of green fluorescent protein. *Proceedings of the National Academy of Sciences USA* 91, 12501-12504.

Hoshi Y, Kondo M, Mori S, Adachi Y, Nakano M, Kobayashi H (2004) Production of transgenic lily plants by *Agrobacterium*-mediated transformation. *Plant Cell Reports* 22, 359-364.

Hoshino Y (2008) Advances in *Alstroemeria* biotechnology. In: Teixeira da Silva JA (Ed) Floriculture, Ornamental and Plant Biotechnology: Advances and Topical Issues, Global Science Books, Isleworth, UK, Volume V, in press.

Hosokawa K, Matsuki R, Oikawa Y, Yamamura S (2000) Production of transgenic gentian plants by particle bombardment of suspension-culture cells. *Plant Cell Reports* 19, 454-458.

Hosoki T, Kajino E (2003) Shoot regeneration from petioles of coral bells (*Heuchera sanguinea* Engelm.) cultured *in vitro*, and subsequent planting and flowering *ex vitro*. *In Vitro Cellular and Developmental Biology Plant* 39, 135-138.

Hsiao P-Y, Sanjaya Su R-C, Teixeira da Silva JA, Chan M-T (2007) Plant native tryptophan synthase beta 1 gene is a non-antibiotic selection marker for plant transformation. *Planta* 225, 897-906.

Hsiao Y-Y, Tsai W-C, Chen W-H, Chen H-H (2008) Floral scent in *Phalaenopsis*. In: Teixeira da Silva JA (Ed) Floriculture, Ornamental and Plant Biotechnology: Advances and Topical Issues, Global Science Books, Isleworth, UK, Volume V, in press.

Jach G, Winter J (2006) Focus on fluorescent proteins. *Studies in Natural Product Chemistry* 33, 3-67.

Joung Y-H, Kamo K (2006) Expression of a polyubiquitin promoter isolated from *Gladiolus*. *Plant Cell Reports* 25, 1081-1088.

Jowkar MM, Sticklen MB (2008) Biotechnological applications in *Narcissus* research. In: Teixeira da Silva JA (Ed) Floriculture, Ornamental and Plant Biotechnology: Advances and Topical Issues, Global Science Books, Isleworth, UK, Volume V, in press.

Khan RS, Chin DP, Nakamura I, Mii M (2006) Production of marker-free transgenic *Nierembergia caerulea* using MAT vector system. *Plant Cell Reports* 25, 914-919.

Kinouchi T, Endo R, Yamashita A, Satoh S (2006) Transformation of carnation with genes related to ethylene production and perception: towards generation of potted carnations with a longer display time. *Plant Cell, Tissue and Organ Culture* 86, 27-35.

Kitahara K, Hibino Y, Ainda R, Matsumoto S (2004) Ectopic expression of the rose *AGAMOUS*-like MADS-box genes '*MASAKO C1* and *D1*' causes similar homeotic transformation of sepal and petal in *Arabidopsis* and sepal in *Torenia*. *Plant Science* 166, 1245-1252.

Knapp JE, Kausch AP, Auer C, Brand MH (2001) Transformation of *Rhododendron* through microprojectile bombardment. *Plant Cell Reports*20, 749-754.

Kojima M, Sparthana P, Teixeira da Silva JA, Nogawa M (2006) Development of *in planta* transformation methods using *Agrobacterium tumefaciens*. In: Teixeira da Silva JA (Ed) Floriculture, Ornamental and Plant Biotechnology: Advances and Topical Issues, Global Science Books, Isleworth, UK, Volume II, pp 41-48.

Kuehnle AR, Sugii N (1992) Transformation of *Dendrobium* orchid using particle bombardment of protocorms. *Plant Cell Reports* 11, 484-488.

Lavy M, Zuker A, Lewinsohn E, Larkov O, Ravid U, Vainstein A, Weiss D (2002) Linalool and linalool oxide production in transgenic carnation flowers expressing the *Clarkia breweri* linalool synthase gene. *Molecular Breeding* 9, 103-111.

Lewinsohn E, Schalechet F, Wilkinson J, Matsui K, Tadmor K, Nam KH, Amar O, Lastochkin E, Larkov O, Ravid U, Hiatt W, Gepstein S, Pichersky E (2001) Enhanced levels of the aroma and flavor compound S-linalool by metabolic engineering of the terpenoid pathway in tomato fruits. *Plant Physiology* 127, 1256-1265.

Li H-Q, Kang P-J, Li M-L, Li M-R (2007) Genetic transformation of *Torenia fournieri* using the PMI/mannose selection system. *Plant Cell, Tissue and Organ Culture* 90, 103-109.

Liau C-H, Lu J-C, Prasad V, Hsiao H, You S-J, Lee J, Yang N-S, Huang H-E, Feng T-Y, Chen W-H, Chan M-T (2003a) The sweet pepper ferredoxin-like protein (*pflp*) conferred resistance against soft rot disease in *Oncidium* orchid. *Transgenic Research* 12, 329-336.

Liau C-H, You S-J, Prasad V, Hsiao H-H, Lu J-C, Yang N-S, Chan M-T (2003a) *Agrobacterium tumefaciens*-mediated transformation of an *Oncidium* orchid. *Plant Cell Reports* 21, 993-998.

Lim JD, Hahn SJ, Yu CY, Chung IM (2005) Expression of the glutathione *S*-transferase gene (*NT107*) in transgenic *Dianthus superbus*. *Plant Cell, Tissue and Organ Culture* 80, 277-286.

Lücker J, Bouwmeester HJ, Schwab W, Blaas J, van der Plas LHW, Verhoeven HA (2001) Expression of *Clarkia S*-linalool synthase in transgenic petunia plants results in the accumulation of S-linalyl-β-D-glucopyranoside. *Plant Journal* 27, 315-324.

MacCartan SA, van Staden J (2003) Micropropagation of the endangered *Kniphofia leucocephala* Baijnath. *In Vitro Cellular and Developmental Biology - Plant* 39, 496-499.

Mahadtanapuk S, Topoonyanont N, Handa T, Sanguansermsri M, Anuntalabhochai S (2006) Genetic transformation of *Curcuma alismatifolia* Gagnep. using retarded shoots. *Plant Biotechnology* 23, 233-237.

Malabadi RB, Teixeira da Silva JA, Nataraja K (2008) Green fluorescent protein as a novel marker during genetic transformation of plants. *Functional Plant Science and Biotechnology* in press.

Marinangeli P, Dalmastro S, Curvetto N (1998) Influence of DB-cAMP, adenosine, forskolin, and traumatic acid, on *in vitro* bulbing of *Lillium longiflorum*. *HortScience* 33, 151-152.

Martens S, Forkmann G, Britsch L, Wellmann F, Matern U, Lukacin R (2003a) Divergent evolution of flavonoid 2-oxoglutarate-dependent dioxygenases in parsley. *FEBS Letters* 544, 93-98.

Martens S, Knott J, Seitz C, Janvari L, Yu S-N, Forkmann G (2003b) Impact of biochemical pre-studies on specific metabolic engineering strategies of flavonoid biosynthesis in plant tissues. *Biochemical Engineering Journal* 14, 227-235.

Men S, Ming X, Wang Y, Liu R, Wei C, Li Y (2003a) Genetic transformation of two species of orchid by biolistic bombardment. *Plant Cell Reports* 21, 592-598.

Men S, Ming X, Liu R, Wei C, Li Y (2003b) *Agrobacterium*-mediated trans-formation of a *Dendrobium* orchid. *Plant Cell, Tissue and Organ Culture* 75, 63-71.

Mercuri A, Sacchetti A, De Benedetti L, Schiva T, Alberti S (2001) Green fluorescent flowers. *Plant Science* 161, 961-968.

Mercuri A, Sacchetti A, De Benedetti L, Schiva T, Alberti S (2002) Green fluorescent flowers. *Plant Science* 162, 647-654 (*erratum*).

Meyer P, Heidemann I, Forkmann G, Saedler H (1987) A new petunia flower colour generated by transformation of a mutant with a maize gene. *Nature* 330, 677-678.

Mirici S, Parmakis Y, Ozcan S, Sancak C, Uranbery S, Sarihan O, Gümüscü A, Gürbüz B, Arslan N (2005) Efficient *in vitro* bulblet regeneration of immature embryos of endangered *Sternbergia fischeriana*. *Plant Cell, Tissue and Organ Culture* 80, 239-246.

Mishiba K, Chin DP, Mii M (2005) *Agrobacterium*-mediated transformation of *Phalaenopsis* by targeting protocorms at an early stage after germination. *Plant Cell Reports* 24, 297-303.

Mishiba K, Nishihara M, Abe Y, Nakatsuka T, Kawamura H, Kodama K, Takesawa T, Abe J, Yamamura S (2006) Production of dwarf potted gentian using wild-type *Agrobacterium rhizogenes*. *Plant Biotechnology* 23, 33-38.

Misra P, Srivastava AK (2004) Molecular investigation in ornamental floricultural plants. *Journal of Plant Biotechnology* 6, 131-143.

Miyoshi K, Asakura N (1996) Callus induction, regeneration of haploid plants chromosome doubling in ovule cultures of pot gerbera (*Gerbera jamesonii*). *Plant Cell Reports* 16, 1-5.

Miyoshi K, Mii M (1998) Stimulatory effects of sodium and calcium hypochlrite, pre-chilling and cytokinins on the germination of *Cypripedium macranthos* seed *in vitro*. *Physiologia Plantarum* 102, 481-486.

Mohapatra A, Rout GR (2005) Study of embryo rescue in floribunda rose. *Plant Cell, Tissue and Organ Culture* 81, 113-117.

Mol J, Grotewold E, Koes R (1998) How genes paint flowers and seeds. *Trends in Plant Sciences* 3, 212-217.

Mol JNM, Holton TA, Koes RE (1995) Floriculture: genetic engineering of commercial traits. *Trends in Biotechnology* 13, 350-355.

Mori S, Kobayashi H, Hoshi Y, Kondo M, Nakano M (2004) Heterologous expression of the flavonoid 3′,5′-hydroxylase gene of *Vinca major* alters flower color in transgenic *Petunia hybrida*. *Plant Cell Reports* 22, 415-421.

Murashige T, Skoog F (1962) A revised medium for rapid growth and bioassays with tobacco tissue cultures. *Physiologia Plantarum* 15, 473-479.

Nielsen K, Deroles SC, Markham KR, Bradley MJ, Podivinsky E, Manson D (2002) Antisense flavonol synthase alters copigmentation and flower color in lisianthus. *Molecular Breeding* 9, 217-229.

Nishihara M, Nakatsuka T, Mizutani-Fukuchi M, Tanaka Y, Yamamura S (2008) Gentians: from gene cloning to molecular breeding. In: Teixeira da Silva JA (Ed) Floriculture, Ornamental and Plant Biotechnology: Advances and Topical Issues, Global Science Books, Isleworth, UK, Volume V, in press.

Ono E, Nakayama T (2007) Molecular breeding of novel yellow flowers by engineering the aurone biosynthetic pathway. *Transgenic Plant Journal* 1, 66-80.

Oud JSN, Schneiders H, Kool AJ, van Grinsven MQJM (1995) Breeding of transgenic orange *Petunia hybrida* varieties. *Euphytica* 84, 175-181.

Pédelacq J-D, Cabantous S, Tran T, Terwilliger TC, Waldo GS (2006) Engineering and characterization of a superfolder green fluorescent protein. *Nature Biotechnology* 24, 79-88.

Petty LM, Harberd NP, Carré IA, Thomas B, Jackson SD (2003) Expression of the *Arabidopsis gai* gene under its own promoter causes a reduction in plant height in chrysanthemum by attenuation of the gibberellin response. *Plant Science* 164, 175-182.

Popowich EA, Firsov AP, Mitiouchkina TY, Filipenya VL, Dolgov SV, Reshetnikov VN (2007) *Agrobacterium*-mediated transformation of *Hyacinthus orientalis* with thaumatin II gene to control fungal diseases. *Plant Cell, Tissue and Organ Culture* 90, 237-244.

Ran Y, Simpson S (2005) *In vitro* propagation of genus *Clivia*. *Plant Cell, Tissue and Organ Culture* 81, 239-242.

Rosati C, Cadic A, Renou JP, Duron M (1996) Regeneration and *Agrobacterium*-mediated transformation of *Forsythia intermedia* "Spring Glory". *Plant Cell Reports* 16, 114-117.

Rosati C, Simoneau P, Treutter D, Poupard P, Cadot Y, Cadic A, Duron M (2003) Engineering of flower color in forsythia by expression of two independently-transformed dihydroflavonol 4-reductase and anthocyanidin synthase genes of flavonoid pathway. *Molecular Breeding* 12, 197-208.

Sacchetti A, Alberti S (1999) Protein tags enhance GFP folding in eukaryotic cells. *Nature Biotechnology* 17, 1046.

Sacchetti A, Ciccocioppo R, Alberti S (2000) The molecular determinants of the efficiency of green fluorescent protein mutants. *Histology and Histopathology* 15, 101-107.

Satoh S, Kosugi Y, Iwazaki Y, Narumi T, Kinouchi T (2006) Genetic engineering of ethylene production and perception in carnation and chrysanthemum. In: Teixeira da Silva JA (Ed) Floriculture, Ornamental and Plant Biotechnology: Advances and Topical Issues, Global Science Books, Isleworth, UK, Volume II, pp 164-171.

Savin KW, Baudinette SC, Graham MW, Michael MZ, Nugent GD, Lu C-Y, Chandler SF, Cornish EC (1995) Antisense ACC oxidase RNA delays carnation petal senescence. *HortScience* 30, 970-972.

Saxena G, Banerjee S, Laiq-ur-Rahman, Verma PC, Mallavarapu GR, Kumar S (2007) Rose-scented geranium (*Pelargonium* sp.) generated by *Agrobacterium rhizogenes* mediated

Ri-insertion for improved essential oil quality. *Plant Cell, Tissue and Organ Culture* 90, 215-223.

Shaw J-F, Chen H-H, Tsai M-F, Kuo C-I, Huang L-C (2002) Extended flower longevity of Petunia hybrida plants transformed with *boers*, a mutated *ERS* gene of *Brassica oleracea. Molecular Breeding* 9, 211-216.

Shirgurkar MV, Naik VB, von Arnold S, Nadgauda RS, Clapham D (2006) An efficient protocol for genetic transformation and shoot regeneration of turmeric (*Curcuma longa* L.) via particle bombardment. *Plant Cell Reports* 25, 112-116.

Siemering KR, Golbik R, Sever R, Haseloff J (1996) Mutations that suppress the thermosensivity of green fluorescent protein. *Current Biology* 6, 1653-1663.

Sjahril R, Mii M (2006) High-efficiency *Agrobacterium*-mediated transformation of *Phalaenopsis* using meropenem, a novel antibiotic to eliminate *Agrobacterium*. *Journal of Horticultural Science and Biotechnology* 81, 458-464.

Sjahril R, Chin DP, Khan RS, Yamamura S, Nakamura I, Amemiya Y, Mii M (2006) Transgenic *Phalaenopsis* plants with resistance to *Erwinia carotovora* produced by introducing wasabi defensin gene using *Agrobacterium* method. *Plant Biotechnology* 23, 191-194.

Smith RH, Norris RE (1983) *In vitro* propagation of African violet chimeras. *HortScience* 18, 436-437.

Springob K, Nakajima J, Yamazaki M, Saito K (2003) Recent advances in the biosynthesis and accumulation of anthocyanins. *Natural Product Reports* 20, 288-303.

Stewart NC Jr (2006) Go with the glow: fluorescent proteins to light transgenic organisms. *Trends in Biotechnology* 24, 155-162.

Stummer BE, Smithb SE, Langridge P (1995) Genetic transformation of *Verticordia grandis* (Myrtaceae) using wild-type *Agrobacterium rhizogenes* and binary *Agrobacterium* vectors. *Plant Science* 111, 51-62.

Tamura M, Togami J, Ishiguro K, Nakamura N, Katsumoto Y, Suzuki K, Kusumi T, Tanaka Y (2003) Regeneration of transformed verbena (*Verbena* × *hybrida*) by *Agrobacterium tumefaciens*. *Plant Cell Reports* 21, 459-466.

Tan R-F, Tao J, Li L (China) Genetic transformation of *Ipomoea purpurea* mediated by *Agrobacterium rhizogenes*. *Floriculture and Ornamental Biotechnology* 1, 131-135.

Tanaka Y, Katsumoto Y, Brugliera F, Mason J (2005) Genetic engineering in floriculture. *Plant Cell, Tissue and Organ Culture* 80, 1-24.

Tanaka Y, Tsuda S, Kusumi T (1998) Metabolic engineering to modify flower color. *Plant and Cell Physiology* 39, 1119-1126.

Tanase K, Onozaki T, Satoh S, Shibata M, Ichimura K (2008) Differential expression levels of ethylene biosynthetic pathway genes during senescence of long-lived carnation cultivars. *Postharvest Biology and Technology* 47, 210-217.

Teixeira da Silva JA (2004) Ornamental chrysanthemums: improvement by biotechnology. *Plant Cell, Tissue and Organ Culture* 79, 1-18.

Teixeira da Silva JA (2005) Simple multiplication and effective genetic transformation (4 methods) of *in vitro*-grown tobacco by stem thin cell layers. *Plant Science* 169, 1046-1058.

Teixeira da Silva JA (Ed) (2006) Floriculture, Ornamental and Plant Biotechnology: Advances and Topical Issues. Global Science Books, Isleworth, UK, Volumes I-IV, 2551 pp.

Teixeira da Silva JA (Ed) (2008) Floriculture, Ornamental and Plant Biotechnology: Advances and Topical Issues. Global Science Books, Isleworth, UK, Volume V, in press.

Teixeira da Silva JA, Fukai S (2003) Four gene introduction methods affect the shoot regeneration and localization of transgene expression in greenhouse stem explants and *in vitro*-grown chrysanthemum stem thin cell layers. *African Journal of Biotechnology* 2, 114-123.

Teixeira da Silva JA, Tanaka M (2008a) Optimization of particle bombardment conditions for *Cymbidium* hybrids. *Transgenic Plant Journal* 2, in press.

Teixeira da Silva JA, Tanaka M (2008b) Effect of initial explant on efficiency of biolistically-transformed *Cymbidium* hybrids. *Orchid Science and Biotechnology* 2, in press.

Teixeira da Silva JA, Tran Thanh Van K, Biondi S, Nhut DT, Altamura MM (2007) Thin Cell Layers: developmental building blocks in ornamental biotechnology. *Floriculture and Ornamental Biotechnology* 1, 1-13.

Torné JM, Rodriguez P, Manich A, Claparols I, Santos MA (1997) Embryogenesis induction in petals of *Araujia sericifera. Plant Cell, Tissue and Organ Culture* 51, 95-102.

Tsuda S, Fukui Y, Nakamura N, Katsumoto Y, Yonekura-Sakakibara K, Fukuchi-Mizutani M, Ohira K, Ueyama Y, Ohkawa H, Holton TA, Kusumi T, Tanaka Y (2004) Flower color modification of *Petunia hybrida* commercial varieties by metabolic engineering. *Plant Biotechnology* 21, 377-386.

Tsugawa H, Kagami T, Suzuki M (2004) High-frequency transformation of *Lobelia erinus* L. by *Agrobacterium*-mediated gene transfer. *Plant Cell Reports* 22, 759-764.

Tzfira T, Citovsky V (2006) *Agrobacterium*-mediated genetic transformation of plants: biology and biotechnology. *Current Opinion in Biotechnology* 17, 147-154.

Vainstein A, Lewinsohn E, Pichersky E, Weiss D (2001) Floral fragrance. New inroads into an old commodity. *Plant Physiology* 127, 1383-1389.

Vinterhalter B, Ninković S, Cingel A, Vinterhalter D (2006) Shoot and root culture of *Hypericum perforatum* L. transformed with *Agrobacterium rhizogenes* A4M70GUS. *Biologia Plantarum* 50, 767-770.

Wang Z-Y, Ge Y (2006) Recent advances in genetic transformation of forage and turf grasses. *In Vitro Cellular and Developmental Biology - Plant* 42, 1-18.

Yang J, Lee HJ, Shin DH, Oh SK, Seon JH, Paek KY, Han K-H (1999) Genetic transformation of *Cymbidium* orchid by particle bombardment. *Plant Cell Reports* 18, 978-984.

Zuker A, Ahroni A, Tzfira T, Ben-Meir H, Vainstein A (1999) Wounding by bombardment yields highly efficient *Agrobacterium*-mediated transformation of carnation (*Dianthus caryophyllus* L.). *Molecular Breeding* 5, 367-375.

Zuker A, Tzfira T, Ben-Meir H, Ovadis M, Shklarman E, Itzhaki H, Forkmann G, Martens S, Neta-Sharir I, Weiss D, Vainstein A (2002) Modification of flower color and fragrance by antisense suppression of the flavanone 3-hydroxylase gene. *Molecular Breeding* 9, 33-41.

Reviewed by D.T. Nhut, Institute of Biology, Vietnam.

In: Transgenic Plants: New Research
Editor: Oliver T. Chan

ISBN 978-1-60692-017-6
© 2009 Nova Science Publishers, Inc.

Chapter 3

TRANSGENIC PLANTS WITH ALTERNATION IN LIGNIN BIOSYNTHESIS

Qing-Hu Ma
Key Laboratory of Photosynthesis and Environmental Molecular Physiology, Institute of Botany, Chinese Academy of Sciences, Beijing 100093, China

ABSTRACT

Lignin is complex, three-dimensional polymer existing in plant cell wall. Lignin biosynthesis has become increasingly highlighted because it plays the important roles in plant growth and development as well as for applications in the human life. In this brief review, the progress in the lignin biosynthesis has been discussed. The genetic engineering approaches on lignin biosynthesis are reviewed. The transgenic plants with alternation on lignin biosynthesis and their applications in agriculture and industry productions are presented, particularly in pulping process, forage digestibility, and bio-fuel. Based on the information generated to date, implications and future research directions are discussed.

Keywords: Lignin biosynthesis; Genetic manipulation; Pulping process; Forage digestibility; Bio-fuel

1. INTRODUCTION

Lignin is a major structural compound of cell wall in vascular plants (pteridophytes, gymnosperms and angiosperms), being incorporated into the cell wall together with the process of thickening plant secondary wall. Distinct from other macromolecules such as protein and cellulose in which all subunits are linked *via* regular bonds, lignin is derived from dehydrogenated polymerization of three main monolignols, named *p*-coumaryl alcohol, coniferyl alcohol and sinapyl alcohol. Those monolignols are all hydroxycinnamyl alcohols different from each other only in the degree of hydroxylation and methylation. Sinapyl alcohol has two methyols, and coniferyl alcohol has one methyol while *p*-coumaryl alcohol has no methyol group. These three monolignols are linked together by polymerization to form high molecular lignin *via* various kinds of intermonomeric linkages (Boerjan et al., 2003). There are 16 kinds of intermonomeric linkages in theory, of which the most frequent inter-unit linkage is the C_β–O–C_4 (beta-aryl ether) linkage, it is also the one most easily cleaved chemically. The other common linkages are C–C aryl linkage, including C_β–C_5, C_β–C_1, C_β–C_β and C_5–C_5, which are all more resistant to chemical degradation. This leads to form hydroxyphenyl (H), guaiacyl (G) and syringyl (S) units in lignin polymers, respectively (Figure 1). Other uncommon phenylpropanoid ramifications, for example, coniferaldehyde, 5-hydroxyconiferaldehyde and coumaraldehyde, are also found to be incorporated into lignin, but the percentage is low (Ralph et al., 1997; Ralph et al., 2001).

Lignin is principally deposited in the secondary wall of certain plant tissues, including xylem, sclerenchyma, phloem fiber and periderm, which are involved in mechanical support of plant stem and hydrophobicity protection of plant cell wall. Some parenchyma cells may also have the lignin deposition in primary wall but the degree of lignifications is low. Lignin is linked to cellulose, hemicellulose and other cell wall polymers to form extracellular matrix, this structure increases the mechanical intensity and supportable ability of plant tissues. It contributes rigidity and compressive strength to plant stem, which is also related to the lodging-resistant phenotype in crop plants such as wheat, rice and maize (Ma, 2007). Because the natural property of hydrophobicity, lignin imparts water impermeable to plant cells and tissues. This function is very important, not only for xylem and phloem to transport water and mineral components but also for the successful colonization of land by plants. In fact, the terrestrial vascular plants are proposed to evolve on the earth by the concomitant evolution of lignin biosynthesis. Therefore, lignin is a substance unique to vascular plants such as pteridophytes, gymnosperms and angiosperms (Peter Neale, 2004). In addition, lignin accumulation in the cell wall forms a physically structural barrier to effectively protect plant from pathogens and lignin synthesis will be induced in response to various kinds of abiotic and biotic stresses such as pathogen infection and wounding (Dixon, 2001).

Second to only cellulose, lignin is the most abundant biopolymer on earth, which accounts for about 30% of carbon (about 1.4×10^{12} kg) fixed into terrestrial plants each year (Battle et al., 2000). Lignin content varies among different kinds of plants. In trees, lignin content represents 27~32% of the plant dry weight, while it only represents approximately 14~25% of dry weight in herbaceous plants. As a bio-undegradable biopolymer in nature, lignin is also intimately related to human life. Lignin content and composition are a limiting factor that affects not only the quality of paper, but also the digestibility of forage crops. In pulping process, lignin is undesirable component that must be removed by expensive and environmentally hazardous processes requiring large amounts of energy and chemicals in the chemical process, which may lead to severe environmental pollution. For the paper-making industry, it would be beneficial to process wood with either less lignin or lignin with an altered chemical reactivity. Lignin content and composition are negatively corrected with forage digestibility for ruminant animals in many forage crops, and improving digestibility is an important goal of forage. Lignin is also an important determinant for bio-fuel production. The value of lignin depends on its final use in energy crops. Due to its high heat value, lignin is desirable for conversion by gasification or pyrolysis. Conversely, enzymatic saccharification of lignocellulose for production of ethanol is inhibited by lignin-derived phenolics (Chen and Dixon, 2007).

Because of important roles of lignin both in plant development and in applications of agriculture and industry, there is considerable attention to lignin biosynthesis in recent decade. As the utilization of molecular biology and genetic manipulation in the area of lignin research, understanding of lignin biosynthesis has become deepened. The transgenic plants with altered lignin metabolism, when combined with molecular breeding process, has shown very promising potential for application in the future.

2. LIGNIN BIOSYNTHESIS

Lignin biosynthesis is a complex biochemical pathway in plant which is involved in deamination of L-phenylalanine into cinnamic acid, then a series of hydroxylation and methylation reactions convert cinnamic acid into a variety of hydroxycinnamic acids. These hydroxycinnamic acids act as precursors not only for lignin but also flavonoids. Activation of hydroxycinnamic acids to their corresponding co-enzyme A (CoA) thioesters, followed by successive reductions produce three monolignols (Figure 1) which is thought to build into the lignin polymer by polymerization. This conventional model of lignin biosynthesis, introduced in 1980's, has much modification after extensive research in biochemistry, molecular biology and genetics. The important revisions have been reviewed below.

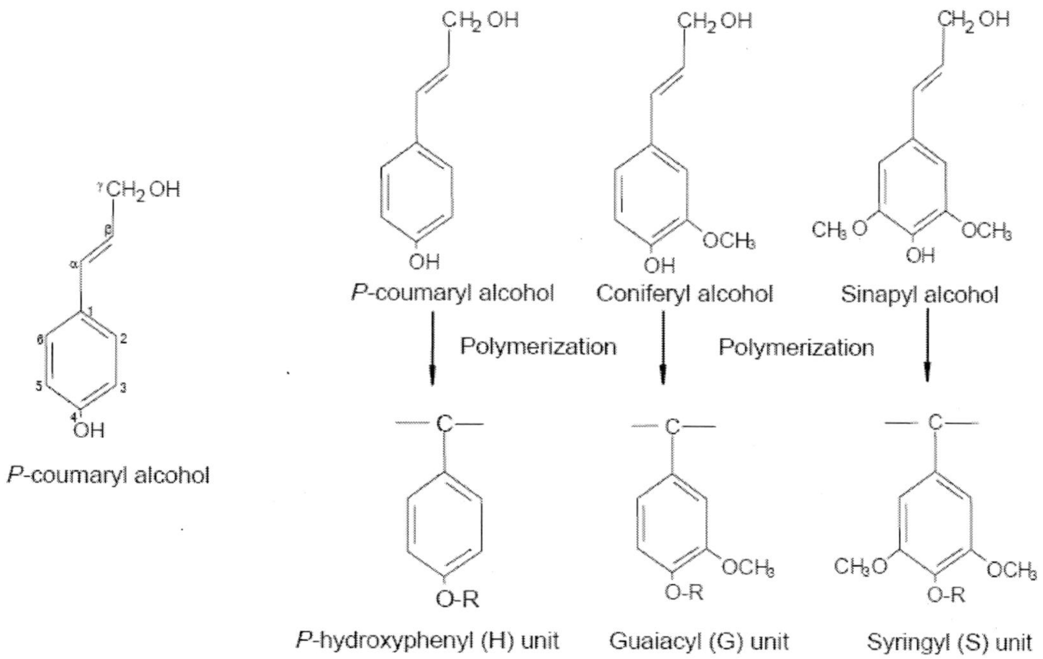

Figure 1. The structure of three monolignols and corresponding *p*-hydroxyphenyl (H), guaiacyl (G) and syringyl (S) units in lignin.

Ferulate 5-hydroxylase (F5H) is a cytochrome-P450-dependent monooxygenase that was originally thought to catalyze the hydroxylation at the C_5 position of ferulic acid to form 5-hydroxyferulic acid, which is the precursor to S lignin. However, a serial of work from the transgenic plants, feeding tests and *in vitro* biochemistry demonstrated that F5H actually used coniferaldehyde and coniferyl alcohol to form 5-hydro-coniferaldehyde and 5-hydro-coniferyl alcohol then syringyl alcohol, and over-expression of F5H in transgenic tobacco and poplar gave rise to lignin that was composed of almost S units (Meyer et al., 1996; Chen et al., 1999; Humphreys et al., 1999; Osakabe et al., 1999; Franke et al., 2000). Therefore, F5H is now also called coniferaldehyde 5-hydroxylase (CAld-5H).

Caffeoyl CoA 3-*O*-methyltransferase (CCoMT) is a new enzyme that has been demonstrated to convert *p*-coumaroyl CoA into feruloyl CoA, and down-regulation of CCoMT in transgenic tobacco lead to reduce S and G monomer production and dramatically lower lignin content, suggesting it is a critical enzyme to function in a route to monolignol synthesis (Zhong et al., 1998). To accompany with the identification of new CCoMT and novel F5H activities, the research has revealed the new functions for caffeic acid/5-hydroxyferulic acid *O*-methyltransferase (COMT). In dicotyledonous angiosperms, both COMT and CCoMT were initially thought to be a bi-specific enzyme that used both caffeic acid and 5-hydroxyferulic acid as substrate (Davin and Lewis, 1992). The biochemical analysis of recombinant COMT proteins and its related enzymes from *Arabidopsis* (Humphreys et al., 1999), sweetgum (Osakabe et al., 1999), poplar (Li et al., 2000), and alfalfa (Parvathi et al., 2001) have been demonstrated that COMT has the higher activity toward 5-hydroxyconiferaldehyde and 5-hydroxyconiferyl alcohol. Therefore, COMT may immediately act after F5H, which is responsible to convert coniferaldehyde and coniferyl alcohol to sinapaldehyde and sinapyl alcohol, respectively. Taken together, F5H and COMT likely act on hydroxylation and methylation at the levels of aldehydes and alcohols, not previously described at free acid level. This result is consistent with transgenic tobacco and poplar in which reduced COMT expression leads to a marked reduction in S lignin content but has little impact on total lignin (Atanassova et al., 1995; Van Doorsselaere et al., 1995), although this has some variation in the transgenic alfalfa (Guo et al., 2001a). It suggests that COMT may also use caffeoyl aldehyde and caffeoyl alcohol that position up F5H. From those results, it is noted that COMT no longer has a function that is redundant with CCoMT. However, a novel multifunctional OMT called AEOMT (acids and esters OMT) has been identified in loblolly pine which has approximately equal substrate preference for caffeic acid and caffeoyl CoA (Li et al., 1997). Our recent work has shown that a COMT from wheat namely TaCM exhibits not only the highest catalyzing efficiency towards caffeoyl aldehyde and 5-hydroxyconiferaldehyde, but also a strong activity towards ester precursors, including caffeoyl-CoA and 5-hydroxyferuloyl-CoA (Ma and Xu, 2008). It supports the notion that COMT has a broad substrate preference and may overlap with other OMT. To date, the biological significance of these potential functions warrants further investigation.

Characterization of *p*-coumarate 3-hydroxylase (C3H) marks a major progress to elucidate the branch pathways to G and S lignin from H lignin. Aid by *Arabidopsis* genomics, a possible C3H candidate, CYP983A, has been identified. Biochemical characterization showed that CYP983A actively converted the *p*-coumaroyl shikimate and quinate into their corresponding caffeoyl shikimic and quinic conjugates instead of acting on free 4-coumarate (Schoch et al., 2001). A mutant named *reduced epidermal fluorescence8* (*ref8*) in *Arabidopsis* has been showed to be CYP98A3 gene encoded the C3H (Franke et al., 2002a; Franke et al.,

2002b). Phenotypic analysis showed that the *ref8* mutant had greatly reduced lignin levels, and that remaining lignin was mainly from *p*-coumaryl alcohol instead of guaiacyl and syringyl alcohols that were generally found in wild-type plants. Down-regulation of C3H in the transgenic alfalfa has also showed to greatly increased the proportion of H units of lignin, suggesting the key role of C3H in regulation on the pathway toward coniferyl and sinapyl alcohols (Ralph et al., 2006). Accompanying with C3H, an enzyme namely quinate/shikimate hydroxycinnamoyl transferase (HCT) has also showed to function in monolignol biosynthesis. Since HCT reversible transfer quinate/shikimate moiety into coumaroyl and caffeoyl units, it has also called "shikimate ester shunt". (Hoffmann et al., 2003). Down-regulation of HCT in transgenic alfalfa resulted in strongly reduced lignin content and striking changes in lignin monomer composition, with predominant deposition of H units in the lignin. These transgenic plants exhibited impairment of vascular structure, significant stunting growth, reduction of biomass and delayed flowering, suggesting the critical role of HCT in lignin biosynthesis and plant development (Shadle et al., 2007). Cinnamate 4-hydroxylase (C4H) catalyzes the conversion of cinnamate into 4-hydroxy-cinnamate, a key reaction of the phenylpropanoid pathway which leads to the synthesis of several kinds of secondary metabolites. C4H genes exist as a multi-gene family in various plant species, each may be responsible for the different sub-branches of the phenylpropanoid pathway. Four C4H cDNAs (PtreC4H) were isolated from *Populus tremuloides* and their ability to convert trans-cinnamate into *p*-coumaric acid was verified by the recombinant PtreC4H proteins in yeast. The expression patterns of individual *Populus* C4H gene members were different from each other, which was associated with cis-acting regulatory elements such as box L, P and H in C4H promoters. It suggests that only some specific C4H isoforms are involved in the biosynthesis of guaiacyl and syringyl monolignols (Lu et al., 2006).

Cinnamoyl-CoA reductase (CCR) is responsible for the CoA ester to aldehyde conversion in monolignol biosynthesis, which diverts phenylpropanoid-derived metabolites into the biosynthesis of lignin. Transgenic approaches have demonstrated that CCR activity has profound influences on lignin synthesis and plant development. CCR down-regulated tobacco plants have a strong reduction in lignin content which is associated with a profound alteration of development (Piquemal et al., 1998). Only one type of CCR protein is found in *Eucalytus* (Lacombe et al., 1997), while at least there are two types of CCR which have the different functions in dicotyledonous (*Medicago truncatula* and *Arabidopsis*) and monocotyledonous (wheat, maize, ryegrass and sugarcane) plants. This can be retracted from the evolution of CCR genes. It is possible that multiple forms of CCR are needed to meet more diversity functions as the evolution from gymnosperms to angiosperms. This is consistent with the recent data for the lignin evolution (Peter and Neale, 2004). A xylem-specific CCR cDNA was cloned from aspen (*Populus tremuloides*) developing xylem tissue. The biochemical analysis of recombinant CCR protein demonstrated that CCR was coupled with CCoMT to catalyze the substrate caffeoyl-CoA ester into coniferaldehyde, suggesting that CCoMT and CCR are neighboring enzymes, although *in vivo* data are still lack (Li et al., 2005). In *Arabidopsis*, the mutant for CCR1 (Jones et al., 2001) or down-regulation of CCR gene expression (Goujon et al., 2003) resulted in the reduced lignin content accompanying with the dramatic influence on plant growth and development. Furthermore, the expression of *At-CCR2*, a CCR gene usually involved in plant defense, was increased in the *At-CCR1* mutants which could account for the lignin synthesis in the CCR1-knockout plants. These mutant plantlets also have three to four-times less sinapoyl malate and accumulate some feruloyl malate as well as more kaempferol glycosides. This suggests that redirection of lignin into flavonoid biosynthesis, reflecting the remarkable plasticity of the phenylpropanoid metabolism. It is also emphasizing the important role of CCR in lignin biosynthesis (Derikvand et al., 2008).

Cinnamyl alcohol dehydrogenase (CAD) is the enzyme responsible for the last catalytic step in the pathway to monolignols, i.e. the reduction of hydroxycinnamaldehydes into the corresponding alcohols. Aid by the progress of plant genomics, CAD has been fully characterized. Nine CAD genes are present in the *Arabidopsis* genome according to bioinformatic studies. These genes are organized in four classes according to phylogenetic analysis with CAD amino acid sequences from other plant species. Only three genes, namely *AtCAD-C*, *AtCAD-D* and *AtCAD-1*, have the major roles in lignin biosynthesis in accordance with their profile of expression (Sibout et al. 2003), substrate specificity (Kim et al. 2004), and mutation analysis (Sibout et al. 2005; Eudes et al., 2006). The data also reminds to be careful to use the database annotation since many genes may have the different functions as annotation labeled (Kim et al., 2007). A full research of CAD gene has also been facilitated by the recently completed rice genome sequence. Blast analysis revealed 12 distinct CAD genes in the rice genome, of which *OsCAD2* has showed to be an orthologue of the CAD gene present at the *bm1* (brown midrib 1) locus of maize. *OsCAD2* expression is associated with vascular tissues and is correlated with the onset of lignification (Tobias and Chow, 2005). This gene has showed to be corresponding to a *Gh2* (*Gold Hull and Internode 2*) mutant. Gh2 mutant showed a dramatic decrease in Klason lignin content (reduced by 5% to 6% compared with the wild-type plants), while the *p*-hydroxyphenyl, guaiacyl, and sinapyl monomers were reduced in almost the same ratio. The purified recombinant OsCAD2 was found to exhibit strong catalytic ability toward coniferaldehyde and sinapaldehyde. Therefore, OsCAD2 (Gh2) may act as a multi-functional CAD to synthesize coniferyl and sinapyl alcohols in rice lignin biosynthesis (Zhang et al., 2006). A novel gene has been identified from poplar namely sinapyl alcohol dehydrogenase (SAD). SAD is similar with CAD but with different substrate specificity and expressing localization. SAD catalyzes the reduction of sinapaldehyde, while CAD is responsible for the reduction of coniferaldehyde. SAD, together with F5H and COMT, is expressed and localized in cells and tissues with S units of lignin, while CAD activity is detected in tissues that has G lignin synthesis (Li et al., 2001), which suggests that SAD may be specifically involved in the reduction of sinapaldehyde. Therefore, the different monolignols are likely to be reduction *via* different pathways and with different enzymes at the last step. However, the CAD-deficient poplar had less conventional S units and beta–*O*–4-bonds and more free phenolic groups in lignin. Furthermore, mainly sinapaldehyde was found to incorporate into the lignin. This suggests that SAD does not play any substantial role in constitutive lignification in poplar (Lapierre et al., 2004). Obviously, the real biological functions of SAD need further investigation.

The polymorphism of CCR and CAD genes has been noticed from recent genomic studies and mutant/transformant analyses. As the CCR and CAD are encoded by small multi-gene families, it is often difficult to use the sequence information to assign substrate specificities to each member of CCR and CAD gene families. However, the expression profiles suggest that both CAD and CCR may have two sub-families: one devoted to developmental lignification, and the other involved in the synthesis of defence-related compounds (Boudet et al., 2004).

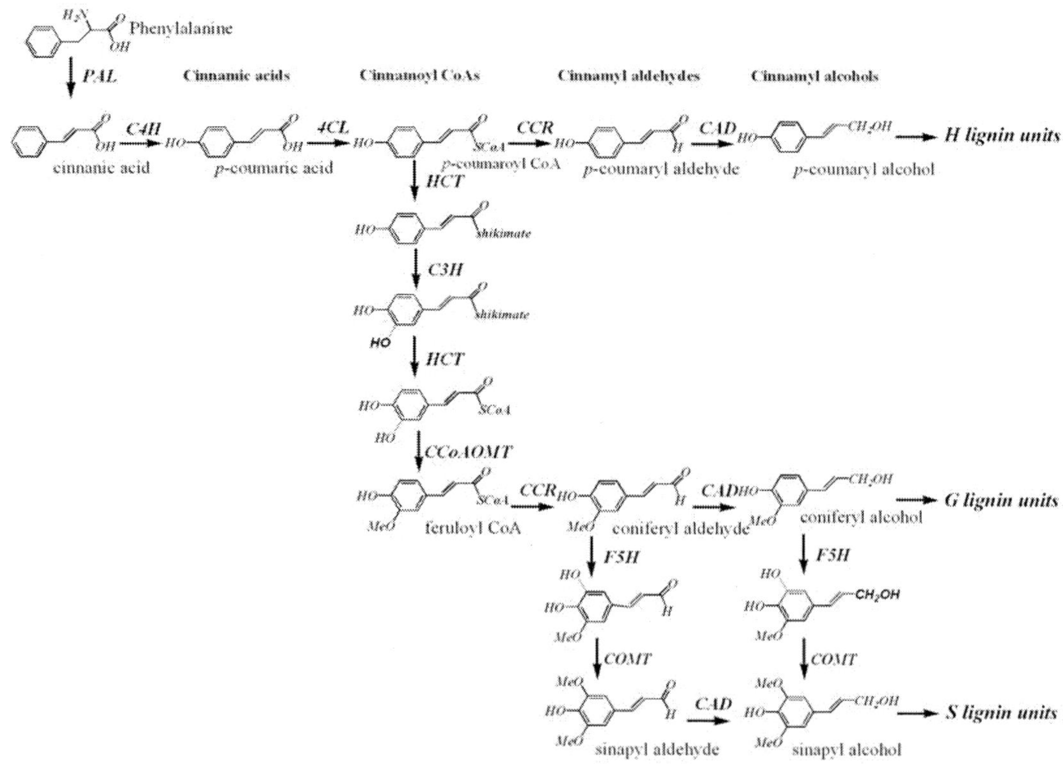

Figure 2. The proposal lignin biosynthesis pathway (data from Dr Richard A. Dixon of Noble Foundation, USA).

Monolignols are synthesized in the cytoplasm, and then transported through the plasma membrane into the cell wall where polymerization of monolignols occurs. As an extraordinary complex process, lignification is proposed to be polymerized radicals initiated by redox enzymes such as peroxidase (POX) and laccase. Direct evidences of involvement of those enzymes in the monolignol polymerization, however, are still limited. The major basic peroxidase from *Zinnia elegans* (ZePrx) suspension cell cultures was purified and cloned. There are two isoforms of ZePrx with a relative molecular mass 34,700 and 33,440, respectively. Four full-length cDNAs coding for these peroxidases were cloned. Enzyme kinetics analysis showed that sinapyl alcohol was the best substrate for ZePrx, among the three monolignols. Western blots showed that one ZePrx was expressed in tracheary elements, roots, and hypocotyls, while another ZePrx was only expressed in roots and young hypocotyls (Gabaldon et al., 2005). It has also reported that Mn^{2+}/Mn^{3+} redox shuttle play an important role in the lignification. *In vitro* synthesis of lignin showed that peroxidase oxidized Mn^{2+} to Mn^{3+}, and Mn^{3+} oxidized monolignols to radicals together with reduction of Mn^{3+} to Mn^{2+}. Peroxidase is never in direct contact with the lignin monolignols. This was called Mn^{2+}/Mn^{3+} redox shuttle. The monolignols with radicals will form high molecular lignin by various kinds of covalent linkages through an automatic process (Önnerud et al., 2002). However, there is other theory that proposes lignin polymerization by non-enzymatic coupling reactions under simple chemical control, since no explicit control of the final configuration of the product is observed. According to this theory, a new class of proteins called dirigent that region- and stereo-specifically couple monolignol radicals to make (optically active) lignans, and this will direct lignification with regular polymers (Gang et al., 1999; Lewis, 1999; Davin and Lewis, 2000). Absolutely, the current theory of lignification remains open to investigate. The currently acceptable lignin biosynthesis pathway is showed in Figure 2.

3. Transgenic Plants and Applications

The transgenic plants with alternation on lignin-related genes have been well documented. The interesting readers can refer the early reports from the comprehensive review papers (Baucher et al., 1998; Anterola and Lewis, 2002; Dixon and Reddy, 2003). Here we just review the recent progress concentrating on application of lignin research. Three main sections are currently extensively been investigated: decrease pollution in pulping process of paper-making industry, improvement of forage quality, and increase production in bio-fuel.

3.1. Pulping

Pulping is the main process in paper-making procedure, which is involved in removing lignin from fiber cell wall and obtaining flexible fibers that mainly consist of cellulose and hemicellulose. All paper is made from plant materials. More specifically, paper consists of fibers obtained from plant materials. Plant fibers in industry term actually include vessel (tracheids), phloem fiber (also called bast fiber), and xylem fiber (also called wood fiber). These fibers are composed of mainly of cellulose, hemicellulose and lignin. Cellulose is a linear polymer of β-(1,4)-D-glucopyranose units synthesized during photosynthesis. Cellulose has good tensile strength, but it is soft and flexible in nature. Hemicellulose is a polymer made up of poly-pentoses and poly-hexoses, which has a similar property in paper-making but the strength is less than cellulose since hemicellulose has a random, amorphous structure in contrast to cellulose that is crystalline. Lignin is a heterogenous hydrophobic phenolic polymer, which is a major structural compound of cell wall in vascular plants (pteridophytes, gymnosperms and angiosperms), being incorporated into the cell wall with cellulose together with the process of thickening plant secondary wall. Lignin is much like an epoxy, with a large molecular weight, random structure, and an enormous capacity to cross-link with other sites on the lignin molecule. The deposition of lignin increases the mechanical intensity and supportable ability of plant tissues. Therefore, the combination of cellulose and lignin makes plant tough and durable material. For the paper-making purpose, however, this structure presents a problem, which is how to remove the lignin and liberate the cellulosic fibers. This process is called pulping. There are two basic methods for pulping: Mechanical pulping uses only mechanical force (and perhaps some heat energy) to pull the raw material apart into fibers with little lignin is removed. Chemical pulping uses heat and chemical energy to preferentially dissolve the lignin from the plant material, leaving only pure cellulosic fibers.

Mechanical pulping has the highest pulp yield, but the resulting pulp is relatively crude, with a large content of damaged fibers, and a high amount of lignin. Therefore, the paper made from this type of the pulp has less strength, limited bleachability and can revert in brightness upon exposure to light (the paper becomes yellow as it ages because of the remaining lignin). In contrast, chemical pulping yields intact fibers that can interact with other fibers via hydrogen bonds, due to removal of the hydrophobic lignin, which insure to make strong and durable paper. Currently, the Kraft process is the most widely used chemical procedure for the paper production. However, this process is gradually being replaced by thermomechanical (TMP) and chemithermomechanical (CTMP) pulping methods that give higher pulp yields and consume less water. In 2004, the world wood pulp production has been estimated at 189,732,000 ton, from which 6% are of mechanical pulp, 16% TMP pulp, 4% CTMP pulp, and 74% chemical pulp (Kraft pulp)(State of the World's Forests 2007, Food and Agriculture Organization of the United Nations, http://www.fao.org/docrep/009/a0773e/a0773e00.htm).

The most widely used pulping method at present is chemical pulp by the Kraft procedure. It uses NaOH and sodium sulfide (Na_2S) as delignifying agents. The wood chips are dissolved into the solution containing above chemical and heated to 160-180 °C for about 2-4 hour. This process is called cooking. During cooking, Na_2S is hydrolyzed into NaOH, NaHS, and H_2S, which react with lignin, forming thio-lignin that are more soluble and easily separated from cellulose. The liquor applied to the chips is called white liquor and the liquor extracted from the reactor that contains the removed compounds is called black liquor, a complex mixture of inorganic and organic compounds.

Generally, the removal of 1000 kg lignin needs 800 kg NaOH and 300 kg Na_2S. The residual chemicals plus lignin in black liquor are very harmful if being directly discharged into environment. Although most of the cooking chemicals from the black liquor can be recovered, the whole pulping procedure is still an expensive and environmentally hazardous process requiring large amounts of energy and chemicals. For the paper-making industry, it would be beneficial to process wood with either less lignin or lignin with an altered chemical reactivity.

The first attempts to alter lignin biosynthesis involved down-regulation of COMT and CAD. A large reduction in CAD activity only slightly reduced lignin content and composition in the transgenic poplar. The lignin structure of the transgenic poplar, however, has been altered which lead to higher extractability in pulping process (Baucher et al., 1996). The transgenic poplar also exhibited red xylem but overall plant growth and development was not affected (Figure 3). Down-regulation of COMT with antisense strategy did not affect the lignin content in transgenic poplar, but induced a dramatic alteration of lignin composition with an increased proportion of G units, together with an increased frequency of resistant interunit bonds, and conversely a lower frequency of S units and labile beta-*O*-4 bonds (Lapierre et al., 1999). The pulping experiments have been performed on these lignin-modified poplars grown in the field (Pilate et al., 2002). Two sites have been selected for these field trials, one in France and one in the United Kingdom. The transgenic trees were harvested after grown for 2 and 4 years, respectively. The analysis data showed that the transgenic trees modified in either COMT or CAD exhibited similar phenotypes in two different field environments with those grown in the greenhouse. The COMT antisense poplars were more resistant to Kraft delignification than control trees, with the higher kappa number, lower pulp brightness, but with similar paper strength as control trees. CAD antisense trees, however, were more easily delignified and had a lower kappa number than control trees. The pulp brightness and paper strength were similar in control and CAD-suppressed poplars. When a range of active alkali charges was used in pulping test, less chemical was needed to reach a given kappa number for the pulp made from CAD antisense poplars and more chemical was needed for COMT antisense poplars compared with the control. Therefore, the transgenic poplars with a suppression of CAD activity have better Kraft pulping properties, with chemicals saving and increase in pulp yield have been estimated at 6% and 2–3%, respectively, without other adverse effects. In contrast, down-regulation of COMT has a negative effect on Kraft pulp properties, due to the more condensed lignin rich in G units is difficult to extract by the Kraft process.

The similar transgenic poplars were obtained by introduction of a sense homologous COMT transgene. A transgenic poplar with COMT activity close to zero in woody tissues was selected. This line showed substantially reduced lignin level (17% decrease) and increased cellulose content in 6-month-old trees. Lignin structure was found to be strongly altered, with two times higher content in condensed bonds, an almost complete lack of S units, buy more G units and 5-hydroxyguaiacyl units. Laboratory-scale kraft-pulping experiments showed that the transgenic poplar trees had the higher pulp yield (10% relative increase), but higher kappa number indicating for a higher residual lignin content. It suggests that the reduction in lignin content has positive effect on pulping, but it may counterbalance by the high condense lignin (Jouanin et al., 2000). From these experiments, it may imply that the tree with less lignin content and less G units will be more suitable for pulping process. Down-regulation of CCoMT gene has created a poplar with such phenotype. Repression of CCoMT expression by the antisense approach in transgenic poplar plants caused a significant decrease in total lignin content (decrease up to 65% of wild-type tree) with a decrease in both G and S units. This resulted in a less condensed and less cross-linked lignin structure in transgenic wood. Unfortunately, no pulping experiment has been tested with this poplar (Zhong et al., 2000). Alternatively, lignin composition may be altered by over-expression of some lignin biosynthetic genes. F5H is a cytochrome P450-dependent monooxygenase that catalyses the hydroxylation of coniferaldehyde and coniferyl alcohol leading to S lignin biosynthesis. Over-expression of F5H under the control of C4H promoter increased S unit lignin, with reduced total lignin content (Franke et al., 2000). Kraft pulping experiments on 2-year-old greenhouse-grown transgenic poplars resulted in a pulp with lower kappa number (decrease 23 kappa units) and an increase in brightness (increases >20 ISO brightness units). These changes were associated with higher ratio of S unit concentrations but no modification in total lignin content and no observed phenotypic differences in the transgenic poplar. Therefore, C4H-F5H-transgenic poplars are expected to increase pulp yields by 60% while concurrently decreasing chemicals employed during processing (Huntley et al., 2003). C3H is another cytochrome P450-dependent monooxygenase that catalyzes the 3'-hydroxylation of *p*-coumaroyl shikimate and *p*-coumaroyl quinate. The transgenic poplar suppressed in C3H expression by RNA interference greatly reduced total lignin content (up to 43% of wild-type poplar) with a significant shift in lignin monomer composition, favoring the generation of H units at the expense of G units while the proportion of S units remained constant. According to results from CAD and COMT transgenic poplar, this type of poplar should have high potential to improve the pulping efficiency but pulping test is needed (Coleman et al., 2008).

Antisense suppression of the enzymes in general phenylpropanoid metabolism, such as hydroxycinnamate CoA ligase (4CL), may also affect lignin biosynthesis. Down-regulation of a specific 4CL gene, which appeared to be involved in directing phenylpropanoids towards lignin biosynthesis, resulted in up to a 45% reduction of lignin without impacting other major branches of phenylpropanoid biosynthesis in the transgenic aspen (*Populus tremuloides*

Michx.). Interesting, the cellulose in the cell wall was increased up to 15% in transgenic aspen probably due to compensate the decrease of lignin in the cell wall. Thus, the cellulose to lignin ratio is higher in the transgenic apsen than in wild-type plants. Furthermore, the transgenic aspens appeared to grow considerably faster than their wild-type counterparts under greenhouse conditions (Hu et al., 1999). These phenotypes, such as high cellulose plus low lignin content, and fast growth, are likely to have very potential commercial importance in pulping and paper industry, providing field trials and commercial pulping tests demonstrate its real values. Suppressed peroxidase (POX), which is involved in the dehydrogenated polymerization of monolignols, on lignin biosynthesis were investigated in transgenic hybrid aspen (*Populus sieboldii X Populus gradidentata*). One transgenic line had lower lignin content and gave a higher pulp yield in kraft pulping than the wild-type aspen. These results indicate that suppression of POX may hold promise for producing woods suited for milder kraft pulping conditions using less chemicals and lower pulping temperature (Tamura et al., 2001).

Figure 3. The debark trunk of transgenic poplar with down-regulated CAD. Inset is the transverse section of the trunk showing the red color confined to developing xylem (Modified from Pilate et al., 2002).

Although most pulping experiments are involved in transgenic poplar, other trees have also been tested. Pulping experiments have been performed with a 12-year-old loblolly pine (*Pinus taeda* L.) mutant deficient in CAD. In soda pulping (26–28% active alkali), the pulping resulted in a lower kappa number than that of the wild type, which indicates the increased lignin reactivity in this mutant of pine. In contrast to the results from the transgenic CAD-suppressed poplars, no improved delignification was obtained using either the kraft-pressure bomb (20–24% active alkali) or the kraft-micro autoclave (17.5–25% active alkali), possibly reflecting the differences in the lignin biosynthesis between gymnosperms and angiosperms. The pulp yield from the mutant pine was lower than that of the wild type. Brightness of the soda pulp of the mutant pine was similar to that of the wild type, but brightness of the kraft pulp from the mutant was slightly lower than that of the wild type with comparable kappa values. This indicates that suppression of CAD in softwood trees may have promise to produce woods suited for "milder" pulping conditions that consume less chemicals and generate less waste (MacKay et al., 1999). RNAi-mediated silencing of the HCT gene in *Pinus radiata* had a strong reduction in lignin content (up to 42%), and remarkable rise of H units from trace amounts in wild-type trees to up to 31% in the transgenic trees. This suggests that HCT also represents the metabolic entry point leading to G lignin biosynthesis in coniferous gymnosperms (Wagner et al., 2007). The pulping experiments are necessary as pine is more widely used in paper-making industry than poplar.

Tobacco is not normally used for pulping, but it has been widely used as a model plant to investigate lignin biosynthesis. Transgenic tobacco with severely suppressed in CCR activity were produced by introduction of a partial sense CCR transgene. Some transgenic lines showed the reduced growth, with significantly decreased lignin content and an increased proportion of free phenolic groups in non-condensed lignin. Direct kraft pulping trials confirmed that more lignin could be removed from the transgenic stems than from wild-type plants at the same alkali charge. Similar transgenic tobacco has been obtained by down-regulated CAD gene. This type of transgenic tobacco had a lower kappa number, similar cellulose degree of polymerization (DP) and a higher brightness for the pulp than those of the control plants (O'Connell et al., 2002). Because the pulping data obtained from CAD-down-regulated tobacco are comparable to those from CAD-suppressed poplar (reduced kappa number and better bleaching with no modification of other pulp properties), therefore tobacco may be an appropriate model plant for rapid evaluation of lignin modifications on pulping. The transgenic tobacco has been obtained with down-regulated 4CL activity. Kraft pulping on a laboratory scale indicated that the xylem of transgenic tobacco exhibited higher delignification efficiency than control, accompanied by a higher pulp yield, and easily bleached with a lower consumption of chemicals than that of the controls. Therefore, this type of plants is useful in pulping process by increased pulping yield and saving of bleaching reagent (Kajita et al., 2002).

Taken together, the data currently available concerning the genetic modification of lignin for improving pulping suggest that CAD and CCR suppression affect lignin structure in ways that improve chemical pulping efficiency by reducing the amount of required chemicals, and thereby reducing the environmental impact from pulping. However, it is likely that CCR-suppression will not be the optimal target for genetic manipulation of pulping character due to the potential association of growth defects in the transgenic plants. COMT suppression has adverse affect on delignification, while CCoMT is likely to facilitate delignification. Other genes, such as F5H, C3H, 4CL, and POX may also be the targets to benefit delignification but warrant further investigations

3.2. Forage Digestibility

Forage crops are a major source of nutrients for domesticated herbivores, particularly ruminant cattle such as cow and sheep. The fiber fraction from forage stem and leaf tissues makes up an important part of the ruminant diet, which originates mainly from plant cell walls that consist of a variety of structural polysaccharides, often cross-linked with proteins and phenolic components, particularly with lignin. The main polysaccharides of the forage cell wall are cellulose, hemicelluloses (arabinoxylans, β-glucan, xyloglucans, arabinogalactans), and pectic polysaccharides (rhamnogalacturonan) (Bach Knudsen, 1997). Lignin cements and anchors the cellulosic microfibrils and other matrix polysaccharides to form the ligninpolysaccharide complex. This structure increases the mechanical intensity and supportable ability of plant tissues thereby protecting them from degradation and physical damage. Lignin is likely the only component in cell wall resistant to bacterial and fungal degradation in the gut of ruminant cattle and therefore lignin association with other matrix components greatly influences the enzymatic degradability of these polysaccharides (Monties, 1991). It has demonstrated that the nutritional value of forage crops is closely related to the lignin composition and content (Casler and Jung. 2006). Forage digestibility, reflecting by *in vitro* dry-matter digestibility (IVDMD), is essential to livestock performance. Relatively small increases in IVDMD typically result in measurable improvements in animal performance. Animal feeding test has shown that as little as 1% increasing IVDMD generally leads to 3.2% average daily live-weight gains in cattle (Casler and Vogel, 1999). Therefore, plant breeding for increasing the nutritional value of forage crops is always extremely interesting to the farmer and rancher.

Genetic variation in lignin characteristics and digestibility was first documented in maize *brown-midrib* (*bm*) mutants in the 1960s and 1970s (Kuc and Nelson, 1964; Kuc et al., 1968; Lechtenberg et al., 1972; Muller et al., 1972). Since then, *bm* mutants of maize and other C_4 grasses (sorghum and millet) have been extensively studied (Barrière et al., 2004). *Bm* plants exhibit a reddish brown pigmentation in the leaf midrib and stalk pith (as their name) when the plants have about five expanded leaves. The molecular characterization of *bm* mutants has demonstrated that *bm1* directly affect expression of the CAD (Halpin et al., 1998), while the *bm3* mutation occurs in the gene encoding COMT (Vignols et al., 1995). However, the increased forage nutritional value in *bm* mutants is often associated with reductions in agricultural fitness traits, such as forage yield and stress tolerance. Therefore, more plant-breeding programs with concomitant selection pressure will be requested in order to apply these genetic resources into real stockbreeding.

Transgenic plants represent a new approach for generating novel grass with improved forage nutritional value. Modification of the gene expression that in the early lignin synthesis pathway such as phenylalanine ammonia lyase (PAL) led to general reductions in lignin concentration and improved digestibility in transgenic tobacco plants, but also with a wide range of negative phenotypes (Sewalt et al., 1997). In this reason, modifying the expression of downstream enzymes in monolignol pathways has more promising. Down-regulation of COMT activity by antisense approach resulted in reduced lignin content with an increased S/

G monomer ratio in transgenic tobacco. Enzymatic digestibility of stem internodes was improved in the transgenic plants and was highly negatively correlated with lignin concentration (r = -0.97) (Sewalt et al., 1997). Down-regulation of CAD in transgenic tobacco showed little change in lignin content but considerable decrease in the S units of lignin, which lead significant increase in digestibility of stem tissues. The changes in lignin composition that were observed in CAD-depressed tobacco were largely similar to those seen in maize and sorghum *bm1* mutants with altered lignification and improved digestibility (Bernard Vailhé et al., 1998). These data suggest that depressing CAD activity may be an effective approach for improving digestibility without reduction of lignin content in forage crop.

Tobacco is not used in ruminant nutrition and has a limited relevance to feeding applications. Alfalfa is an important forage crop widely grown throughout the world and its primary use is for dairy production, followed by beef, horse, sheep, and goat, and it is sometimes also used for human consumption. The down-regulation of CAD by an antisense approach resulted a lower S/G ratio (reduced by less than 50%), mainly because of a decreased amount of S units, but the lignin content remained unchanged in the transgenic alfalfa. An increase in *in situ* disappearance of dry matter and of cell wall residue was detected in the transgenic plants grown both in the greenhouse and the field. An increase in forage digestibility in cannulated sheep was recorded for these transgenic plants (Baucher et al., 1999). Down-regulation of COMT reduced lignin content, accompanied by near total loss of S units with remarkable decrease in S/G ratio, whereas down-regulation of CCoMT reduced lignin content but to a reduction in G units without reduction in S units (with a modest increase in S/G ratio) in the transgenic alfalfa. Analysis of rumen digestibility of the transgenic alfalfa in fistulated steers revealed improved digestibility of forage from COMT down-regulated plants, but a greater improvement in digestibility in CCoMT down-regulated plants, with increase in *in vitro* true digestibility (IVTD) of up to 4.4% and of in-rumen digestibility of up to 2.8–6% (Figure 4) (Guo et al., 2001b). Further analysis showed that the alfalfa deficient in CCoMT exhibited a structurally similar lignin to the control plant but with a dramatic decrease in lignin content (20%) and modest increase in cellulose (10%), therefore reflecting a 30% increase in cellulose to lignin ratio. This compositional change will potentially enhance utilization of alfalfa as a major forage crop (Marita et al., 2003).

The relationship between lignin content and composition with forage digestibility has been investigated using isogenic alfalfa created by the transgenic approach. Down-regulation three specific cytochrome P450 enzymes in the lignin pathway, C4H, C3H and F5H, generated the transgenic alfalfa lines with a range of differences in lignin content and composition. C4H down-regulated alfalfa had a greatly reduced lignin without significant impact on composition, C3H down-regulated alfalfa had lignin-rich in H units, while F5H down-regulated alfalfa had lignin rich in G units with reduced S units. Together with the transgenic alfalfa down-regulated COMT and CCoMT, analyses from these five types of alfalfa demonstrate that a strong negative relationship between lignin content and rumen digestibility, but no relationship between lignin composition and digestibility, which is quite contrast to the situation that modification in lignin composition increases paper-pulping efficiency in trees (Reddy et al., 2005). It seems unlikely that increased condensation of lignin (with more G units as expected in the COMT down-regulate plants) is a serious factor affecting forage digestibility. Therefore, lignin content may be the overriding factor influencing cell wall digestibility although lignin composition is also affected. This conclusion need to be further validated in other grass species.

Figure 4. Feeding test with transgenic alfalfa of down-regulation of COMT and CCoMT (data from Dr Richard A. Dixon of Noble Foundation, USA).

Down-regulation of the genes early in the phenylpropanoid pathway may limit the overall flux of metabolites into lignin, therefore modifying lignin content. Down-regulation of HCT in transgenic alfalfa resulted in strongly reduced lignin content (around 50%) and striking changes in lignin monomer composition, with predominant deposition of H units in the lignin. The transgenic alfalfa exhibited significant stunting, reduction of biomass and delayed flowering. Although modification of lignin in HCT transgenic alfalfa greatly improve forage digestibility, accompanying effects on plant development need to be overcome (Shadle et al., 2007). The similar situation may exist in C3H and C4H down-regulated alfalfa lines. Down-regulation of C3H provided the largest improvements in digestibility yet seen in a forage crop (with up 18% increase in IVDMD), some lines exhibited delayed flowering phenotype. C4H down-regulated alfalfa not only showed a loss of flower coloration, but also possessed a different floral scent, which may have significant effects on pollination by honey bees (Reddy et al., 2005).

Besides alfalfa, other forage crops have also been tested for improving digestibility by modifying lignin biosynthesis. Transgenic maize (*Zea mays*) plants were generated with a maize COMT cDNA in the antisense orientation under the control of the maize alcohol dehydrogenase promoter, which localized the gene expression in vascular tissues and lignifying sclerenchyma. One transgenic line displayed a significant reduction in COMT activity (15%–30% residual activity) and a strong decrease in Klason lignin content (25% to 30% lower as compared with controls) at the flowering stage, accompanied by a decrease in S units and the occurrence of unusual 5-OH guaiacyl units. These characteristics of lignin metabolism are similar with what have found in maize *bm3* mutant, as well as in COMT down-regulated alfalfa. The transgenic maize was more digestible than controls as indicated by its higher *in vitro* neutral detergent fiber digestibility value (IVNDFD) (84% in transgenic line *versus* 77% in control), suggesting beneficial for digestibility (Piquemal et al., 2002). Down-regulation of maize COMT by an antisense sorghum COMT driven by the maize ubiquitin-1 promoter resulted a brown-midrib phenotype with significantly lower COMT activity (up to 60% reduction) in transgenic plants compared with controls. The transgenic maize had 20% less lignin in stems and 12% less lignin in leaves compared with controls. On a whole-plant basis, lignin was reduced by an average of 17% with the greatest reduction up to 31%. Digestibility was significantly improved in transgenic plants by 2% in leaves and 7% in stems. Mean whole-plant digestibility increased from 72% to 76% (He et al., 2003). Down regulation of COMT by sense co-suppression in transgenic tall fescue (*Festuca arundinacea*) resulted in a strong decrease in lignin content, a reduction in total S and G units but final S/G ratio was increased. IVDMD of the transgenic tall fescue, as measured by a filter bag method, showed a significant increase (9.8-10.8%) as comparison with control plants (Chen et al., 2004).

3.3. Bio-fuel

Due to the progressive depletion of conventional fossil fuels (The world's oil reserves will be depleted in about 40 years according to Energy Information Administration estimates in 2006, Energy Information Administration. 2006. World oil markets. p. 25–35. In International Energy Outlook 2006. Available at http://www.eia.doe.gov/oiaf/archive/ieo06/pdf/oil.pdf. USDOE, Energy Information Administration, Washington, DC.), renewable energy sources have gained great interest and rapidly applications in this field are taking place in recent years. Biomass energy is an attractive renewable fuel. Biomass, as the solar energy stored in chemical form in plant and animal materials, is among the most precious and versatile resources on earth. The components of biomass include cellulose, hemi-cellulose, lignin, lipid, protein, simple sugar, starch, water, ash, and other compounds. However, most important biomass for energy utilization is starch, cellulose, hemi-cellulose and lignin. Biomass may be used to meet a wide variety of energy needs, including generating electricity, providing heat for industry and home, and fuelling vehicles. The conversion of biomass to such useful forms of energy, also called bio-fuel or bio-energy, can be achieved using a number of different technology but basically they can be classed into two basic categories, namely thermochemical process and biochemical/biological process. The energy crops and lignocellulosic residues are two represented biomasses with the global potentiality. Conversion of these biomasses into bio-fuel is an important choice for the exploitation of alternative energy sources and reduction of polluting gases. Particularly, the plantation of energy crops will decrease the CO_2 emissions and alleviate greenhouse effect. Therefore, the utilization of bio-fuel has important ecological and social effects (Snchez and Cardona, 2007).

Bio-ethanol is the most widely used bio-fuel either as a fuel or as a gasoline enhancer at present. Many countries have implemented or are implementing programs for addition of ethanol to gasoline. Bio-ethanol production has increased remarkably in recent year because many countries hope to reduce oil imports and improve air quality. The world bio-ethanol production has reached about 51,000 mill liters in 2006 (Renewable Fuels Association, 2007, Renewable Fuels Association. 2007. Industry statistics. <http://www.ethanolrfa.org/industry/statistics>). The main biomass to produce bio-ethanol is sucrose-containing feedstocks (mainly sugar cane) and starchy materials (mainly maize). Because the competition with food production, producing bio-ethanol in large scale has lead the global increase in food price which may be one reason for the inflation since 2007 in the world. Moreover, recent analysis from a worldwide agricultural model has doubt maize-based ethanol production on the effects to reduce greenhouse gases in the long term (over 30 to 167 years) (Scharlemann and Laurance, 2008; Searchinger et al., 2008). This result raises concerning about the importance of lignocellulosic biomass as a feedstock for large-scale of bio-fuel production. Lignocellulosic biomass is the most abundant biopolymer in the Earth, which comprises about 50% of world biomass and annual production estimate in 10–50 billion ton (Claassen et al., 1999). Many technologies have been developed to test for bio-ethanol production from various types of lignocellulosic materials. Basically, these technologies include three main steps: (1) pretreatment of lignocellulosic biomass to separate cellulose from lignin; (2) hydrolysis of cellulose; and (3) fermentation of biomass hydrolyzates into ethanol. Among these steps, separation of cellulose from lignocellulosic biomass is the main limiting factor which is complicated, energy-consuming and non-completely developed. This situation is some extent similar with what in pulping process. Therefore, transgenic approach has been tested to reduce or eliminate the need for pretreatment of lignocellulosic biomass by modification of lignin biosynthesis.

Six different transgenic alfalfa lines expressing antisense construct for down-regulating lignin biosynthesis were used to test modification of lignin on bio-ethanol production, these lines including C4H, HCT, C3H, CCoMT, F5H, and COMT transgenic plants. Lignin content from these transgenic alfalfa stems decreased in the order: F5H and control (highest lignin level)>COMT and CCoMT>C4H, C3H and HCT (lowest lignin level at 50% of the wild-type value). Lignin S/G ratios varied from about 0.3 to 1.0 depending on the targeted gene, with a high proportion of H units in HCT and C3H transgenic plants. The data showed that recalcitrance to both acid pretreatment and enzymatic digestion is directly proportional to lignin content. C3H and HCT transgenic lines yielded nearly twice as much sugar from cell walls as wild-type plants (the enzymatic saccharification efficiencies of acid-pretreated cell walls is 67–79% in C3H compared to 43% in controls). In addition, both lignin content and composition might affect cellulose substrate-enzyme interactions during enzymatic hydrolysis for untreated stem material, whereas lignin composition might not be a major factor affecting hydrolysis for pretreated stems (Chen and Dixon, 2007). This provides the first glimpse of what can be done to improve bio-fuel production by modification of lignin biosynthesis.

The relation between lignin and bio-fuel production, however, is more complicated and warrant further investigations. High contents of lignin in biomass are desirable in some situations, especially when they are used as solid bio-fuel, for two main reasons. Firstly, lignin has a high heating value due to the high content of carbon in lignin (about 64%). Secondly, highly lignified plants can stand upright with low water contents. Therefore, the plants can dry 'on the stem' and harvest at lower water contents for improved biomass quality.

The energy content of different composition in biomass can be expressed by using a relative scale of combustion degree, the R value, calculated from the degree of reduction of the carbon in an organic compound which is related to the percentage of carbon, hydrogen, and oxygen (Spoer and Milner, 1949). By assigning carbon dioxide has an R value of zero, while methane, the most reduced carbon compound, has an R value as 100, then R values of other organic compounds can be calculated. Thus, the calculated R value for lipids, lignin, and cellulose plus starch is 70, 60 and 30, respectively. Therefore, lignin represents the component in lignocellulosic biomass with highest energy potentiality and a higher level of lignin will increase the intrinsic heat content of biomass (Baucher et al. , 1998). Actually, the higher heating value (HHV) of different biomasses (sunflower shell, almond shell, hazelnut shell, wood bark, olive husk, hazelnut kernel husk, and walnut shell) are highly correlated with their lignin contents, with a correlation coefficient 0.9734, but there are no direct relations between HHV and the holocellulose in the lignocellulosic fuels (Demirbas, 2003). Moreover, the correlation coefficients between the lignin content and the oil plus char yields were about 0.898 for non-catalytic aqueous liquefaction and 0.883 for catalytic aqueous liquefaction in the thermochemical conversion of lignocellulosic materials, indicating a strong correlation. With both processes of aqueous liquefaction, by increasing lignin content, the oil yield decreased, but the char yield increased (Demirbas, 2000). Obviously, this warrants more investigations, particularly with transgenic approach and breeding program to explore the potentiality of modifying lignin biosynthesis in order to increase energy efficiency of lignocellulosic biomass.

4. Perspective

Lignin as the complex, heterogeneous, cell wall bound polymer is synthesized by almost all land plants, where it plays important roles in plant growth and development. The progress to understand lignin biosynthesis is constant and rapid. The combination of biochemical approach with the molecular biology, particularly transgenic approach, to investigate the biosynthetic pathway has contributed to recent progress in this research area. The interests in reducing pulping pollution, improving forage digestibility, and increasing bio-fuel yield for industrial applications, have greatly promoted the extensive research of the transgenic plants with the modified lignin.

After completing the genome sequence in *Arabidopsis* and rice, the genome of poplar has been completely sequenced recently, which is the first tree and the third plant genome to be sequenced (Tuskan et al., 2006). The large-scale of expressed sequence tags (ESTs) from xylem tissues has generated for loblolly pine, poplar, and Zinnia (Hertzberg et al., 2001). The lignin transcriptional profiling using micro-array technology has identified genes encoding the enzymes involved in lignin, hemicellulose and cellulose biosynthesis as well as other potent regulators of plant cell wall (Demura et al., 2002). However, our understanding of the lignin biosynthesis is still not complete. Firstly, many models about lignin biosynthesis pathway have been proposed at present, two best known as randomness theory (Sederoff et al, 1999) and independent pathways (Dixon et al., 2001), but still these models can not completely explained all aspects in lignin biosynthesis. Secondly, the lignin content, composition, structure and metabolic pathway are different from each other in different species, tissues and organs, and at different developmental stages. The reasons for these differences are largely open to investigate. Thirdly, the regulating mechanism of lignin biosynthesis is still very poorly understood and much has still to be learned about how to control and regulate the lignin synthesis at transcriptional and biochemical levels. Moreover, lignin constitutes a large portion of plant dry mass (up to 30%) and its biosynthesis is a metabolically costly process that requires large quantities of carbon skeletons and bio-energy (Amthor, 2003). Once lignin is deposited in plant tissues, it will not be converted or degraded.

Therefore, it is essential to accurately regulate lignin synthesis in order to balance to fit physiological functions and metabolite cost (Patzlaff et al., 2003). This makes the regulation on lignin biosynthesis essential not only for its applications but also for basic physiology in planta. Finally, the transgenic plants with modifying lignin biosynthesis should combine with breeding program in order to select the trees, forage crops and energy crops not only with better lignin characters but also fine agricultural performance. This will solve the problems that frequently meet in the present experiments like lowering lignin to increase forage digestibility but generally accompanied by a reduction in biomass. It is expected that lignin research and the transgenic plants with modifying lignin biosynthesis will have the exciting potentiality in agriculture and industry which will be greatly benefit for human life and environment in the near future.

ACKNOWLEDGMENTS

This work was supported by grants from the National High Technology and Research Development Program of P.R. China (“863” program, No. 2007AA10Z101), the National Natural Science Foundation of China (No.30570133, No30671043), the Chinese National Special Foundation for Transgenic Plant Research and Commercialization (No. 2008ZX08002-003), and Innovation Project of Chinese Academy of Sciences. The author sincerely thanks Dr. Richard A. Dixon and Fang Chen of Noble Foundation, USA, and Dr Claire Halpin of University of Dundee, UK for providing photo data.

REFERENCES

Amthor J. S., 2003. Efficiency of lignin biosynthesis: a quantitative analysis. *Ann. Bot.* 91: 673 - 695.

Anterola A.M., Lewis N.G. 2002.Trends in lignin modification: a comprehensive analysis of the effects of genetic manipulations/mutations on lignification and vascular integrity. *Phytochem.* 61: 221–294.

Atanassova R., Favet N., Martz F., Chabbert B., Tollier M., Monties B,. Fritig B., Legrand M. 1995. Altered lignin composition in transgenic tobacco expressing *O*-methyltransferase sequences in sense and antisense orientation. *Plant J.* 8: 465-477.

Bach Knudsen K.E. 1997. Carbohydrate and lignin contents of plant materials used in animal feeding. *Animal Feed Sci. Technol.* 67: 319–338.

Barrière Y., Ralph J., Méchina V., Guillaumie S., Grabber J.H., Argillier O., Chabbert B., Lapierre C. 2004. Genetic and molecular basis of grass cell-wall degradability. II. Lessons from *brown-midrib* mutants. *C. R. Biologies* 327: 847–860.

Battle M., Bender M.L., Tans P.P., White J.W.C., Ellis J.T., Conway T., Francey R.J. 2000. Global carbon sinks and their variability inferred from atmospheric O_2 and $\delta^{13}C$. Science 287: 2467-2470.

Baucher M., Bernard-Vailhe M.A., Chabbert B., Besle J.M., Opsomer C., Van Montagu M., Botterman J. 1999. Down-regulation of cinnamyl alcohol dehydrogenase in transgenic alfalfa (*Medicago sativa* L.) and the effect on lignin composition and digestibility. Plant Mol. Biol. 39: 437–447.

Baucher M., Chabbert B., Pilate G., Van Doorsselaere J., Tollier M.-T., Petit-Conil M., Cornu D., Monties B., Van Montagu M., Inze D., Jouanin L., Boerjan W. 1996. Red xylem and higher lignin extractability by down-regulating a cinnamyl alcohol dehydrogenase in poplar (*Populus tremula x P. alba*). Plant Physiol. 112: 1479–1490.

Baucher M., Monties B., Montagu M.V., Boerjan W. 1998. Biosynthesis and genetic engineering of lignin. *Crit. Rev. Plant Sci.* 17: 125-197.

Bernard Vailhé M.A., Besle J.M., Maillot M.P., Cornu A., Halpin C., Knight M. 1998. Effect of down-regulation of cinnamyl alcohol dehydrogenase on cell wall composition and on degradability of tobacco stems. *J. Sci. Food Agric.* 76: 505–514.

Boerjan W., Ralph J., Baucher M. 2003. Lignin biosynthesis. *Annu. Rev. Plant Biol.* 54: 519–46.

Boudet A.M., Hawkins S., Rochange S. 2004. The polymorphism of the genes/enzymes involvedin the last two reductive steps of monolignol synthesis: what is the functional significance? *C. R. Biologies* 327: 837–845.

Casler M.D., Jung H. J. G., 2006. Relationships of fibre, lignin, and phenolics to *in vitro* fibre digestibility in three perennial grasses. *Animal Feed Sci. Technol.* 125: 151–161.

Casler M.D., Vogel K.P. 1999. Accomplishments and impact from breeding for increased forage nutritional value. *Crop Sci.* 39: 12–20.

Chen F., Dixon R.A. 2007. Lignin modification improves fermentable sugar yields for biofuel production. *Nature Biotechnol.* 25: 759-761.

Chen F., Yasuda S., Fukushima K. 1999. Evidence for a novel biosynthetic pathway that regulates the ratio of syringyl to guaiacyl residues in lignin in the differentiating xylem of Magnolia kobus DC. *Planta* 1207: 597-603.

Chen L., Auh C., Dowling P., Bell J., Lehmann D., Wang Z.Y. 2004. Transgenic down-regulation of caffeic acid *O*-methyltransferase (COMT) led to improved digestibility in tall fescue (*Festuca arundinacea*). *Funct. Plant Biol.* 31: 235-245.

Claassen P.A.M., van Lier J.B., Lopez Contreras A.M., van Niel E.W.J., Sijtsma L., Stams A.J.M., de Vries S.S., Weusthuis R.A. 1999. Utilisation of biomass for the supply of energy carriers. *Applied Microbiol. Biotechnol.* 52: 741–755.

Coleman H.D., Park J.Y., Nair R., Chapple C., Mansfield S.D. 2008. RNAi-mediated suppression of *p*-coumaroyl-CoA 3'-hydroxylase in hybrid poplar impacts lignin deposition and soluble secondary metabolism. *Proc. Natl. Acad. Sci.* USA 105: 4501–4506.

Davin L.B., Lewis N.G. 1992. Phenylpropanoid metabolism: biosynthesis of monolignols, lignans, neolignans, lignins and suberins. *Rec. Adv. Phytochem.* 26: 325-375.

Davin L.B., Lewis N.G. 2000. Dirigent proteins and dirigent sites explain the mystery of specificity of radical precursor coupling in lignan and lignin biosynthesis. *Plant Physiol.* 123: 453–461.

Demirbas A. 2000. Effect of lignin content on aqueous liquefaction products of biomass. *Energy Conversion & Management* 41: 1601-1607.

Demirbas A. 2003. Relationships between heating value and lignin, fixed carbon, and volatile material contents of shells from biomass products. *Energy Sources* 25: 629–635.

Derikvand M.M., Sierra J. B., Ruel K., Pollet B., Do C.T., Thévenin J., Buffard D., Jouanin L., Lapierre C. 2008. Redirection of the phenylpropanoid pathway to feruloyl malate in *Arabidopsis* mutants deficient for cinnamoyl-CoA reductase 1. *Planta* 227: 943–956.

Demura T., Tashiro G., Horiguchi G., Kishimoto N., Kubo M., Matsuoka N., Minami A., Nagata-Hiwatashi M., Nakamura K., Okamura Y., Sassa N., Suzuki S., Yazaki J., Kikuchi S., Fukuda H. 2002. Visualization by comprehensive microarray analysis of gene expression programs during transdifferentiation of mesophyll cells into xylem cells. *Proc. Natl. Acad. Sci.* USA 99: 15794–15799.

Dixon R.A. 2001. Natural products and disease resistance. *Nature* 411 843-847.

Dixon R.A., Chen F., Guo D., Parvathi K. 2001. The biosynthesis of monolignols: a "metabolic grid", or independent pathways to guaiacyl and syringyl units? *Phytochem.* 57: 1069-1084.

Dixon R.A., Reddy M.S.S. 2003. Biosynthesis of monolignols. Genomic and reverse genetic approaches. *Phytochem. Reviews* 2: 289–306.

Eudes A., Pollet B., Sibout R., Do C., Séguin A., Lapierre C., Jouanin. L. 2006. Evidence for a role of AtCAD 1 in lignification of elongating stems of *Arabidopsis thaliana.* Planta 225: 23–39.

Franke R., Hemm M.R., Denault J.W., Ruegger M.O., Humphreys J.M., Chapple C. 2002a. Changes in secondary metabolism and deposition of an unusual lignin in the *ref8* mutant of *Arabidopsis*. *Plant J.* 30: 47–59.

Franke R., Humphreys J.M., Hemm M.R., Denault J.W., Ruegger M.O., Chapple C. 2002b. The *Arabidopsis* REF8 gene encodes the 3-hydroxylase of phenylpropanoid metabolism. *Plant J.* 30: 33–45.

Franke R., McMichael C.M., Meyer K., Shirley A.M., Cusumano J.C., Chapple C. 2000. Modified lignin in tobacco and poplar plants overexpressing the *Arabidopsis* gene encoding ferulate 5-hydroxylase. *Plant J.* 22: 223-234.

Gabaldon C., Lopez-Serrano M., Pedreno M. A., Barcelo A. R. 2005. Cloning and molecular characterization of the basic peroxidase isoenzyme from *Zinnia elegans*, an enzyme involved in lignin biosynthesis. *Plant Physiol.* 139: 1138 - 1154.

Gang D.R., Costa M.A., Fujita M., Dinkova-Kostova A.T., Wang H.B., Burlat V., Martin W., Sarkanen S., Davin L.B., Lewis N.G. 1999. Regiochemical control of monolignol radical coupling: a new paradigm for lignin and lignan biosynthesis. *Chem. Biol.* 6:143–151.

Goujon T., Ferret V., Mila I., Pollet B., Ruel K., Burlat V., Joseleau J.P., Barriere Y., Lapierre C., Jouanin L. 2003. Down-regulation of the AtCCR1 gene in *Arabidopsis thaliana*: effects on phenotype, lignins and cell wall degradability. *Planta* 217: 218–228.

Guo D.J., Chen F., Inoue K, Blount J.W., Dixon R.A. 2001a. Downregulation of caffeic acid 3-*O*-methyltransferase and caffeoyl CoA 3-*O*-methyltransferase in transgenic alfalfa: Impacts on lignin structure and implications for the biosynthesis of G and S lignin. *Plant Cell* 13: 73–88.

Guo D.J., Chen F., Wheeler J., Winder J., Selman S., Peterson M., Dixon R.A. 2001b. Improvement of in-rumen digestibility of alfalfa forage by genetic manipulation of lignin *O*-methyltransferases. *Transgenic Research* 10: 457–464.

He X., Hall M.B., Gallo-Meagher M., Smith R.L. 2003. Improvement of forage quality by downregulation of maize *O*-methyltransferase. *Crop Sci.* 43:2240–2251.

Hertzberg M., Aspeborg H., Schrader J., Andersson A., Erlandsson R., Blomqvist K., Bhalerao R., Uhlen M., Teeri T.T., Lundberg J., Sundberg B., Nisson P., Sandberg G.

2001. A transcriptional roadmap to wood formation. *Proc. Natl. Acad. Sci.* USA 98: 14732–14737.

Halpin C., Holt K., Chojecki J., Oliver D., Chabbert B., Monties B., Edwards K., Barakats A., Foxon G. A. 1998. *Brown-midrib* maize (*bm1*) – a mutation affecting the cinnamyl alcohol dehydrogenase gene. *Plant J.* 14: 545-553.

Hoffmann L., Maury S., Martz F., Geoffroy P., Legrand M. 2003. Purification, cloning, and properties of an acyltransferase controlling shikimate and quinate ester intermediates in phenylpropanoid metabolism. *J. Biol. Chem.* 278: 95–103.

Hu W.J., Harding S.A., Lung J., Popko J.L., Ralph J., Stokke,D.D., Tsai C.J., Chiang V.L. 1999. Repression of lignin biosynthesis promotes cellulose accumulation and growth in transgenic trees. *Nature Biotechnol.* 17: 808-812.

Humphreys J.M., Hemm M.R., Chapple C. 1999. New routes for lignin biosynthesis defined by biochemical characterization of recombinant ferulate 5-hydroxylase, a multifunctional cytochrome P450-dependent monooxygenase. *Proc. Natl. Acad. Sci.* USA 96: 10045-10050.

Huntley S.K., Ellis D., Gilbert M., Chapple C., Mansfield S.D. 2003. Significant increases in pulping efficiency in C4H-F5H-transformed poplars: improved chemical savings and reduced environmental toxins. *J. Agric. Food Chem.* 51: 6178 -6183

Jones L., Ennos A.R., Turner S.R. 2001. Cloning and characterization of irregular xylem4 (irx4): a severely lignin-deficient mutant of *Arabidopsis*. *Plant J.* 26: 205–216.

Jouanin L., Goujon T., de Nadaı V., Martin M.T., Mila I., Vallet C., Pollet B., Yoshinaga A., Chabbert B., Petit-Conil M., Lapierre C. 2000. Lignification in transgenic poplars with extremely reduced caffeic acid *O*-methyltransferase activity. *Plant Physiol.* 123: 1363–1373.

Kajita S., Ishifuji M., Ougiya H., Hara S., Kawabata H., Morohoshi N., Katayama Y. 2002. Improvement in pulping and bleaching properties of xylem from transgenic tobacco plants. *J. Sci. Food Agric.* 82: 1216–1223.

Kim S.J., Kim M.R., Bedgar D., Moinuddin S.G.A., Cardenas C.L., Davin L.B., Kang C.C.K., Lewis N.G. 2004. Functional reclassification of the putative cinnamyl alcohol dehydrogenase multigene family in *Arabidopsis*. *Proc. Natl. Acad. Sci.* USA 101: 1455–1460.

Kim S.J., Kim K.W., Cho M.H., Franceschi V.R., Davin L.B., Lewis N.G.. 2007. Expression of cinnamyl alcohol dehydrogenases and their putative homologues during *Arabidopsis thaliana* growth and development: Lessons for database annotations? *Phytochem.* 68: 1957–1974.

Kuc J., Nelson O.E. 1964. The abnormal lignins produced by the *brown-midrib* mutants of maize. *Arch. Biochem. Biophys.* 105: 103–113. [80]

Kuc J., Nelson O.E., Flanagan P. 1968. Degradation of abnormal lignins in the *brown-midrib* mutants and double mutants of maize. *Phytochem.* 7: 1435–1436.

Lacombe E., Hawkins S., Doorsselaers J.V., Piquemal J., Goffner D., Poeydomege, O., Boudet A.M., Grima-Pettenati J. 1997. Cinnamoyl-CoA reductase, the first committed enzyme of lignin branch biosynthetic pathway: cloning, expression and phylogenetic relationships. *Plant J.* 11: 429-441.

Lapierre C., Pilate G., Pollet B., Mila I., Leple J. C., Jouanin L., Kim H., Ralph J. 2004. Signatures of cinnamyl alcohol dehydrogenase deficiency in poplar lignins. *Phytochem.* 65: 313–321.

Lapierre C., Pollet B., Petit-Conil M., Toval G., Romero J., Pilate G., Leple J.C., Boerjan W., Ferret V., De Nadai V., Jouanin L. 1999. Structural alterations of lignins in transgenic poplars with depressed cinnamyl alcohol dehydrogenase or caffeic acid *O*-methyltransferase activity have opposite impact on the efficiency of industrial Kraft pulping. *Plant Physiol.* 119:153–163.

Lechtenberg V.L., Muller L.D., Bauman L.F., Rhykerd C.L., Barnes R.F. 1972. Laboratory and in vitro evaluation of inbred and F_2 populations of *brown midrib* mutants of *Zea mays* L. *Agron.* J. 64: 657–660.

Lewis N.G. 1999. A 20th century roller coaster ride: a short account of lignification. Current Opin. *Plant Biol.* 2: 153–162.

Li, L., Cheng, X., Leshkevich, J., Umezawa, T., Harding, S.A., Chiang, V.L. 2001. The last step of syringyl monolignol biosynthesis in angiosperms is regulated by a novel gene encoding sinapyl alcohol dehydrogenase. *Plant Cell* 13: 1567–1585.

Li L., Cheng X., Lu S., Nakatsubo T., Umezawa T., Chiang V. L. 2005. Clarification of cinnamoyl co-enzyme A reductase catalysis in monolignol biosynthesis of aspen. *Plant Cell Physiol.* 46: 1073–1082.

Li L., Popko J.L., Umezawa T., Chiang V.L. 2000. 5-Hydroxyconiferyl aldehyde modulates enzymatic methylation for syringyl monolignol formation, a new view of monolignol biosynthesis in angiosperms. *J. Biol. Chem.* 275: 6537-6545.

Li L., Popko J.L., Zhang X.H., Osakabe K., Tsai C.J., Joshi C.P., Chiang V.L. 1997. A novel multifunctional *O*-methyltransferase implicated in a dual methylation pathway associated with lignin biosynthesis in loblolly pine. *Proc. Natl. Acad. Sci.* USA 94: 5461–5466.

Lu S., Zhou Y., Li L., Chiang V.L. 2006. Distinct roles of cinnamate 4-hydroxylase genes in Populus. *Plant Cell Physiol.* 47: 905–914.

Ma Q.H. 2007. Characterization of a cinnamoyl-CoA reductase that is associatedwith stem development in wheat. *J. Exp. Bot.* 58: 2011-2021.

Ma Q.H., Xu Y. 2008. Characterization of a caffeic acid 3-*O*-methyltransferase from wheat and its function in lignin biosynthesis. *Biochimie* 90: 515-524.

MacKay J., Presnell T., Jameel H., Taneda H., O'Malley D., Sederoff R. 1999. Modified lignin and delignification with a CAD-deficient loblolly pine. *Holzforschung* 53: 403–410.

Marita J.M., Ralph J., Hatfield R.D., Guo D.J., Chen F., Dixon R.A. 2003. Structural and compositional modifications in lignin of transgenic alfalfa down-regulated in caffeic acid 3-*O*-methyltransferase and caffeoyl coenzyme A 3-*O*-methyltransferase. *Phytochem.* 62: 53–65.

Meyer K., Cusumano J.C., Somerville, C., Chapple, C.C.S. 1996. Ferulate-5-hydroxylase from *Arabidopsis thaliana* defines a new family of cytochrome P450-dependent monooxygenases. *Proc. Natl. Acad. Sci.* USA 93: 6869-6874.

Monties B., 1991. Plant cell walls as fibrous lignocellulosic composites–Relations with lignin structure and function. *Anim.Feed Sci. Technol.* 32: 159–175.

Muller L.D., Lechtenberg V.L., Bauman L.F., Barnes R.F., Rhykerd C.L. 1972. *In vivo* evaluation of a *brown midrib* mutant of *Zea mays*. *J. Anim. Sci.* 35: 883–889.

O'Connell A., Holt K., Piquemal J., Grima-Pettenati J., Boudet A., Pollet B., Lapierre C., Petit-Conil M., SchuchW., Halpin C. 2002. Improved paper pulp from plants with suppressed cinnamoyl-CoA reductase or cinnamyl alcohol dehydrogenase. *Transgenic Res.* 11: 495–503.

Önnerud H., Zhang L., Gellerstedt G., Henriksson G. 2002. Polymerization of monolignols by redox shuttle–mediated enzymatic oxidation: a new model in lignin biosynthesis *I. Plant Cell* 14: 1953-1962.

Osakabe K., Tsao C.C., Li L., Popko J.L., Umezawa T., Carraway D.T., Smeltzer R.H., Joshi C.P., Chiang V.J. 1999. Coniferyl aldehyde 5-hydroxylation and methylation direct syringyl lignin biosynthesis in angiosperms. *Proc. Natl. Acad. Sci.* USA 96: 8955-8960.

Parvathi K., Chen F., Guo D., Blount J.W., Dixon R.A. 2001. Substrate preferences of *O*-methyltransferases in alfalfa suggest new pathways for 3-O-methylation of monolignols. *Plant J.* 25: 193-202.

Patzlaff A., McInnis S., Courtenay A., Suman C., Newman L.J., Smith C., Bevan M.W., Mansfield S., Whetten R.W., Sederoff R.R., Campbell M.M. 2003. Characterisation of a pine MYB that regulates lignification. *Plant J.* 36: 743–754

Peter G., Neale D. 2004. Molecular basis for the evolution of xylem lignification. *Curr. Opin. Plant Biol.* 7: 737–742.

Pilate G., Guiney E., Holt K., Petit-Conil M., Lapierre C., Leple J.C., Pollet B., Mila I., Webster E.A., Marstorp H.G., Hopkins D.W., Jouanin L., Boerjan W., Schuch W., Cornu D., Halpin C. 2002. Field and pulping performances of transgenic trees with altered lignification. *Nature Biotechnol.* 20: 607–612.

Piquemal J., Chamayou S., Nadaud I., Beckert M., Barrière Y., Mila I., Lapierre C., Rigau J., Puigdomenech P., Jauneau A., Digonnet C., Boudet A.M., Goffner D., Pichon M.. 2002. Down-regulation of caffeic acid *O*-methyltransferase in maize revisited using a transgenic approach. *Plant Physiol.* 130: 1675–1685.

Piquemal J., Lapierre C., Myton K., O'Connell A., Schuch W., Grima-Pettenati J., Boudet A.M. 1998. Down-regulation of cinnamoyl-CoA reductase induces significant changes of lignin profiles in transgenic tobacco plants. *Plant J.* 13: 71-83.

Ralph J., Lapierre C., Lu F., Marita J.M., Pilate G., Van Doorsselaere J., Boerjan W., Jouanin L. 2001. NMR evidence for benzodioxane structures resulting from incorporation of 5-hydroxyconiferyl alcohol into lignins of *O*-methyltransferase-deficient plants. *J. Agric. Food Chem.* 49: 86–91.

Ralph J., Akiyama T., Kim H., Lu F., Schatz P.F., Marita J.M., Ralph S. A., Reddy M. S.S., Chen F., Dixon R.A. 2006. Effects of coumarate 3-hydroxylase down-regulation on lignin structure. *J. Biol. Chem.* 281: 8843–8853.

Ralph J., MacKay J.J., Hatfield R.D., O'Malley D.M., Whetten R.W., Sederoff R.R. 1997. Abnormal lignin in a loblolly pine mutant. *Science* 277: 235–239.

Reddy M.S.S., Chen F., Shadle G., Jackson L., Aljoe H., Dixon R.A. 2005. Targeted down-regulation of cytochrome P450 enzymes for forage quality improvement in alfalfa (*Medicago sativa* L.). *Proc. Natl. Acad. Sci.* USA 102: 16573–16578.

Scharlemann J.P.W., Laurance W.F. 2008. How green are biofuels? *Science* 319: 43-44.

Schoch G., Goepfert S., Morant M., Hehn A., Meyer D., Ullmann P., Werck-Reichhart D. 2001. CYP98A3 from *Arabidopsis thaliana* is a 3'-hydroxylase of phenolic esters, a missing link in the phenylpropanoid pathway. *J. Biol. Chem.* 276: 36566-36574.

Searchinger T., Heimlich R., Houghton A., Dong F., Elobeid A., Fabiosa J., Tokgoz S., Hayes D., Yu T.H. 2008. Use of U.S. croplands for biofuels increases greenhouse gases through emissions from land-use change. *Science* 319: 1238-1240.

Sederoff R.R., MacKay J.J., Ralph J., Hatfield R,D. 1999. Unexpected variation in lignin. *Curr. Opin. Plant Biol.* 2: 145–152.

Sewalt V.J.H., Ni W., Jung H.G., Dixon R.A. 1997. Lignin impact on fiber degradation: Increased enzymatic digestibility of genetically engineered tobacco (*Nicotiana tabacum*) stems reduced in lignin content. *J. Agric. Food Chem.* 45: 1977–1983.

Shadle G., Chen F., Reddy M.S.S., Jackson L., Nakashima J., Dixon R.A. 2007. Down-regulation of hydroxycinnamoyl CoA: Shikimate hydroxycinnamoyl transferase in transgenic alfalfa affects lignification, development and forage quality. *Phytochem.* 68: 1521–1529.

Sibout R., Eudes A., Mouille G., Pollet B., Lapierre C., Jouanin L., Séguin A. 2005. Routing sinapyl and coniferyl alcohol pathway though the last reduction step in *Arabidopsis thaliana* using a double CAD mutant. *Plant Cell* 17: 2059–2076.

Sibout R., Eudes A., Pollet B., Goujon T., Mila I., Granier F., Séguin A., Lapierre C., Jouanin L. 2003. Expression pattern of two paralogs encoding cinnamyl alcohol dehydrogenases in *Arabidopsis*. Isolation and characterization of the corresponding mutants. *Plant Physiol.* 132:848–860.

Snchez O.J., Cardona C.A. 2007. Trends in biotechnological production of fuel ethanol from different feedstocks. *Bioresource Technol.* doi:10.1016/j.biortech.2007.11.013.

Spoer H.A., Milner H.W. 1949. The Chemical composition of *Chlorella*; effect of environmental conditions. *Plant Physiol.* 24:120-149.

Tamura T., Morohoshi N., Yasuda S. 2001. Suitability of peroxidase-suppressed transgenic hybrid aspen (*Populus sieboldii X Populus gradidentata*) for pulping. *Holzforschung* 55: 335-339.

Tobias C.M., Chow E.K. 2005. Structure of the cinnamyl-alcohol dehydrogenase gene family in rice and promoter activity of a member associated with lignification. *Planta* 220: 678–688.

Tuskan G.A., DiFazio S., Jansson S., Bohlmann J., Grigoriev I., Hellsten U., Putnam N., Ralph S., Rombauts S., Salamov A., Schein J., Sterck L., Aerts A., Bhalerao R.R., Bhalerao R.P., Blaudez D., Boerjan W., Brun A., Brunner A., Busov V., Campbell M., Carlson J., Chalot M., Chapman J., Chen G.L., Cooper D., Coutinho P.M., Couturier J., Covert S., Cronk Q., Cunningham R., Davis J., Degroeve S., Déjardin A., dePamphilis C., Detter J., Dirks B., Dubchak I., Duplessis S., Ehlting J., Ellis B., Gendler K., Goodstein D., Gribskov M., Grimwood J., Groover A., Gunter L., Hamberger B., Heinze B., Helariutta Y., Henrissat B., Holligan D., Holt R., Huang W., Islam-Faridi N., Jones S., Jones-Rhoades M., Jorgensen R., Joshi C., Kangasjärvi J., Karlsson J., Kelleher C., Kirkpatrick R., Kirst M., Kohler A., Kalluri U., Larimer F., Leebens-Mack J., LepléJ.C., Locascio P., Lou Y., Lucas S., Martin F., Montanini B., Napoli C., Nelson D.R., Nelson C., Nieminen K., Nilsson O., Pereda V., Peter G., Philippe R., Pilate G., Poliakov A., Razumovskaya J., Richardson P., Rinaldi C., Ritland K., Rouzé P., Ryaboy D., Schmutz J., Schrader J., Segerman B., Shin H., Siddiqui A., Sterky F., Terry A., Tsai C.J., Uberbacher E., Unneberg P., Vahala J., Wall K., Wessler S., Yang G., Yin T., Douglas C., Marra M., Sandberg G., Van de Peer Y., Rokhsar D.. 2006. The Genome of black cottonwood, *Populus trichocarpa* (Torr. & Gray). *Science* 313: 1596 – 1604.

Van Doorsselaere J., Baucher M., Chognot E., Chabbert B., Tollier M., Petit-Conil M., Leplé J., Pilate G., Cornu D., Monties B., Van Montagu M., Inzé D., Boerjan W., Jouanin L. 1995. A novel lignin in poplar trees with a reduced caffeic acid/5-hydroxyferulic acid *O*-methyltransferase activity. *Plant J.* 8: 855-864.

Vignols F., Rigau J., Torres M.A., Capeliades M., Puigdomenech, P. 1995. The *brown midrib3* (*bm3*) mutation in maize occurs in the gene encoding caffeic acid *O*-methyltransferase. *Plant Cell* 7: 407-416.

Wagner A., Ralph J., Akiyama T., Flint H., Phillips L., Torr K., Nanayakkara B., Kiri L.T. 2007. Exploring lignification in conifers by silencing hydroxycinnamoyl-CoA:shikimate hydroxycinnamoyltransferase in *Pinus radiate*. *Proc. Natl. Acad. Sci.* USA 104: 11856–11861.

Zhang K., Qian Q., Huang Z., Wang Y., Li M., Hong L., Zeng D., Gu M., Chu C.,, Cheng Z. 2006. GOLD HULL AND INTERNODE2 encodes a primarily multifunctional cinnamyl-alcohol dehydrogenase in rice. *Plant Physiol.* 140: 972–983.

Zhong R., Morrison W.H., Himmelsbach D.S., Poole F.L., Ye Z.H. 2000. Essential role of caffeoyl coenzyme A *O*-methyltransferase in lignin biosynthesis in woody poplar plants. *Plant Physiol*. 124: 563–577.

Zhong R., Morrison W.H., Negrel J., Ye Z.H. 1998. Dual methylation pathways in lignin biosynthesis. *Plant Cell* 10:2033-2045.

In: Transgenic Plants: New Research
Editor: Oliver T. Chan

ISBN 978-1-60692-017-6
© 2009 Nova Science Publishers, Inc.

Chapter 4

NOVEL APPROACHES TO IMPROVING PLANT TRANSFORMATION

A. Boyko and I. Kovalchuk
Department of Biological Sciences, University of Lethbridge,
Lethbridge, AB, T1K 3M4, Canada

ABSTRACT

Agrobacterium tumefaciens-mediated plant transformation represents a powerful biotechnological technique for introducing desirable traits into commercially important plant species. Multiple approaches were undertaken during the last decade in order to improve the efficiency of plant transformation. Unfortunately, a relatively low frequency of stable integration events, high-copy-number insertions, a need for using antibiotic based selection markers, and a number of plant species recalcitrant to *Agrobacterium* transformation vote for continuous search for improved transformation procedures. The work of our group was directed on elucidating novel factors that influence the efficiency of tissue regeneration and transgenesis. The integration of foreign DNA into the plant genome is believed to be strongly dependent on the activity of the host DNA repair machinery and, therefore, can be stimulated via upregulation of homologous recombination rate (RR) in host cells just before transformation. This should enable the manipulation of a frequency and perhaps, quality of integration events. Our group was able to develop several unique growth medium compositions that allow a significant increase in RR without causing a negative effect on plants. The medium composition that exhibited no genotoxic influence was further tested for its effect on stable transformation in tobacco. The results obtained demonstrated that the application of a novel medium composition allows us to achieve higher transformation rates, as compared to those obtained while using conventional media. Plants grown on the modified medium generally looked healthier and produced more transgenic calli per amount of tissue transformed. We demonstrated that the application of this novel medium resulted in callus induction and plant regeneration as well as an increase in transformation rates by 2–3-fold, as compared to the application a conventional medium. Moreover, it resulted in the quality of integration events similar to that for the conventional medium. Currently, we continue searching for other factors that can be successfully used to improve plant transformation efficiency. Several chemical agents displaying a positive effect on *Agro*-transformation have been successfully identified. The possibility of using these factors alone and in combination is currently under the scope of our study. Overall, we believe that our approach represents a low-cost and highly efficient strategy for achieving a high number of stable transformants in a single round of plant transformation with *Agrobacterium*.

INTRODUCTION

During the past decade, *Agrobacterium*-mediated genetic transformation became one of the dominant technologies used to produce a variety of genetically modified transgenic plants. The successful introduction of this technology can be partially explained by the fact that *Agrobacterium tumefaciens* used for donor DNA delivery is a typical plant pathogen present in soil. However, in contrast to many other plant pathogens, it has a unique ability of trans-kingdom DNA transfer (reviewed in Gelvin 2003). The natural ability to transform a host cell with the bacterial DNA observed in *Agrobacterium* triggered an extensive search for other plant pathogens with similar properties. To date, three non-*Agrobacterium* species, all sharing the ability of plant genetic transformation, have been identified. This short list includes *Rhizobium sp.* NGR234, *Sinorhizobium meliloti* and *Masorhizobium loti* (Broothaerts *et al.*, 2005; reviewed in Chung *et al.*, 2006).

However, *Agrobacterium* remains a favoured tool used for transgenesis. This is reflected by a growing number of *Agrobacterium*-related patents being claimed every year (Roa-Rodriguez and Nottenburg, 2003), and by a continuously increasing range of host species that currently includes various plant species, yeast, mushrooms, filamentous and phytopathogenic fungi, and even human cells (Tzfira and Citovsky, 2003; reviewed in Lacroix *et al.*, 2006). The ability to transform human cells offers great prospects for using *Agrobacterium* not only in plant biotechnology but also in human and animal gene therapy.

Despite the continuously increasing popularity of *Agrobacterium*-mediated genetic transformation, the improvement of efficiency and precision of transgene integration is still necessary. Moreover, a number of agriculturally important plant species remains highly recalcitrant to *Agrobacterium* transformation. Understanding of biology of host-*Agrobacterium* interactions may help to solve these problems. Specifically, manipulating of *Agrobacterium* and host to enhance the virulence and susceptibility respectively may help to overcome the existing obstacles and lead to the major advancements in a field of plant transformation.

In this chapter we will discuss the various host manipulation approaches that can be undertaken to improve the transformation efficiency. We will demonstrate, how modifying plant growth medium composition allows to increase the frequency of transgene integration events and a number of transgenic plants produced during one transformation round.

THE BASIC MECHANISM OF *AGROBACTERIUM*-MEDIATED GENETIC TRANSFORMATION

Agrobacterium performs genetic transformation of recipient species by transfer and integration into the host genome the bacterium tumor-inducing plasmid (Ti) region (approximately 10 to 30kbp in size). The transferred DNA region is known as the T-DNA, and it usually represents less than 10% of the Ti plasmid. Borders of the T-DNA region are well-defined by the so-called border sequences that are represented by the 25 bp direct long repeats, flanking the T-DNA from the right and the left borders. Generally, there is only one T-DNA region present on the Ti plasmid, however the Ti plasmids containing multiple T-DNA sequences are also known. T-regions sizes of some naturally occurring Ti plasmids can reach approximately 23 kpb, allowing the introduction of many genes into the T-region (reviewed in Gelvin 2003, and Tzfira *et al.*, 2004a).

In the wild-type *Agrobacterium* strains, the T-DNA encodes a set of oncogenes and opine-catabolism genes, and its expression in a plant cell results in neoplastic growth and production of opines. In contrast, biotechnology relies on using the so-called recombinant or disarmed *Agrobacterium* strains that have an entire sequence of the native T-DNA replaced with any genes of interest, and therefore are capable of plant transformation without causing tumour development (Tzfira and Citovsky, 2006 and references within). Interestingly, recombinant *Agrobacterium* was shown to transfer large T-DNAs (up to 150 kbp) (Hamilton *et al.*, 1996) and 200 kbp (Miranda *et al.*, 1992).

The Ti plasmid also contains a set of virulence (*vir*) genes. These genes together with chromosomal virulence (*chv*) genes mediate processing of T-DNA from the Ti plasmid and its subsequent export from the bacterium to the host cell. Briefly, VirD1 and VirD2 act as the site-specific endonucleases that recognize the T-DNA border sequences and release a single stranded (ss)T-DNA molecule from the Ti plasmid. The resulted ssT-DNA with one molecule of VirD2 attached to the 5' end represents an immature T-complex that is exported by the VirB/D4 type IV secretion system into the plant cell along with several other Vir proteins, including VirE2, VirE3 and VirF. Once inside the host cell cytoplasm, the T-DNA covered with numerous VirE2 molecules can form a "telephone cord"-like structure of a mature T-complex, which is then actively transported to the nucleus using a cell dynein motor (Tzfira, 2006). A mature T-complex enters the nucleus through a nucleus pore complex (NPC), and is believed to be directed to double strand break (DSB) sites, where T-DNA is stripped of accompanying proteins and integrated into the host genome (reviewed in Tzfira *et al.*, 2004b; Tzfira and Citovsky, 2006).

It is noteworthy that on various stages of T-DNA delivery and integration (including the immature T-complex delivery to the host cell cytoplasm, transport through the cytoplasm and nuclear import, intracellular transport, T-DNA uncoating and integration), the activity and factors of various host cellular mechanisms are actively utilized by *Agrobacterium* to complete transformation (reviewed in Citovsky *et al.*, 2007). The fact that *Agrobacterium* depends greatly on host factors for the successful T-DNA integration suggests a variety of promising strategies for improving plant transformation efficiency that are based on various manipulations with host factors participating in T-DNA integration.

Host Systems Involved in *Agrobacterium* Infection

Among host factors used by *Agrobacterium* on its way to the nucleus are plant cell surface receptors. The attachment of *Agrobacterium* to the host cell is an absolute requirement for transformation (McCullen and Binns, 2006). Recent studies made it possible to isolate several putative receptors and cell-attachment factors localized on the host cell surface and utilized by *Agrobacterium*. Among those are a vitronectin-like protein, a rhicadhesin-binding protein, a cellulose synthase-like gene and several VirB2-interacting proteins (BTIs) (reviewed in Citovsky *et al.*, 2007).

Once inside the host cytoplasm, the mature T-complex has to be delivered to the nucleus. A large size of the complex (~ 15.7 nm outer diameter (Abu-Arish *et al.*, 2004) and a dense structure of cytoplasm contradicts passive transport of the complex. In contrast, a model, based on the available experimental evidences, that suggested an active mechanism mediating transport of this complex to the nucleus was recently developed (Salman *et al.*, 2005; Tzfira, 2006). Tzfira (2006) suggested that active transport is performed by a dynein-like *Arabidopsis* protein DLC3, which interacts with another host factor, a VirE2-interacting protein 1 (VIP1), mediating the recognition of VirE2 by the nuclear import machinery (Citovsky *et al.*, 2007).

A larger size of the T-complex (as compared to the NPC (~ 9 nm) implies that a kind of an active mechanism should exist in a plant cell that could provide the T-complex delivery to the nucleus. The host nuclear-import machinery itself is the best tool for doing this. In consistence with this hypothesis, both proteins interacting with T-DNA in the T-complex, VirD2 and VirE2 proteins, can also interact with host importin/karyopherin α (Ballas and Citovsky, 1997) and VIP1 (Tzfira *et al.*, 2001) proteins, respectively. While VirD2 is directly recognized by the nuclear-import machinery and initiates nuclear import, VIP1 along with another bacterial factor, a VirE3 protein, plays a role of a molecular adaptor between VirE2 and host karyopherin α (reviewed in Tzfira and Citovsky, 2006). Consistent with importin α family role in the T-complex import, *Arabidopsis* plants deficient in one of the importin α genes are resistant to *Agrobacterium* infection (Zhu *et al.*, 2003a). Interestingly, a VirE3 protein may play a role of a bacterial host range factor (Hirooka and Kada, 1986) that compensates the absence or low levels of VIP1 in some plants and non-plant species (discussed in Citovsky *et al.*, 2007). Indeed, VirE3 promotes nuclear accumulation of VirE2 in mammal cells that lack VIP1 and in plants with suppressed expression of VIP1 (Lacroix *et al.*, 2005). Deng *et al.* (1998) reported interactions between VirD2 and host cyclophilins that serve as molecular chaperons and maintain the proper conformation of VirD2 during its transport to the nucleus.

Another group of host factors, identified in *Arabidopsis* and alfalfa, that influence nuclear import of the T-complex, is represented by a nuclear cyclin-dependent kinase-activating kinase (CAK2Ms) (Bakó *et al.*, 2003), and by a type 2C serine/threonine phosphatase DIG3 (Tao *et al.*, 2004). While the former factor phosphorylates VirD2 and promotes its nuclear import, the latter reverses phosphorylation and negatively regulates nuclear import of the T-complex. Interestingly, the recent report by Djamei *et al.* (2007) suggested that *Agrobacterium* is also able to use a part of a mitogen-activated protein kinase (MAPK) defence signalling pathway, the MPK3 kinase, for phosphorylation of VIP1 that leads to its nuclear localization (reviewed in Dafny-Yelin *et al.*, 2008).

Delivery of the T-complex to its integration site in chromatin also requires the participation of host factors, particularly CAK2M, a TATA-box binding protein (TBP), VIP1 and core histones (reviewed in Citovsky *et al.*, 2007). The first two factors, CAK2M and TBP, bind to VirD2 (Bakó *et al.*, 2003). Moreover, CAK2M interacts with a RNA polymerase II largest subunit, thus permitting recruitment of TBP that allows to control transcription and transcription-coupled repair. Overall, interactions between CAK2M, TBP and VirD2 target the T-complex to the host chromatin (Bakó *et al.*, 2003). Considering the involvement of VIP1 in chromatin decondensation, its interactions with VirE2 may play a role similar to that of the T-complex targeting to chromatin (Tzfira *et al.*, 2001). In consistence with this model, VIP1 was shown to interact *in planta* with *Arabidopsis* core histone H2A (Li *et al.*, 2005a; Loyter *et al.*, 2005). An *Arabidopsis* histone H2A mutant was found to be deficient in T-DNA integration (Mysore *et al.*, 2000a). Finally, the involvement of the host transcription machinery in delivery of T-complexes to insertion sites can be further supported by finding of Alonso *et al.* (2003), who demonstrated that T-DNA insertions into regulatory regions of plant genes occur at high frequencies.

During the integration stage, the interaction of the host DNA repair machinery with T-DNA requires uncoating T-DNA of its cognate proteins. Uncoating is completed via the plant-targeted proteolysis machinery. VirF, a bacterial F-box protein that functions as a subunit of E3 ubiquitin ligase forms a ternary complex with VIP1 and VirE2. This results in their destabilization by targeted proteasomal degradation (Tzfira *et al.*, 2004b). The foregoing is consistent with a nuclear localisation of VirF and a negative effect of proteasomal inhibitors on early T-DNA expression (Tzfira *et al.*, 2004b). Dispensability of VirF for transformation of some plant species indicates that unidentified plant F-box proteins may be also involved in this process (Regensburg-Tuink and Hooykaas, 1993).

The final step of transformation, the actual integration of T-DNA into plant genomes, is almost completely host-dependent. It mainly determines the precision of transgene integration (Tzfira *et al.*, 2004a). Unfortunately, there are still some blank spots and data controversy regarding mechanisms and factors employed during this stage. Two mutually exclusive models that could explain T-DNA integration were originally proposed (reviewed in Tzfira *et al.*, 2004a). The first, the double-strand-break repair (DSBR) model, postulates that T-DNA preferentially integrates into DSB sites. Integration via the DSBR model requires conversion of ssT-DNA molecules into dsT-DNA molecules prior to integration. In contrast, the single-strand-gap repair (SSGR) model suggests that ssT-DNA integrates into a nicked DNA and then provides a template for complimentary DNA strand synthesis that substitutes an original host DNA sequence within the integration site. Both models were later modified to incorporate new experimental evidences (reviewed in Tzfira *et al.*, 2004a).

The increasing number of experimental evidences, however suggests targeted integration of T-DNA into DSB sites via double stranded intermediates (Chilton and Que, 2003; Tzfira *et al.*, 2003). Hence, ssT-DNA that enters the nucleus is being converted into a double stranded intermediate that later is directed to naturally occurring DSBs. Once at DSB site, T-DNA is integrated in the host genome using the host DNA repair machinery. This implies that host factors are required for conversion of a single stranded T-DNA to a double stranded T-DNA, production of DSBs in the host genome, and finally for incorporation (ligation) of T-DNA molecule into these breaks.

Strand Break Repair Pathways Potentially Involved in Transgene Integration

DNA DSBs, and thus T-DNA integration can be processed using one of two alternative DNA repair pathways, non-homologous end-joining (NHEJ) or homologous recombination (HR). Both pathways differ in their fidelity of DNA repair (Gorbunova and Levy, 1999; Puchta, 2005) that may affect the intactness of T-DNA sequence during integration. NHEJ is a relatively inaccurate DNA repair pathway and is frequently associated with small and large-scale deletions (from 1bp up to >1kb), insertions (up to 1.2kp long) or point mutations (Gorbunova and Levy, 1999). In contrast, HR is considered to be a more precise DNA repair pathway with a 100% accuracy, if a perfectly homologous template is used to prime repair synthesis. This could allow a better quality of integration events and generation of site-specific transgene insertions and gene targeting (GT) (Vergunst and Hooykaas, 1999; Puchta 2002; Reiss 2003).

Unfortunately, the majority of DSBs in plant cells are repaired via NHEJ, and the contribution of HR is very low (Gorbunova and Levy, 1999; Puchta, 2005). This leads to a major drawback in the development of technology for a site-specific transgene integration in plant cells. A number of studies conducted in the past several years were focused on: identification of host DNA repair factors involved in T-DNA integration, revealing mechanisms that control the DSB repair pathway choice in a cell, and manipulating these mechanisms during T-DNA integration in the host genome (reviewed in Tzfira *et al.*, 2004a; Tzfira and Citovsky, 2006; and Shrivastav *et al.*, 2008).

Most of experimental evidences supporting the role of host DNA repair factors in T-DNA integration were obtained using yeast as a model system. Using this system, it was first demonstrated that simultaneous inactivation of HR and NHEJ by mutations in RAD52 and KU70 genes inhibits T-DNA integration completely (van Attikum and Hooykaas, 2003), proving that the DSB repair machinery of a cell is absolutely required for successful transformation. Consistently, mutations inactivating the RAD52 or KU70 gene lead to preferential integration of T-DNA via the NHEJ (van Attikum and Hooykaas, 2003) or HR (van Attikum *et al.*, 2001) pathway, respectively. This suggests Rad52 and Ku70 enzymes to play a key role in the determination of a T-DNA integration mode in yeast. Using different yeast mutants deficient in DNA repair genes, van Attikum and Hooykaas (2003) demonstrated that integration of T-DNA via the NHEJ pathway requires the activity of Ku70, Rad50, Mre11, Xrs2, Lig4 and Sir4. In contrast, integration via the HR pathway is independent of Rad50, Mre11, Xrs2, Lig4 and Ku70 and requires the activity of Rad52 and Rad51 (van Attikum *et al.*, 2001).

Unfortunately, these results can not be directly applied to plants. In yeast, HR is the predominant mode of DNA repair. In contrast, plants similar to other higher eukaryotes predominantly use the NHEJ pathway. The difference is also reflected at the molecular level: vertebrates and plants have three critical NHEJ-related factors that are missing in yeast, including DNA-PKcs, BRCA1 and Artemis (Bleuyard *et al.*, 2006 and references within; Reidt *et al.*, 2006; reviewed in Shrivastav *et al.*, 2008). The presence of DNA-PKcs in higher eukaryotes may play a critical role in the regulation of NHEJ and could account for its domination over HR (reviewed in Shrivastav *et al.*, 2008). Since a low HR frequency in vertebrate cells coexists with the presence of a yeast Rad52 homologue, the absence of Rad52 in plants (Iyer *et al.*, 2002) can not be considered to be the main cause of a low HR activity.

One of the key factors involved in T-DNA integration in plants is a Ku80 protein that in complex with Ku70 and DNA-PKcs proteins represents the NHEJ pathway. The role of Ku80 was supported by genetic studies using *Arabidopsis* insertional mutants in the KU80 gene that were found to be deficient in T-DNA integration in somatic cells (Li *et al.*, 2005b). Immunoprecipitation of Ku80-dsT-DNA complexes from *Agrobacterium*-infected plants supported a physical interaction between Ku80 and T-DNA in the nucleus (Li *et al.*, 2005b). Consistently, overexpression of Ku80 in plant somatic cells resulted in an increased susceptibility to *Agrobacterium* infection (Li *et al.*, 2005b). In contrast to somatic cells, data on the involvement of Ku80 in T-DNA integration in germ-line cells remain controversial. While Friesner and Britt (2003) reported Ku80 to be essential for transformations, the study of Gallego *et al.* (2003) demonstrated its dispensability. Similarly, another NHEJ factor, *Arabidopsis* LIG4 ligase, was dispensable for transformation in somatic cells (van Attikum *et al.*, 2003) and required in germ-line cells (Friesner and Britt, 2003). Interestingly, some *Arabidopsis* mutants, like *rat1,* that show resistance to *Agrobacterium* somatic transformation (Zhu *et al.*, 2003b) also display susceptibility to germ line transformation (Mysore *et al.*, 2000b). This phenomenon may indicate the difference in a spectrum of host factors required for transformation in somatic (Zhu *et al.*, 2003b) and germ-line cells (Ye *et al.*, 1999).

Experimental evidences provided by studies of Mysore *et al.* (2000a) demonstrated that core histones represent another important factor influencing the last step of plant transformation. Indeed, histone H2A expression in various plant tissues was positively correlated with their susceptibility to *Agrobacterium* infection (Yi *et al.*, 2002). Moreover, inactivation of the H2A gene in the *Arabidopsis rat5* mutant leads to deficiency in T-DNA integration (Mysore *et al.*, 2000a). Overall, histone H2A may promote T-DNA integration via relaxation of a host chromatin structure (Mysore *et al.*, 2000a). Consistently, the importance of chromatin structure in T-DNA integration was further supported using *fas1* and *fas2* mutant plants (Endo *et al.*, 2006). FAS1 and FAS 2 represent two of three subunits of Chromatin Assembly Factor 1 (CAF-1) (Kaya *et al.*, 2001; Henning *et al.*, 2003) that mediates deposition of H3/H4 histones onto replicating DNA, thereby promoting chromatin condensation (Tagami *et al.*, 2004). Indeed, reduced heterochromatin in *fas* mutants (Schonrock *et al.*, 2006) was consistent with an enhanced frequency of T-DNA integration into a plant genome (Endo *et al.*, 2006).

The significance of host chromatin structure for T-DNA integration is consistent with a genome-wide T-DNA insertional analysis that includes more than 80 000 independent insertions and demonstrates preferential integration of T-DNA in intergenic regions, including promoters, 5' and 3' UTRs, where active transcription usually occurs (Alonso *et al.*, 2003). However, it should be taken into consideration that the authors who reported preferential integration of T-DNA into transcriptionally active regions of the host genome (Alonso *et al.*, 2003; Tzfira *et al.*, 2003) used an active selection during callus regeneration step. It could result in a preferential selection of actively expressed T-DNA insertions located in transcriptionally active regions, thereby discriminating the selection of transgenic plants containing T-DNA insertions within heterochromatin or regions with a low transcriptional activity. In fact, recent studies that used non-selective conditions supported the idea that T-DNA integration occurs more randomly in a genome, and integration events can be found also in heterochomatic regions at relatively high frequencies (Francis and Spiker, 2005; Kim *et al.*, 2007).

MANIPULATION OF A HOST TO IMPROVE TRANSFORMATION EFFICIENCY

The existence of multiple host factors required for *Agrobacterium*-mediated transformation suggests a promising strategy directed on transformation improvement. Specifically, the various manipulations of host factors can be performed to increase its susceptibility to genetic transformation. In fact, gaining better control over the activity of host DNA repair pathways may help direct T-DNA integration preferentially via the HR pathway, thus increasing a precision and site-specificity of transgene integration.

Different strategies can be suggested for improving plant transformation via manipulation of host factors (Tzfira and Citovsky, 2006). The first strategy, which is probably the most used nowadays, is overexpression of host factors that has a positive effect on plant transformation. Indeed, overexpression of histone H2A in wild-type *Arabidopsis* plants greatly increased their susceptibility to *Agrobacterium* infection (Mysore *et al.*, 2000a). Similarly, overexpression of BTI in *Arabidopsis* (Hwang and Gelvin, 2004) and VIP1 in tobacco (Tzfira *et al.*, 2002) increased their susceptibility to genetic transformation. Overexpression of DNA repair proteins derived from different organisms in plant cells may increase the frequency of site-specific integration events. In fact, expression of the *E.coli RecA* gene in transgenic tobacco plants stimulated HR between sister chromatids by 2.4-fold over wild type plants (Reiss *et al.*, 2000). In a similar way, expression of the bacterial resolvase RuvC resulted in more that a 10-fold induction of somatic crossover and intrachromosomal recombination in transgenic tobacco plants (Shalev *et al.*, 1999). Nevertheless, despite its positive effect on HR, overexpression of RecA and RuvC in plant cells had no significant effect on the GT frequency.

In contrast, expression of the yeast *RAD54* gene, a member of the SWI2/SNF2 superfamily of chromatin remodelling genes in *Arabidopsis,* allowed increasing the GT frequency by one-two orders of magnitude (Shaked *et al.*, 2005). This specific induction is consistent with reduced rates of GT previously reported in yeast, mouse stem cells and chicken DT40 cells with the inactivated *RAD54* gene or its homologs (Bezzubova *et al.*, 1997; Essers *et al.*, 1997). An important role of chromatin remodelling factors in controlling the frequency of HR was also supported by early studies of Hanin *et al.* (2000) who identified the *Arabidopsis MIM* gene encoding for a protein belonging to the SMS (structure maintenance of chromosomes) family. Overexpression of a MIM protein in plant cells resulted in a 2-fold increase in the intrachromosomal HR frequency relative to the wild type (Hanin *et al.*, 2000). Mysore *et al.* (2000a) demonstrated that overexpression of the *RAT5* histone H2A gene in wild-type *Arabidopsis* plants increases transformation efficiency.

Inactivation of host genes that negatively influence T-DNA transformation and the HR pathway may represent an alternative to the strategy of overexpression. In fact, mutation in the *Arabidopsis RAD50* gene that encodes a protein involved in the NHEJ pathway stimulated HR and caused a hyper-recombinant phenotype (Gherbi *et al.*, 2001). Similarly, other host factors that negatively control T-DNA integration, like a protein phosphotase DIG3 that inhibits a nuclear import of VirD2 (Tao *et al.*, 2004), can be considered to be possible targets for down-regulation during genetic transformation. In fact, the studies of Ninomiya *et al.* (2004) using *Neurospora* strains deficient in human *KU70* and *KU80* homologs demonstrated that the rate of HR and GT can be greatly increased by suppression or blocking the NHEJ repair pathway. Similarly, *Arabidopsis fas1* and *fas2* mutant plants deficient in the activity of a CAF-1 protein involved in nucleosome assembly and chromatin condensation were hypersusceptible to *Agrobacterium* transformation (Endo *et al.*, 2006). Moreover, *fas* mutants displayed a hyper-recombinant phenotype which was consistent with enhanced transcription of the *AtRad51* and *AtRad54* genes (Endo *et al.*, 2006).

In plant species recalcitrant to *Agrobacterium,* transient overexpression of specific host factors during co-transformation can be problematic. Hence, it can be suggested to use an *Agrobacterium*-independent transformation technique (like biolistics) for achieving transient overexpression of specific host factors during inoculation with *Agrobacterium* (Tzfira and Citovsky, 2006). Alternatively, it would be interesting to express specific host factors fused to a bacterial export signal in *Agrobacterium* itself, so they could be delivered to host cells in a manner similar to other Vir proteins (Tzfira and Citovsky, 2006). Successful expression in *Agrobacterium* and the following export to a plant cell of a chimeric Cre recombinase fused to a VirF protein export signal during transformation (Vergunst *et al.*, 2005) supports the application of this strategy for enhancing transformation efficiency on a routine basis.

A choice of plant tissues for subsequent transformation can also greatly influence transformation efficiency. Early reports by Grevelding *et al.* (1993) demonstrated that transformation of *Arabidopsis* root tissues yields plants with fewer T-DNA insertions than that of leaf disks. Similarly, tissue-specific variations in histone H2A expression in *Arabidopsis* (Yi *et al.*, 2002) underline the importance of a tissue type selected for transformation. The age of plant explants chosen for transformation was also reported to be an important factor that influenced transformation frequency in cauliflower (Chakrabarty *et al.*, 2002) and wheat (Pastori *et al.*, 2001). In both cases, explants taken from young donors performed better compared to those obtained from older plants (Pastori *et al.*, 2001; Chakrabarty *et al.*, 2002). Overall, the creation of complex profiles describing tissue- and stage-specific expression of various host factors that contribute to transformation could greatly facilitate a search for optimal conditions for genetic modifications.

Proliferation rates in tissues used for transformation, cell types or cell cycle phases can represent another important component for improving transformation efficiency. In fact, early data from the synchronized petunia cell population indicated that entry in the S-phase of the cell cycle was essential for stable genetic transformation (Villemont *et al.*, 1997). Similarly, transformation of synchronized tobacco protoplasts during S–M phase resulted in increased recovery of selection-resistant colonies (Meyer *et al.*, 1985; Okada *et al.*, 1986). The important role of active cell division in plant transformation was also supported by higher transformation efficiency observed in maize cells expressing a modified version of viral replication-associated protein (RepA) that stimulates cell division (Gordon-Kamm *et al.*, 2002). Finally, a very low frequency of HR in vertebrate somatic cells that represent a serious obstacle for efficient GT was overcome by choosing cell types that initially displayed high HRFs (Yamazoe *et al.*, 2004) and actively suppressed NHEJ (Orii *et al.*, 2006). This allowed the successful application of chicken B lymphocyte cells (D-40 system) and mouse embryonic stem cells for genetic transformation and GT (Yamazoe *et al.*, 2004).

Another important consideration on the way of improving plant transformation is that even disarmed *Agrobacterium tumefaciens* used for transformation can activate host defence mechanisms (Djamei *et al.*, 2007; reviewed in Dafny-Yelin *et al.*, 2008). A number of economically important plants like maize and soybean are susceptible to transient transformation, but stable transformants are extremely hard to obtain (Gelvin 2003 and references within). This can be explained by activation of the host defence system triggered by bacterium infection, and it may lead to apoptotic responses and multiple necroses, thereby drastically decreasing the yield of stable transformants of the agronomically important plants like grape (Pu and Goodman, 1992; Deng *et al.*, 1995) and maize (Hansen, 2000). This obstacle was successfully overcome by using antioxidants that allowed to significantly increase the number of regenerated stable transformants of grape, rice, maize and soybean (Gelvin, 2003 and references within). Similarly, overexpression of the baculovirus cell death suppressor genes *p35* and *iap* greatly reduced necrotic responses and endogenous DNA cleavage in cells transformed by *Agrobacterium* (Gelvin, 2003 and references within).

Among other types of host defence mechanisms activated during *Agrobacterium* infection is a perception of bacterial pathogen-associated molecular patterns (PAMPs) that lead to the enhanced production of reactive oxygen species and increased callose deposition that reinforces cell walls (reviewed in Nürnberger *et al.*, 2004). A negative effect of PAMP perception on genetic transformation can be removed via mutation of the *EFR* plant gene that encodes a receptor kinase for perception of the bacterial EF-Tu PAMP. In fact, inactivation of the *EFR* receptor kinase results in super-susceptibility of the host to transformation (Zipfel *et al.*, 2006).

DNA DSBs and Transformation Improvement

Taking into consideration that the number of DSBs available in the host genome represents a limiting factor for T-DNA integration (Salomon and Puchta, 1998), transient induction of DSB by X-rays or chemical agents may be considered as a possible strategy to increase transformation efficiency (Kohler *et al.*, 1989). The main disadvantage of this approach is a negative effect DSBs have on genome stability and cell viability (Evans *et al.*, 1993; Critchlow and Jackson, 1998; Shrivastav *et al.*, 2008). Hence, the large scale induction of DSBs may affect the final yield and quality of transgenic plants. A good alternative would be a directed site-specific generation of DSBs in genetic loci, in which T-DNA integration is desirable. There are several strategies that use site-specific DSBs for improving precision of transformation.

The first system is based on using rare cutting endonucleases (I-*Sce*I, I-*Ceu*I), which can be transiently expressed in cells during co-transformation with two *Agrobacterium* strains that carry T-DNA containing the genes of interest and T-DNA containing the endonuclease gene (Tzfira *et al.*, 2003). Two major drawbacks of this approach are: a rare and random dispersion of endonuclease recognition sites through the plant genome (Puchta, *et al.*, 1993; Li *et al.*, 2007) that could drastically decrease precision of integration, and a need for the previous generation of transgenic plants that have a recognition site in a desired location.

The second approach that results in effective targeting of T-DNA into the desired loci is represented by the Cre-*lox* system (reviewed in Gilbertson, 2005). This bacteriophage derived system consists of the Cre recombinase and the 34 bp *lox* site. Cre binds to 13 bp long inverted repeats in the *lox* sites and catalyzes a crossover in the 8 bp spacer region of two *lox* sites. This system can be effectively used for both site-specific transgene integration and selectable marker gene removal (reviewed in Gilbertson, 2005). However, this system also has several drawbacks. Despite the absence of wild type *lox* sequences in plant genomes, they still contain pseudo-*lox* sites, which display more than 50% similarity with *lox* sites and can still be involved in recombination (Schmidt *et al.*, 2000; Silver and Livingston, 2001). Moreover, it was demonstrated that high expression of *cre* results in chromosomal rearrangements in mammalian cells (Schmidt *et al.*, 2000; Silver and Livingston, 2001) and produces aberrant phenotypes in plants including chlorosis, leaf deformation, stunted growth and delayed flowering (Que *et al.*, 1998; Coppoolse *et al.*, 2003). High levels of *cre* expression trigger recombination in chloroplasts between the *lox* site and AT-reach sequences located about 500 bp away from it (Hajdukiewicz *et al.*, 2001). Finally, Cre-*lox* as well as the I-*Sce*I system require preceding integration of *lox* sites and *Sce*I recognition sites in a location selected for site-specific transgene integration.

The third approach uses zinc-finger nucleases (ZFNs) that are specifically designed to recognise and introduce DSBs into predicted locations within the host genome (Lloyd *et al.*, 2005; Wright, *et al.*, 2005). This ZFN is a dimmer of two ZFNs, each consisting of a set of zinc-finger motifs allowing specificity in recognition of target DNA sequences, and a non-specific endonuclease (reviewed in Wright, *et al.*, 2005). Each pair of three-finger ZFNs has a recognition sequence of 18 pb, which should occur in a genome every 4^{18} bp. It means that three-finger ZFNs could provide enough specificity in genomes of a size of about 10^{10} bp (Lloyd *et al.*, 2005). Moreover, recognition specificity can be further increased by using four-finger ZFEs, which have a recognition sequence of 24 bp. Overall, this strategy has great potential for plant transgenesis and GT technology. Its routine application in biotechnology has currently been the subject of many ongoing studies (reviewed in Wu *et al.*, 2007).

ALTERNATIVE HOST-MANIPULATION STRATEGIES TO IMPROVE TRANSFORMATION EFFICIENCY

In contrast to direct overexpression, inactivation or co-delivery of host factors during transformation, indirect stimulation or preparation of the host for future transformation can be viewed as a possible approach for improving transformation efficiency. Reports in the early nineties suggested some promising ideas for transformation enhancement. It was demonstrated that supplementation of growth media of the future host with various chemicals can influence DNA repair. Waldman and Waldman (1991) grew mammalian cells in the continuous presence of 2 mM of 3-methoxybenzamide, a competitive inhibitor of PARP (Waldman and Waldman, 1990). Authors demonstrated a 3–4-fold induction of intrachromosomal homologous recombination (Waldman and Waldman, 1991). Since PARP facilitates NHEJ repair pathway, the application of PARP inhibitor promoted HR pathway.

Similarly, the use of another PARP inhibitor, niacinamide, for biolistic-mediated wheat transformation increased the number of low-copy transgene integration events (De Block *et al.*, 1997). Moreover, in the case of multiple integrations in the host genome, these copies segregated in the next generation indicating their unlinked nature (De Block *et al.*, 1997).

After its successful application for transformation, the niacinamide treatment was patented (De Block, US Patent No. 6074876). Transient expression of a reporter gene was increased by several folds by performing bombardment in the presence of silver thiosulfate and calcium nitrate (Perl *et al.*, 1992). Lowering temperatures during co-cultivation with *Agrobacterium* (Li *et al.*, 2003a) and chilling during a regeneration step (Immonen, 1996) also promoted transformation efficiency.

High transformation efficiency does not only depend on high frequency of T-DNA integration events; it also requires the establishment of optimal conditions for efficient regeneration of transgenic material from transformed tissues. In most of the cases, regeneration of transformed cells represents a major challenge in transformation of many agronomically important plants that are susceptible to T-DNA integration and recalcitrant to somatic embryogenesis (SE) (Takumi and Shimada, 1997; Li *et al.*, 2003b). To date, a number of chemicals promoting SE and diminishing negative effects during a tissue culture stage have been reported. Among the most interesting are the following: various methods of increasing cytosolic levels of calcium that can promote SE (Racusen and Schiavone, 1990; Chugh and Khurana, 2002; Malabadi and Staden, 2006); the application of silver that prevents cellular necrosis caused by ethylene (Dias and Martins, 1999; Sahrawat, *et al.*, 2003); promotion of SE in wheat by zinc deficiency (He *et al.*, 1991; Kothari, *et al.*, 2004) and various nitrogen sources (Immonen, 1996); promotion or prevention of shout regeneration by cupric sulphate (Saharawat, *et al.*, 2003; Kothari, *et al.*, 2004) and high EDTA (Kothari, *et al.*, 2004). Similarly, supplementation of regeneration media with spermidine was shown to improve the recovery of wheat transformants by more than 3-folds (Khanna and Daggard, 2003).

Ammonium and nitrate levels were reported to promote SE in various plant species (He, *et al.*, 1989; Grimes and Hodges, 1990; Mordhorst and Lorz, 1993; Choi *et al.*, 1998; Jiménez, *et al.*, 2001; Kothari, *et al.*, 2004). High concentrations of ammonium nitrate in regeneration media greatly improved SE and increased the yield of somatic embryos in ginseng (Choi *et al.*, 1998). In rice, the relative ratio of ammonium to nitrate could affect the sensitivity of immature embryos to auxin; the ratio of 1:1 produced the greatest insensitivity (Grimes and Hodges, 1990). In contrast, in barley microspore culture, it was shown that the ratio of two ions (except when it is extreme) had no influence upon SE yields (Mordhorst and Lorz, 1993). However, a link between the total concentration of inorganic nitrogen and the yield of somatic embryos has been established (Mordhorst and Lorz, 1993). In wheat however, it was not possible to show a strong connection between either the ratio or the total content of nitrogen species and SE yield (Menke-Milczarek and Zimny, 2001).

Modification of Host Growth Medium with Factors Enhancing the Activity of HR Can Improve Transformation Efficiency

Our previous studies demonstrated that the activity of HR in plant somatic cells can be effectively modified by changing the composition of plant growth medium. We found that changing the concentration of chemicals in MS medium (Murashige and Skoog, 1962) and supplementing MS medium with chemicals that are missing from its original composition can increase the frequency of HR events in plants. This suggests that growing plants on the modified medium that increases the activity of HR can also improve the efficiency of subsequent transformation with *Agrobacterium*. Our approach was somewhat similar to previously reported successful modifications of media used for embryo regeneration (He *et al.*, 1991; Maës, *et al.*, 1996; Choi *et al.*, 1998; Kothari, *et al.*, 2004). However, in contrast to the aforementioned reports, our approach was directed on increasing the frequency of transgene integration rather than enhancing SE.

In brief, our approach requires the identification of chemicals and medium modifications that can increase the activity of HR in somatic cells without causing substantial negative effects on the physiology of the future host. Next, the effects of the modified growth medium on transformation efficiency can be tested. Finally, the quality of integration events can be compared between the conventional and the modified media. If modification is successful, an active concentration range of chemicals selected can be adjusted for using during *Agrobacterium*-mediated transformation of various agriculturally important species that may differ in their growth requirements and general physiology.

Using this strategy, we were able to identify several compounds that could increase the activity of HR in somatic cells and lead to a higher frequency of stable transgene integrations. This allowed us to develop the growth medium composition that improves plant transformation (Boyko and Kovalchuk, 2006, US Patent No. 11/466184). Currently, we are characterizing the effects of other experimental medium compositions and preparing a public disclosure for their application. For this reason, we are unable to reveal the nature of chemical modifications made to the growth media discussed in this chapter. However, we will provide experimental evidence supporting an important role of growth medium composition for the improvement of transformation.

Identification of Chemicals and Medium Modifications that Can Increase the Activity of HR

The composition of MS medium was developed several decades ago, and it does not incorporate species-specific requirements for macro and micro nutrients for a broad range of plants used for transgenesis. In fact, a majority of recent studies conducted in the field of plant SE indicated the importance of the individualized growth medium composition for many species that are currently used for genetic manipulations. We hypothesized that the identification of growth medium composition enhancing the activity of HR in plant somatic cells can significantly improve transformation efficiency with *Agrobacterium*.

The effect of various chemicals on the activity of HR was tested using transgenic *Arabidopsis thaliana* line #11 plants (Swoboda, *et al.*, 1994; Ilnytskyy, *et al.*, 2004) grown on various types of a modified medium. The presence of a specially designed recombination substrate in the plant genome enabled sensitive measurements of the frequency of HR events in somatic cells. The recombination substrate consisted of two non-functional, truncated, and partially overlapping copies of the GUS gene cloned in direct orientation.

Upon a recombination event in the region of homology, the GUS gene is restored, and its activity can be visualized using histochemical staining (Jefferson, 1987). Cells containing the active GUS gene appear as blue spots under a dissecting microscope. Each blue spot represents an independent HR event. Next, the number of spots per plant can be calculated to obtain the frequency of HR. To account for a difference in plant sizes and growth rates on different media, the frequency of HR is related to the total number of haploid genomes (i.e. recombination substrate copies) in a single plant, as previously described (Boyko *et al.*, 2005). This value represents RR and is used to compare the effect of different media on the HR activity.

In general, two different approaches can be undertaken to determine key factors in MS medium that influence the activity of HR. MS medium can be depleted in one or several components. Alternatively, MS medium can be enriched with chemical(s) that are already available in the medium or are novel.

Using a depletion method, we developed and tested several experimental medium compositions for their effect on RR in *Arabidopsis* (Figure 1). Modified medium compositions were named A, B, C, D and E for simplicity's sake; and standard MS medium served as a control. The RR was measured after 3 weeks of continuous growth on modified media. Four out of five compositions being tested significantly increased the RR by 1.5 – 3.5 fold (Student's test, $\alpha=0.05$). In contrast, *Composition D* drastically reduced the RR, suggesting a critical role of a missing element in the activity of HR in plants. Enriching MS growth medium with this element might have a positive effect on recombination.

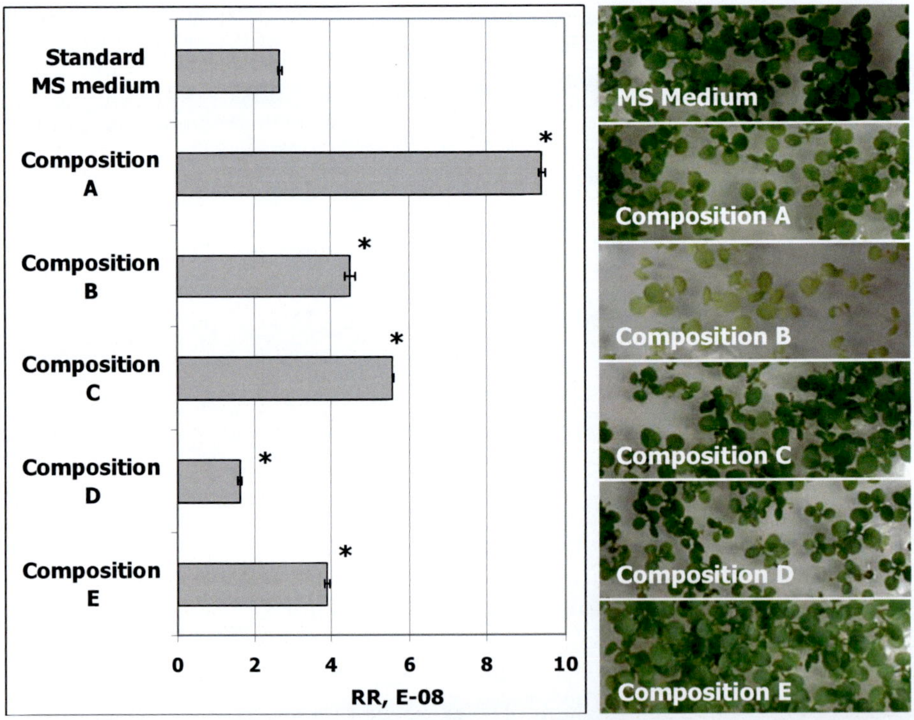

Figure 1. The effects of "depleted" MS medium on RR in *Arabidopsis*.
Plants were grown for 3 weeks on control and modified MS medium depleted in various MS components. The values represent the mean RR ± s.d. Asterisks show statistically significant difference from plants grown on control MS medium (Student's test, $\alpha=0.05$). The pictures show phenotype of plants grown on different media types. Pictures were taken on 22nd day post germination.

Despite a positive influence on RR, *Composition B* negatively affected plant physiology, as it was indicated by a plant phenotype (Figure 1). In contrast, *Composition E* improved plant growth and yielded larger plants, which supports an idea of growth medium composition being individually adjusted to selected plant species. Finally, plants grown on *Composition C* and *Composition A* media were not significantly different from control plants in their appearance. Overall, quick screening of medium components allowed identifying three different elements that can be manipulated to induce the HR activity in plants prior to a transformation step.

As an alternative to depletion of one particular component from the growth medium, we performed enrichment of the medium with chemicals that were either parts of the original MS medium composition or were novel. Bellow, we will discus two most successful modifications, namely *Composition F* and *Composition G*. Both compositions caused a substantial dose-dependent increase in RR in plants.

In brief, a 5-fold increase in the concentration of chemical *F* in the *Composition F* medium caused a 4.0-fold increase in RR in *Arabidopsis* plants (Figure 2). However, it also resulted in moderate growth inhibition caused by excessive amounts of chemical *F*. In contrast, a 2.5-fold increase in the concentration of chemical *F* improved plant growth and increased RR by 2.2-fold (Figure 2).

Similarly, the presence of chemical G in the *Composition G* medium had a positive effect on RR (Figure 3). The concentration of G chemical, named "1x" and "1.5x", had no significant negative effect on plant appearance, and increased RR by 1.5- and 3.0-folds, respectively (Figure 3). In contrast, the concentration "2x" of chemical G caused a 3.9-fold increase of RR and moderate growth inhibition.

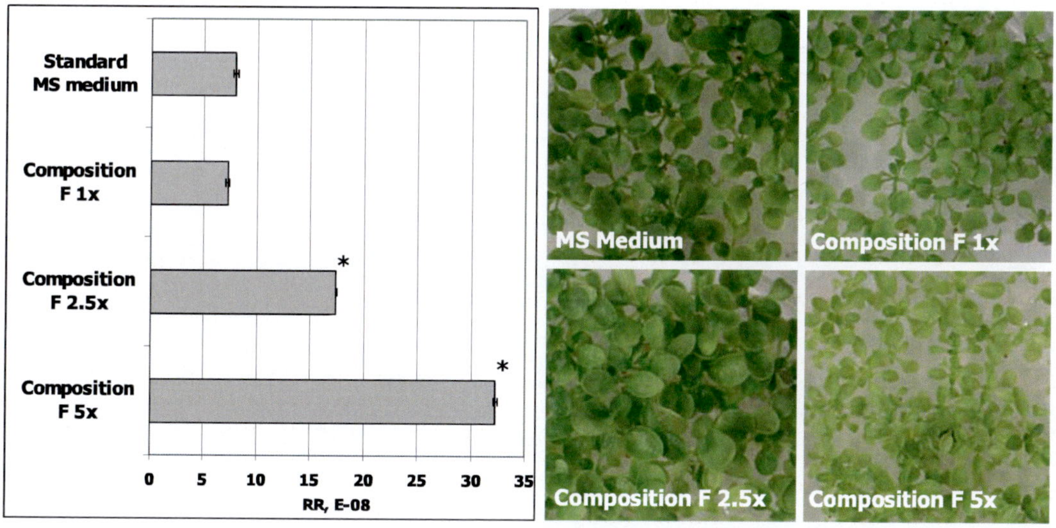

Figure 2. The effects of *Composition F* growth medium on RR in *Arabidopsis*.
Plants were grown for 3 weeks on control and modified MS medium supplemented with various amounts of chemical F. 'x' – shows the base level of chemical F selected for the experiment. The values represent the mean RR ± s.d. Asterisks show statistically significant difference from plants grown on control MS medium (Student's test, $\alpha=0.05$). The pictures show plants appearance on different types of the media. Pictures were taken on 22nd day post germination.

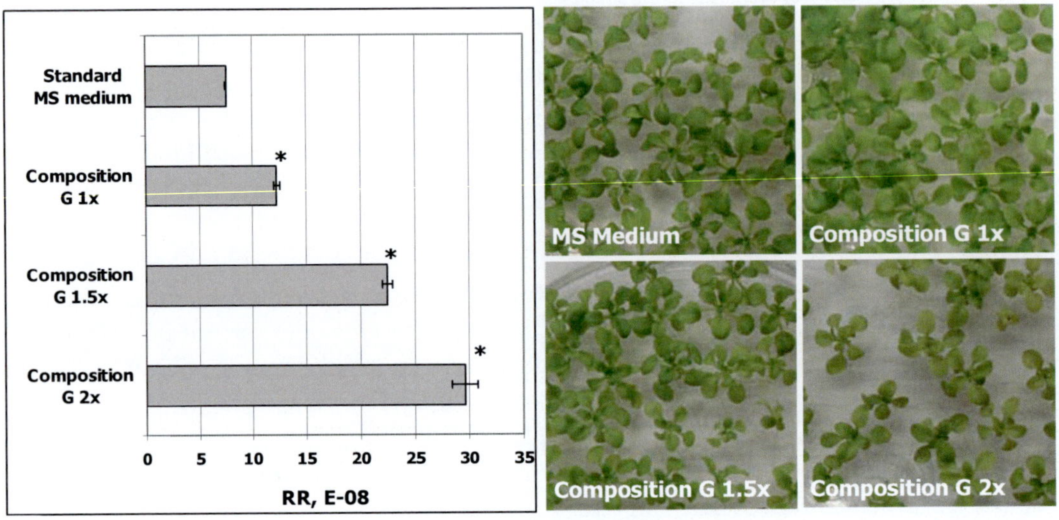

Figure 3. The effects of *Composition G* growth medium on RR in *Arabidopsis.*
Plants were grown for 3 weeks on control and modified MS medium supplemented with various amounts of chemical G. 'x' – shows the base level of chemical G selected for the experiment. The values represent the RR mean ± s.d. Asterisks show statistically significant difference from plants grown on control MS medium (Student's test, $\alpha=0.05$). The pictures show plants appearance on different types of the media. Pictures were taken on 22nd day post germination.

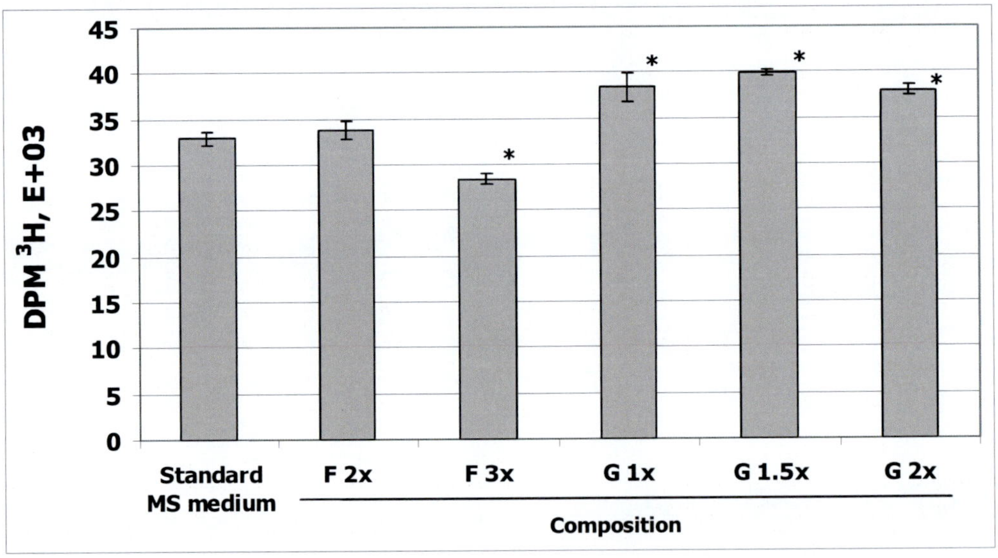

Figure 4. The level of DNA DSBs ($DPM\,^3H$) in plants grown on MS media supplemented with various amounts of chemical F and G.
Number of DNA DSBs was measured in 3-week-old *Arabidopsis* plants as previously described (Basnakian and James, 1996). Plants were grown on control and modified MS medium supplemented with various amounts of chemical F and G. 'x' – shows the base level of these chemicals selected for the experiment. The values represent the mean number of strand breaks (in arbitrary units) ± s.d. Asterisks show statistically significant difference from plants grown on control MS medium (Student's test, $\alpha=0.05$).

Major differences in the mechanism that mediated the effect of these two chemicals on the HR activity were observed upon the comparison of DNA DSBs levels. Whereas high amounts of chemical F did not induce DNA damage, a statistically significant increase in the number of DNA DSBs was observed for all types of *Composition G* media (Student's test $\alpha=0.05$) (Figure 4). This suggests key differences in the mechanisms mediating the HR activity increase by *Composition F* and *Composition G* media. While the presence of chemical G in the medium indirectly induced the activity of the DNA repair machinery via extensive DNA damage, chemical F was likely to increase RR via direct induction of factors involved in HR by unknown mechanisms. Moreover, the presence of 3x amount of chemical F in the growth medium significantly reduced the level of DNA DSBs (Figure 4). Our study illustrates need for an assessment of genotoxic effects caused by the application of selected chemicals in growth media. Indeed, to prevent negative effects on genome stability of host cells and regenerated stable transformants, medium composition modifications should be limited to those that do not induce extensive DNA damage.

EFFECTS OF MODIFIED GROWTH MEDIA ON THE FREQUENCY OF STABLE TRANSFORMATION OF *N. TABACUM* WITH *AGROBACTERIUM TUMEFACIENS*

As it was already emphasized, attempts to increase transformation efficiency should rely on chemicals that can induce the activity of HR without causing negative effects on plant physiology. The aforementioned experiments involving *Composition F* and *Composition G* media clearly demonstrated how two possible different mechanisms can mediate induction of the HR activity. Considering the foregoing, the *Composition G* medium was selected for the next round of experiments that included transformation of *N. tabacum* wild type plants with *Agrobacterium* and regeneration of stable transformants.

We hypothesise that growing plants on the medium that allows an increase in RR can potentially increase either the frequency of T-DNA integration or quality of integration events. Indeed, after testing the effect of chemical F on the transformation frequency, we found a strong positive correlation ($r=0.97$, $p<0.05$) between the number of transgenic plants produced and the amount of chemical F included in the modified growth medium (Figure 5). Plant leaves harvested from the *Composition F* medium showed better regeneration of transgenic calli, as compared to leaves obtained from the control MS medium. Overall, using two modified media enriched with various quantities of chemical F, namely 2x and 3x, led to a significant increase in the total number of transgenic plants produced during one transformation round by 1.9- and 2.7-folds, respectively (Student's test, $\alpha=0.05$) (Figure 5).

Importantly, we did not observe any significant difference in the copy number insertion and intactness of T-DNA/genomic DNA junctions among transgenic plants produced from the modified or control MS media. Overall, using conditions that promote the activity of HR substantially improved the frequency of transgene integration in *N. tabacum*. The application of this modification for genetic transformation of other plant species is also considered to be possible. Additional studies, however, are required to determine an active concentration range optimal for selected species.

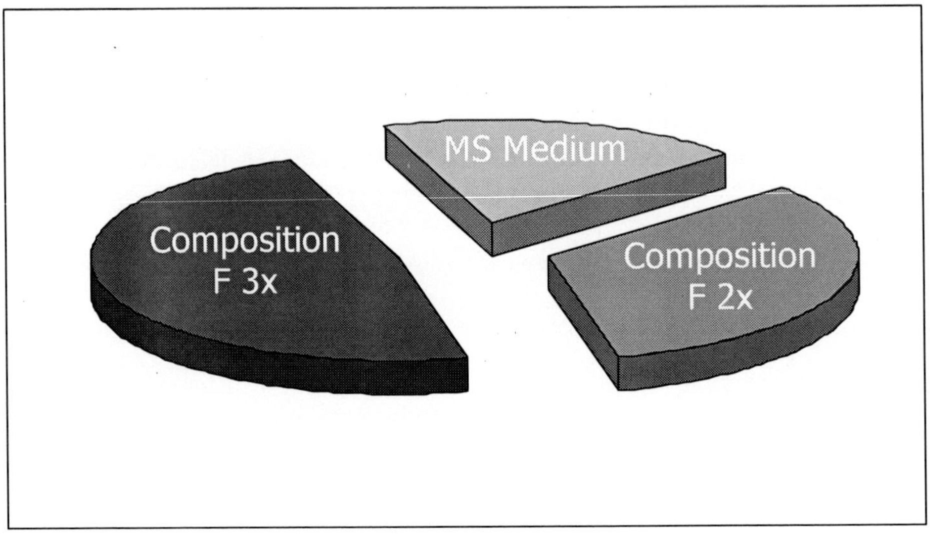

Figure 5. The effects of *Composition F* growth medium on stable transformation efficiency in *N. tabacum* plants.
The diagram shows the total yield of transgenic plants obtained from *Agro*-transformation of leaf disks harvested from *N. tabacum* plants grown on control MS medium and MS medium supplemented with 2x and 3x quantities of chemical F. The same amount of plant material was harvested for *Agro*-transformation from each medium. Transformed tobacco leaf disks were regenerated under the active selection conditions (hygromycin, 25 mg/L). Stable integration events were confirmed via detection of transgene (*luciferase*) expression in CCD camera as previously described (Ilnytskyy, *et al.*, 2004).

CONCLUSION

Genetic transformation is a complex process that relies on a rather complicated network of *Agrobacterium*-plant interactions. Understanding cross-talks between *Agrobacterium* and its host could help improve existing transformation protocols and obtain control over precision of transgene integration in the plant genome. Manipulating host factors and host growth environment represent promising strategies directed toward transformation improvement. Enhancing plant transformation with factors inducing the activity of HR in host somatic cells is a low-cost efficient approach that can be used for *Agro*-transformation of many plant species. Further studies should be directed toward identification and characterization of other chemical factors and physical conditions that could promote T-DNA integration in a variety of agronomically important plant species.

REFERENCES

Abu-Arish A, Frenkiel-Krispin D, Fricke T, Tzfira T, Citovsky V, Grayer Wolf S, Elbaum M (2004) Three-dimensional reconstruction of *Agrobacterium* VirE2 protein with single-stranded DNA. *J Biol Chem* 279:25359-25363.

Alonso JM, Stepanova AN, Leisse TJ, Kim CJ, Chen H, Shinn P, Stevenson DK, Zimmerman J, Barajas P, Cheuk R, Gadrinab C, Heller C, Jeske A, Koesema E, Meyers CC, Parker H, Prednis L, Ansari Y, Choy N, Deen H, Geralt M, Hazari N, Hom E, Karnes M, Mulholland C, Ndubaku R, Schmidt I, Guzman P, Aguilar-Henonin L, Schmid M, Weigel D, Carter DE, Marchand T, Risseeuw E, Brogden D, Zeko A, Crosby WL, Berry CC, Ecker JR (2003) Genome-wide insertional mutagenesis of *Arabidopsis thaliana*. *Science* 301:653-657.

Bakó L, Umeda M, Tiburcio AF, Schell J, Koncz C (2003) The VirD2 pilot protein of *Agrobacterium*-transferred DNA interacts with the TATA box-binding protein and a nuclear protein kinase in plants. *Proc Natl Acad Sci* U S A 100:10108-10113.

Ballas N, Citovsky V (1997) Nuclear localization signal binding protein from *Arabidopsis* mediates nuclear import of *Agrobacterium* VirD2 protein. *Proc Natl Acad Sci* U S A 94:10723-10728.

Basnakian AG, James SJ (1996) Quantification of 3'OH DNA breaks by random oligonucleotideprimed synthesis (ROPS) assay. *DNA Cell Biol* 15(3):255-262.

Bezzubova O, Silbergleit A, Yamaguchi-Iwai Y, Takeda S, Buerstedde JM (1997) Reduced X-ray resistance and homologous recombination frequencies in a RAD54-/- mutant of the chicken DT40 cell line. *Cell* 89:185-193.

Bleuyard JY, Gallego ME, White CI (2006) Recent advances in understanding of the DNA double-strand break repair machinery of plants. *DNA Repair* (Amst) 5(1):1-12.

Boyko A, Filkowski J, Kovalchuk I (2005) Homologous recombination in plants is temperature and day-length dependent. *Mutat Res* 572(1-2):73-83.

Boyko A, Kovalchuk I. U.S. Patent No 11/466184 "Composition and method for enhancing plant transformation and homologous recombination". August 22nd, 2006.

Broothaerts W, Mitchell HJ, Weir B, Kaines S, Smith LM, Yang W, Mayer JE, Roa-Rodríguez C, Jefferson RA (2005) Gene transfer to plants by diverse species of bacteria. *Nature* 433(7026):629-633.

Chakrabarty R, Viswakarma N, Bhat SR, Kirti PB, Singh BD, Chopra VL (2002) *Agrobacterium*-mediated transformation of cauliflower: optimization of protocol and development of Bt-transgenic cauliflower. *J Biosci* 27(5):495-502.

Chilton M-DM, Que Q (2003) Targeted integration of T-DNA into the tobacco genome at double-strand breaks: new insights on the mechanism of T-DNA integration. *Plant Physiol* 133:956-965.

Choi Y-E, Yang D-C, Choi K-T (1998) Induction of somatic embryos by macrosalt stress from mature zygotic embryos of *Panax ginseng*. *Plant Cell, Tissue Culture and Organ Culture* 52:117-181.

Chugh A, Khurana P (2002) Gene expression during somatic embryogenesis – recent advances. *Curr Sci* 83(6):715-729.

Chung SM, Vaidya M, Tzfira T (2006) *Agrobacterium* is not alone: gene transfer to plants by viruses and other bacteria. *Trends Plant Sci* 11(1):1-4.

Citovsky V, Kozlovsky SV, Lacroix B, Zaltsman A, Dafny-Yelin M, Vyas S, Tovkach A, Tzfira T (2007) Biological systems of the host cell involved in *Agrobacterium* infection. *Cell Microbiol* 9(1):9-20.

Coppoolse ER, de Vroomen MJ, Roelofs D, Smit J, van Gennip F, Hersmus BJ, Nijkamp HJ, van Haaren MJ (2003) Cre recombinase expression can result in phenotypic aberrations in plants. *Plant Mol Biol* 51(2):263-279.

Critchlow E, Jackson P (1998) DNA end-joining: from yeast to man. *Trends Biochem Sci* 23:394-402.

Dafny-Yelin M, Levy A, Tzfira T (2008) The ongoing saga of *Agrobacterium*-host interactions. *Trends Plant Sci* 13(3):102-105.

De Block M, Debrouwer D, Moens T (1997) The development of a nuclear male sterility system in wheat. Expression of the barnase gene under the control of tapetum specific promoters. T*heor Appl Genet* 95:125-131.

De Block M. U.S. *Patent No 6074876* "Genetic transformation using a PARP inhibitor". June 13th, 2000.

Deng W, Chen L, Wood DW, Metcalfe T, Liang X, Gordon MP, Comai L, Nester EW (1998) *Agrobacterium* VirD2 protein interacts with plant host cyclophilins. *Proc Natl Acad Sci* U S A 95:7040-7045.

Deng W, Pu X-A, Goodman RN, Gordon MP, Nester EW (1995) T-DNA genes responsible for inducing a necrotic response on grape vines. *Mol Plant-Microbe Interact* 8:538-548.

Dias JS, Martins MG (1999) Effect of silver nitrate on anther culture embryo production of different *Brassica oleracea* morphotypes. *Sci Hortic* 82:299-307.

Djamei A, Pitzschke A, Nakagami H, Rajh I, Hirt H (2007) Trojan horse strategy in *Agrobacterium* transformation: abusing MAPK defense signaling. *Science* 318(5849):453-456.

Endo M, Ishikawa Y, Osakabe K, Nakayama S, Kaya H, Araki T, Shibahara K, Abe K, Ichikawa H, Valentine L, Hohn B, Toki S (2006) Increased frequency of homologous recombination and T-DNA integration in *Arabidopsis CAF-1* mutants. *Europ Mol Biol Org J* 25(23):5579-5590.

Essers J, Hendriks RW, Swagemakers SM, Troelstra C, de Wit J, Bootsma D, Hoeijmakers JH, Kanaar R (1997) Disruption of mouse *RAD54* reduces ionizing radiation resistance and homologous recombination. *Cell* 89:195-204.

Evans HH, Ricanati M, Horng M-F, Jiang Q, Mencl J, Olive P (1993) DNA double-strand break rejoining deficiency in TK6 and other human B-lymphoblast cell lines. *Radiat Res* 134:307-315.

Francis KE, Spiker S (2005) Identification of *Arabidopsis thaliana* transformants without selection reveals a high occurrence of silenced T-DNA integrations. *Plant* J 41(3):464-477.

Friesner J, Britt AB (2003) Ku80- and DNA ligase IV-deficient plants are sensitive to ionizing radiation and defective in T-DNA integration. *Plant J* 34:427-440.

Gallego ME, Bleuyard JY, Daoudal-Cotterell S, Jallut N, White CI (2003) Ku80 plays a role in non-homologous recombination but is not required for T-DNA integration in *Arabidopsis*. *Plant J* 35:557-565.

Gelvin SB (2003) *Agrobacterium*-mediated plant transformation: the biology behind the 'gene-jockeying' tool. *Microbiol Mol Biol Rev* 67:16-37.

Gherbi H, Gallego ME, Jalut N, Lucht JM, Hohn B, White CI (2001) Homologous recombination *in planta* is stimulated in the absence of *Rad50*. *Europ Mol Biol Org Rep* 2(4):287-291.

Gilbertson L (2005) Cre-lox recombination: Cre-activated tools for plant biotechnology *Trends Biotechnol* 21(12):550-555.

Gorbunova V, Levy AA (1999) How plants make ends meet: DNA double strand break repair. *Trends Plant Sci* 4:263-269.

Gordon-Kamm W, Dilkes BP, Lowe K, Hoerster G, Sun X, Ross M, Church L, Bunde C, Farrell J, Hill P, Maddock S, Snyder J, Sykes L, Li Z, Woo YM, Bidney D, Larkins BA (2002) Stimulation of the cell cycle and maize transformation by disruption of the plant retinoblastoma pathway. *Proc Natl Acad Sci* U S A 99(18):11975-11980.

Grevelding C, Fantes V, Kemper E, Schell J, Masterson R (1993) Single-copy T-DNA insertions in *Arabidopsis* are the predominant form of integration in root-derived transgenics, whereas multiple insertions are found in leaf discs. *Plant Mol Biol* 23:847-860.

Grimes HD, Hodges TK (1990) The inorganic NO_3^- : NH_4^+ ratio influences plant regeneration and auxin sensitivity in primary callus derived from immature embryos of indica rice (*Oryza sativa L.*). *J Plant Physiol* 136:362-367.

Hajdukiewicz PT, Gilbertson L, Staub JM (2001) Multiple pathways for Cre/lox-mediated recombination in plastids. *Plant J* 27(2):161-170.

Hamilton CM, Frary A, Lewis C, Tanksley SD (1996) Stable transfer of intact high molecular weight DNA into plant chromosomes. *Proc Natl Acad Sci* U S A 93:9975-9979.

Hanin M, Mengiste T, Bogucki A, Paszkowski J (2000) Elevated levels of intrachromosomal homologous recombination in *Arabidopsis* overexpressing the *MIM* gene. *Plant J* 24(2):183-189.

Hansen G (2000) Evidence for *Agrobacterium*-induced apoptosis in maize cells. *Mol Plant-Microbe Interact* 13:649-657.

He D-G, Yang Y-M, Scott KJ (1989) The effect of macroelements in the induction of embryogenic callus from immature embryos of wheat (*Triticum aestivum L.*). *Plant Sci* 64:251-258.

He D-G, Yang Y-M, Scott KJ (1991) Zinc deficiency and the formation of white structures in immature embryo cultures of wheat (*Triticum aestivum L.*). *Plant Cell Tissue Organ Cult* 24(1):9-12.

Henning L, Taranto P, Walser M, Scho¨nrock N, Gruissem W (2003) *Arabidopsis* MSI1 is required for epigenetic maintenance of reproductive development. *Development* 130:2555-2565.

Hirooka T, Kado CI (1986) Location of the right boundary of the virulence region on *Agrobacterium tumefaciens* plasmid pTiC58 and a host specifying gene next to the boundary. *J Bacteriol* 168:237-243.

Hwang HH, Gelvin SB (2004) Plant proteins that interact with VirB2, the *Agrobacterium tumefaciens* pilin protein, mediate plant transformation. *Plant Cell* 16:3148-3167.

Ilnytskyy Y, Boyko A, Kovalchuk I (2004) *Luciferase*-based transgenic recombination assay is more sensitive than *beta-glucoronidase*-based. *Mutat Res* 559(1-2):189-197.

Immonen AST (1996) Influence of media and growth regulators on somatic embryogenesis and plant regeneration for production of primary triticales. *Plant Cell Tissue Organ Cult* 44(1):45-52.

Iyer LM, Koonin EV, Aravind L (2002) Classification and evolutionary history of the single-strand annealing proteins, RecT, Redbeta, ERF and RAD52. *BMC Genomics* 3(1):8.

Jefferson R (1987) Assaying chimeric genes in plants: the GUS gene fusion system, *Plant Mol Biol Rep* 5:387-405.

Jiménez JM (2001) Regulation of *in vitro* somatic embryogenesis with emphasis on the role of endogenous hormones. *Revista Brasileira de Fisiologia Vegetal* 13(2):196-223.

Kaya H, Shibahara KI, Taoka KI, Iwabuchi M, Stillman B, Araki T (2001) *FASCIATA* genes for chromatin assembly factor-1 in *Arabidopsis* maintain the cellular organization of apical meristems. *Cell* 104:131-142.

Khanna HK, Daggard GE (2003) *Agrobacterium tumefaciens*-mediated transformation of wheat using a superbinary vector and a polyamine-supplemented regeneration medium. *Plant Cell Rep* 21(5):429-436.

Kim S-I, Veena, Gelvin SB (2007) Genome-wide analysis of *Agrobacterium* T-DNA integration sites in the *Arabidopsis* genome generated under non-selective conditions. *Plant J* 51:779-791.

Kohler F, Cardon G, Pohlman M, Gill R, Schider O (1989) Enhancement of transformation rates in higher plants by low-dose irradiation: are DNA repair systems involved in incorporation of exogenous DNA into the plant genome? *Plant Mol Biol* 12:189-199.

Kothari SL, Agarwal K, Kumar S (2004) Inorganic nutrient manipulation for highly improved *in vitro* plant regeneration in finger millet - *Elusine coracana (L.) gaertn. In Vitro Cell Dev Biol - Plant* 40:515-519.

Lacroix B, Tzfira T, Vainstein A, Citovsky V (2006) A case of promiscuity: *Agrobacterium's* endless hunt for new partners. *Trends Genet* 22:29-37.

Lacroix B, Vaidya M, Tzfira T, Citovsky V (2005) The VirE3 protein of *Agrobacterium* mimics a host cell function required for plant genetic transformation. *Europ Mol Biol Org J* 24:428-437.

Li DD, Shi W, Deng XX (2003a) Factors influencing *Agrobacterium*-mediated embryogenic callus transformation of Valencia sweet orange (*Citrus sinensis*) containing the pTA29-barnase gene. *Tree Physiol* 23(17):1209-1215.

Li J, Hsia A-P, Schnable PS (2007) Recent advances in plant recombination. *Curr Opin Plant Biol* 10:131-135.

Li J, Krichevsky A, Vaidya M, Tzfira T, Citovsky V (2005a) Uncoupling of the functions of the *Arabidopsis* VIP1 protein in transient and stable plant genetic transformation by *Agrobacterium. Proc Natl Acad Sci* U S A 102:5733-5738.

Li J, Vaidya M, White C, Vainstein A, Citovsky V, Tzfira T (2005b) Involvement of KU80 in T-DNA integration in plant cells. *Proc Natl Acad Sci* U S A 102(52):19231-19236.

Li W, Ding C-H, Hu Z, Lu W, Guo G-Q (2003b) Relationship between tissue culture and agronomic traits of spring wheat. *Plant Sci* 164:1079-1085.

Lloyd A, Plaisier CL, Carroll D, Drews GN (2005) Targeted mutagenesis using zinc-finger nucleases in *Arabidopsis. Proc Natl Acad Sci* U S A 102(6):2232-2237.

Loyter A, Rosenbluh J, Zakai N, Li J, Kozlovsky SV, Tzfira T, Citovsky V (2005) The plant VirE2 interacting protein 1. A molecular link between the *Agrobacterium* T-complex and the host cell chromatin? *Plant Physiol* 138:1318-1321.

Maës OC, Chibbar RN, Caswell K, Leung N, Kartha KK (1996) Somatic embryogenesis from isolated scutella of wheat: effects of physical, physiological and genetic factors. *Plant Sci* 121:75-84.

Malabadi RB, Staden JV (2006) Cold enhanced somatic embryogenesis in *Pinus patula* is mediated by calcium. *S Afr J Bot* 72(4):613-618.

McCullen CA, Binns AN (2006) *Agrobacterium tumefaciens* and plant cell interactions and activities required for interkingdom macromolecular transfer. *Annu Rev Cell Dev Biol* 22:101-127.

Menke-Milczarek I, Zimny J (2001) NH_4^+ and NO_3^- requirements for wheat somatic embryogenesis. *Acta Physiol Plant* 23(1):37-42.

Meyer P, Walgenbach E, Bussmann K, Hombrecher G, Saedler H (1985) Synchronized tobacco protoplasts are efficiently transformed by DNA. *Mol Gen Genet* 201:513-518.

Miranda A, Janssen G, Hodges L, Peralta EG, Ream W (1992) *Agrobacterium tumefaciens* transfers extremely long T-DNAs by a unidirectional mechanism. *J Bacteriol* 174:2288-2297.

Mordhorst AP, Lorz H (1993) Embryogenesis and development of isolated barley (*Hordeum vulgare L.*) microspores are influenced by the amount and composition of nitrogen sources in culture media. *J Plant Physiol* 142:485-492.

Murashige T, Skoog F (1962) A revised medium for rapid growth and bioassays with tobacco tissue cultures. *Physiol Plant* 15:473-497.

Mysore KS, Kumar CT, Gelvin SB (2000b) *Arabidopsis* ecotypes and mutants that are recalcitrant to *Agrobacterium* root transformation are susceptible to germ-line transformation. *Plant* J 21:9-16.

Mysore KS, Nam J, Gelvin SB (2000a) An *Arabidopsis* histone H2A mutant is deficient in *Agrobacterium* T-DNA integration. *Proc Natl Acad Sci* U S A 97:948-953.

Ninomiya Y, Suzuki K, Ishii C, Inoue H (2004) Highly efficient gene replacements in *Neurospora* strains deficient for nonhomologous end-joining. *Proc Natl Acad Sci* U S A 101(33):12248-12253.

Nürnberger T, Brunner F, Kemmerling B, Piater L (2004) Innate immunity in plants and animals: striking similarities and obvious differences. *Immunol Rev* 198:249-266.

Okada K, Takebe I, Nagata T (1986) Expression and integration of genes into highly synchronized plant protoplasts. *Mol Gen Genet* 205:398-403.

Orii KE, Lee Y, Kondo N, McKinnon PJ (2006) Selective utilization of nonhomologous end-joining and homologous recombination DNA repair pathways during nervous system development. *Proc Natl Acad Sci* U S A 103:10017-10022.

Pastori GM, Wilkinson MD, Steele SH, Sparks CA, Jones HD, Parry MA (2001) Age-dependent transformation frequency in elite wheat varieties. *J Exp Bot* 52(357):857-863.

Perl A, Kless H, Blumenthal A, Galili G, Galun E (1992) Improvement of plant regeneration and GUS expression in scutellar wheat calli by optimization of culture conditions and DNA-microprojectile delivery procedures. *Mol Gen Genet* 235(2-3):279-284.

Pu X-A, Goodman RN (1992) Induction of necrosis by *Agrobacterium tumefaciens* on grape explants. *Physiol Mol Plant Pathol* 41:245-254.

Puchta H (2002) Gene replacement by homologous recombination in plants. *Trends Plant Sci* 1:340-348.

Puchta H (2005) The repair of double-strand breaks in plants: mechanisms and consequences for genome evolution. *J Exp Bot* 56:1-14.

Puchta H, Dujon B, Hohn B (1993) Homologous recombination in plant cells is enhanced by *in vivo* induction of double strand breaks into DNA by a site-specific endonuclease. *Nucleic Acids Res* 21(22):5034-5040.

Que Q, Wang HY, Jorgensen RA (1998) Distinct patterns of pigment suppression are produced by allelic sense and antisense chalcone synthase transgenes in petunia flowers. *Plant J* 13(4):401-409.

Racusen RH, Schiavone FM (1990) Positional cues and differential gene expression in somatic embryos of higher plants. *Cell Differ Dev* 30:159-169.

Regensburg-Tuïnk AJ, Hooykaas PJ (1993) Transgenic *N. glauca* plants expressing bacterial virulence gene virF are converted into hosts for nopaline strains of *A. tumefaciens*. *Nature* 363(6424):69-71.

Reidt W, Wurz R, Wanieck K, Chu HH, Puchta H (2006) A homologue of the breast cancer-associated gene *BARD1* is involved in DNA repair in plants. *Europ Mol Biol Org J* 25(18):4326-4337.

Reiss B (2003) Homologous recombination and gene targeting in plant cells. *Int Rev Cytol* 228:85-139.

Reiss B, Schubert I, Kopchen K, Wendeler E, Schell J, Puchta H (2000) RecA stimulates sister chromatid exchange and the fidelity of double-strand break repair, but not gene targeting, in plants transformed by *Agrobacterium*. *Proc Natl Acad Sci* U S A 97:3358-3363.

Roa-Rodriguez C, Nottenburg C (2003) *Agrobacterium*mediated transformation in plants. CAMBIA (http://www.bios.net/Agrobacterium).

Sahrawat AK, Becker D, Lütticke S, Lörz H (2003) Genetic improvement of wheat via alien gene transfer, an assessment. *Plant Sci* 165:1147-1168.

Salman H, Abu-Arish A, Oliel S, Loyter A, Klafter J, Granek R, Elbaum M (2005) Nuclear localization signal peptides induce molecular delivery along microtubules. *Biophys J* 89(3):2134-2145.

Salomon S, Puchta H (1998) Capture of genomic and T-DNA sequences during double strand break repair in somatic plant cells. *Europ Mol Biol Org J* 17:6086-6095.

Schmidt EE, Taylor DS, Prigge JR, Barnett S, Capecchi MR (2000) Illegitimate Cre-dependent chromosome rearrangements in transgenic mouse spermatids. *Proc Natl Acad Sci* U S A 97(25):13702-71370.

Schonrock N, Exner V, Probst A, Gruissem W, Hennig L (2006) Functional genomic analysis of *CAF-1* mutants in *Arabidopsis thaliana*. *J Biol Chem* 281:9560-9568.

Schuermann D, Molinier J, Fritsch O, Hohn B (2005) The dual nature of homologous recombination in plants. *Trends Genet* 21(3):172-181.

Shaked H, Melamed-Bessudo C, Levy AA (2005) High frequency gene targeting in *Arabidopsis* plants expressing the yeast *RAD54* gene. *Proc Natl Acad Sci* U S A 102:12265-12269.

Shalev G, Sitrit Y, Avivi-Ragolski N, Lichtenstein C, Levy AA (1999) Stimulation of homologous recombination in plants by expression of the bacterial resolvase *ruvC*. *Proc Natl Acad Sci* U S A 96:7398-7402.

Shrivastav M, De Haro LP, Nickoloff JA (2008) Regulation of DNA double-strand break repair pathway choice. *Cell Res* 18(1):134-147.

Silver DP, Livingston DM (2001) Self-excising retroviral vectors encoding the Cre recombinase overcome Cre-mediated cellular toxicity. *Mol Cell* 8(1):233-243.

Swoboda P, Gal S, Hohn B, Puchta H (1994) Intrachromosomal homologous recombination in whole plants. *Europ Mol Biol Org J* 13:484-489.

Tagami H, Ray-Gallet D, Almouzni G, Nakatani Y (2004) Histone H3.1 and H3.3 complexes mediate nucleosome assembly pathways dependent or independent of DNA synthesis. *Cell* 116:51-61.

Takumi S, Shimada T (1997) Variation in transformation frequencies among six common wheat cultivars through particle bombardment of scutellar tissues. *Genes Genet Syst* 72:63-69.

Tao Y, Rao PK, Bhattacharjee S, Gelvin SB (2004) Expression of plant protein phosphatase 2C interferes with nuclear import of the *Agrobacterium* T-complex protein VirD2. *Proc Natl Acad Sci* U S A 101:5164-5169.

Tzfira T (2006) On tracks and locomotives: the long route of DNA to the nucleus. *Trends Microbiol* 14(2):61-63.

Tzfira T, Citovsky V (2003) The *Agrobacterium*-plant cell interaction. Taking biology lessons from a bug. *Plant Physiol* 133(3):943-947.

Tzfira T, Citovsky V (2006) *Agrobacterium*-mediated genetic transformation of plants: biology and biotechnology. *Curr Opin Biotechnol* 17(2):147-154.

Tzfira T, Frankman R, Vaidya M, Citovsky V (2003) Site-specific integration of *Agrobacterium* T-DNA via double-stranded intermediates. *Plant Physiol* 133:1011-1023.

Tzfira T, Li J, Lacroix B, Citovsky V (2004a) *Agrobacterium* T-DNA integration: molecules and models. *Trends Genet* 20:375-383.

Tzfira T, Vaidya M, Citovsky V (2001) VIP1, an *Arabidopsis* protein that interacts with *Agrobacterium* VirE2, is involved in VirE2 nuclear import and *Agrobacterium* infectivity. *Europ Mol Biol Org J* 20:3596-3607.

Tzfira T, Vaidya M, Citovsky V (2002) Increasing plant susceptibility to *Agrobacterium* infection by overexpression of the *Arabidopsis VIP1* gene. *Proc Natl Acad Sci* U S A 99:10435-10440.

Tzfira T, Vaidya M, Citovsky V (2004b) Involvement of targeted proteolysis in plant genetic transformation by *Agrobacterium. Nature* 431:87-92.

van Attikum H, Bundock P, Hooykaas PJJ (2001) Non-homologous end-joining proteins are required for *Agrobacterium* T-DNA integration. *Europ Mol Biol Org J* 20:6550-6558.

van Attikum H, Bundock P, Overmeer R, Lee L-Y, Gelvin B, Hooykaas PJJ (2003) The *Arabidopsis AtLIG4* gene is required for the repair of DNA damage, but not for the integration of *Agrobacterium* T-DNA. *Nucleic Acids Res* 31:4247-4255.

van Attikum H, Hooykaas PJJ (2003) Genetic requirements for the targeted integration of *Agrobacterium* T-DNA in *Saccharomyces cerevisiae*. *Nucleic Acids Res* 31:826-832.

Vergunst AC, Hooykaas PJJ (1999) Recombination in the plant genome and its application in biotechnology. *Crit Rev Plant Sci* 18:1-31.

Vergunst AC, van Lier MC, den Dulk-Ras A, Grosse Stuve TA, Ouwehand A, Hooykaas PJJ (2005) Positive charge is an important feature of the C-terminal transport signal of the VirB/D4-translocated proteins of *Agrobacterium*. *Proc Natl Acad Sci* U S A 102:832-837.

Villemont E, Dubois F, Sangwan RS, Vasseur G, Bourgeois Y, Sangwan-Norreel BS (1997) Role of the host cell cycle in the *Agrobacterium*-mediated genetic transformation of *Petunia*: evidence of an S-phase control mechanism for T-DNA transfer. *Planta* 201:160-172.

Waldman AS, Waldman BC (1991) Stimulation of intrachromosomal homologous recombination in mammalian cells by an inhibitor of poly(ADP-ribosylation). *Nucleic Acids Res* 19(21):5943-5947.

Waldman BC, Waldman AS (1990) Illegitimate and homologous recombination in mammalian cells: differential sensitivity to an inhibitor of poly(ADP-ribosylation). *Nucleic Acids Res* 18(20):5981-5988.

Wright DA, Townsend JA, Winfrey RJ Jr, Irwin PA, Rajagopal J, Lonosky PM, Hall BD, Jondle MD, Voytas DF (2005) High-frequency homologous recombination in plants mediated by zinc-finger nucleases. *Plant J* 44(4):693-705.

Wu J, Kandavelou K, Chandrasegaran S (2007) Custom-designed zinc finger nucleases: what is next? *Cell Mol Life Sci* 64(22):2933-2944.

Yamazoe M, Sonoda E, Hochegger H, Takeda S (2004) Reverse genetic studies of the DNA damage response in the chicken B lymphocyte line DT40. *DNA Repair* (Amst) 3:1175-1185.

Ye GN, Stone D, Pang SZ, Creely W, Gonzalez K, Hinchee M (1999) *Arabidopsis* ovule is the target for *Agrobacterium in planta* vacuum infiltration transformation. *Plant J* 19:249-257.

Yi H, Mysore KS, Gelvin SB (2002) Expression of the *Arabidopsis* histone *H2A-1* gene correlates with susceptibility to *Agrobacterium* transformation. *Plant J* 32:285-298.

Zhu Y, Nam J, Carpita NC, Matthysse AG, Gelvin SB (2003a) *Agrobacterium*-mediated root transformation is inhibited by mutation of an *Arabidopsis* cellulose synthase-like gene. *Plant Physiol* 133:1000-1010.

Zhu Y, Nam J, Humara JM, Mysore K, Lee LY, Cao H, Valentine L, Li J, Kaiser A, Kopecky A (2003b) Identification of *Arabidopsis rat* mutants. *Plant Physiol* 132:494-505.

Zipfel C, Kunze G, Chinchilla D, Caniard A, Jones JD, Boller T, Felix G (2006) Perception of the bacterial PAMP EF-Tu by the receptor EFR restricts *Agrobacterium*-mediated transformation. *Cell* 125:749-760.

In: Transgenic Plants: New Research
Editor: Oliver T. Chan

ISBN 978-1-60692-017-6
© 2009 Nova Science Publishers, Inc.

Chapter 5

Marker-free Technologies for Transgenic Plant Research

Renlin Xu[1] and Shu Wei[2]

[1]Eastern Cereal and Oilseed Research Centre,
Agriculture and Agri-Food Canada, Ottawa, ON, Canada
[2]Saskatoon Research Center, Agriculture and Agri-Food Canada, Saskatoon, SK, Canada

Abstract

Genetic engineering of higher plants typically involves stable introduction of antibiotic or herbicide resistance genes as selective markers in order to obtain transgenic plants recovered from rarely occurred transformation events against a background of a large number of non-transformants. Over the past several years, concerns over the environmental and human health risks associated with antibiotic- and herbicide-resistance genes present in transgenic crops have been raised. Developing technologies to eliminate the marker genes and other superfluous T-DNA fragments from transgenic plants would be helpful not only for public acceptance of transgenic crops but also for studying plant gene functions and for generation of 'super' transgenic crop varieties with improved complex traits. In this paper, we reviewed the emerged techniques for removing marker genes from transgenic plants in the following aspects: (1) intra-genomic relocation of transgenes via transposable elements, (2) site-specific recombination systems, (3) intra-chromosomal recombination; and (4) co-transformation segregation approach. We put our emphasis on the application of site-specific recombination systems for use in plant gene transformation and discuss some future directions and improvements of this technology.

INTRODUCTION

Agrobacterium and biolistic mediated plant gene transformation technologies provide a feasible means to modify crops genetically. The significance of economic, environmental and healthy benefits using transgenic or genetically modified (GM) crops have been well documented (Christou et al., 2004; Ferry et al., 2006). The use of the selectable marker gene systems facilitates the gene transformation process and allows the relatively straightforward recovery of transgenic crop plants. Without the systems, the few plant cells that take up and stably integrate the foreign DNA would simply be lost in an ocean of wild-type cells, which would certainly overgrow these transformed cells in the absence of an effective selection against them. Two main classes of such genes, antibiotic or herbicide resistance genes have been widely applied for this purpose. A comprehensive review on selectable marker genes in transgenic plants has been published (Miki and McHugh, 2004).

However, over the use of such marker genes there have been public concerns, which fall into two categories: environmental hazards and human health risks. Issues frequently raised with respect to antibiotic resistance genes include: (1) antibiotic resistance gene products in transgenic crops or products might render clinically important therapeutic antibiotics ineffective; (2) the potential for horizontal transfer of the antibiotic resistance marker genes to intestinal and soil microorganisms, in other words, when antibiotic selection markers having wide clinical and veterinary application are used, the marker gene could be transferred into human and animal gut microorganisms, which could make such antibiotics functionally useless, although there is no evidence to suggest that the currently used antibiotic resistance markers, such as *npt*II, pose any risks to humans, animals or the environment. For herbicide resistance genes, the issues are somewhat different: (1) gene flow – the potential for the marker genes in the engineered crops spread by natural out-crossing to wild or weedy relatives (Poppy, 2004); (2) weediness – the potential for a crop or its sexually compatible wild relatives to become a new aggressive weed as a result of newly introduced genes (Ramessar et al., 2007). Therefore, to alleviate public concerns, elimination of all antibiotic resistance genes from GM plants has been recommended.

Additionally, one of the practical goals for plant biotechnological studies is to create 'super' cultivars by introducing multiple transgenes into the genomes of elite crop varieties so that several agronomy traits of the cultivars will be collectively and significantly improved. While crosses will still be useful to pyramid transgenes, in many cases, there will also be the situation where repeated transformation of genes into cultivars will be more effective. However, current transgenic protocols severely limit the number of genes that can be introduced simultaneously. Therefore, re-transformation of a single line is a feasible and important approach towards selective introduction of multiple genes for complex traits such as productivity, broad pathogen resistance or tolerance to abiotic stress. Unfortunately, the presence of a particular marker gene in a transgenic plant necessarily precludes the use of the same marker in subsequent transformation round or event. Thus, any technique that can eliminate a selection marker gene in transgenic crops is highly desirable. Moreover, the marker free transgenic crops could also lower the costs for developing and marketing of genetically modified products and might speed up the commercial release of new products (Kuiper et al., 2001). Furthermore, it is also desirable in plant molecular biology studies to eliminate the marker genes and superfluous foreign T-DNA fragment in transgenic plants in order to exclude their possible interference on gene function elucidation. Considering these issues, the production of marker-free transgenic plants has become a major objective for plant biologists and the plant biotech industry.

A variety of strategies have been developed to remove or eliminate marker genes after the selection phase of plant production to create marker-free transgenic plants (Review by Afolabi, 2007; Baszczynski et al., 2003; Hare and Chua, 2002; Hohn et al., 2001; Puchta, 2003). These are: 1) intra-genomic relocation of transgenes via transposable elements; 2) site-specific recombination systems; 3) intra-chromosomal recombination; and 4) co-transformation. Among them the site-specific excision of a transgenic DNA sequence containing the marker gene is most commonly used. Here, we highlight recent advances in these approaches with an emphasis on site-specific recombination systems and their applications in plant gene transformation. Possible development and improvements of this technology in future are discussed as well.

Transposition-mediated Elimination of Marker Gene in Transgenic Plants

Transposable Elements: Overview

Transposable elements (transposons) were discovered in corn by McClintock more than five decades ago. Her remarkable discoveries were made without the knowledge of the structure of DNA and without modern molecular techniques. Transposons were rediscovered some 20 years later in prokaryotes and they are having been identified in virtually all organisms examined. Transposable elements can be described as discrete DNA segments that are able to move between different, non homologous, genomic loci (Craig, 1997). Transposition of an element is thus a recombination reaction involving transposase and three separate sites: the two transposon ends and the new target locus. In autonomous transposable elements, the transposase is encoded by a gene located within the element. Transposon ends often contain inverted repeated sequences on which the transposase binds specifically (Sadowski, 1993).

The models describing transposition mechanisms fall into two categories: 1) replicative model, and 2) cut-and-paste model. The replicative (or cointegrate) model, which was proposed to explain replication of the mutagenic bacteriophage Mu, postulated that the ends of the transposable element are nicked and joined to a staggered break in the target sequence and that replication through the element creates a “cointegrate” structure that contains two inverted copies of the element (Shapiro, 1993; Sherratt et al., 1995). In the non-replicative or ‘cut-and-paste’ model the entire transpososn is excised from the donor site via double-strand breaks and joined to the staggered break induced in the recipient DNA. No replication of the element occurs and the donor DNA suffers a potentially lethal double-strand break (Hallet and Sherratt, 1997; Sadowski, 1993). This mode of transposition results in the simple insertion of the element in the target DNA. The gapped donor molecule resulting from the transposon excision is either degraded or repaired by homologous recombination using a second copy of the donor as a template (Bender et al., 1991). Both modes of transposition generate short target sequence duplications of a characteristic length (2–14 bp) that form direct repeats flanking the element in its new locus (Hagemann and Craig, 1993).

In plant genomes the most extensively studied transposable element is maize (*Zea mays*) *Activator* (*Ac*) and its nonautonomous *Dissociation* (*Ds*) element. *Ac* is an autonomous transposable element of 4565 bp in length, coding for a single gene product, the transposase (*Tpa*) with 807 amino acids. The *Tpa* is a sequence-specific DNA binding protein that mediates the initiation of transposition by specifically binding to multiple motifs positioned near the terminus of the element (Kunze and Starlinger, 1989). The *Ac* element has two functions essential for transposition which can be physically separated, a transposase gene and inverted repeat termini. The *Ds* elements lack the transposase function and are stable in the absence of *Ac* but can be trans-activated by introducing of *Ac* transposase coding sequence in *trans*. *Ac* is generally believed to transpose via cut-and-paste DNA recombination (Fedoroff, 1989) by which the target DNA is presumably cleaved by 8-bp staggered cuts prior to insertion because of the presence of an 8-bp target site duplication on each side of the transposed *Ac*/*Ds* elements. Excision is initiated by the assembly of an active synaptic complex in which the two ends of the element are paired and held together by bound *Ac*-transposase subunits (Gorbunova and Levy, 2000). A notable feature of *Ac* in maize is its preference for insertion into target sites that are linked to the donor site from which the element transposed, although insertions to both unlinked and linked targets are observed (Craig, 1997). In general, transposition occurs to linked and unlinked sites in roughly equal frequencies in maize and transformed dicots (Jones et al., 1990).

Elimination of Marker Gene Using *As/Dc* Transposition System

Transposable elements can be used to obtain marker-free transgenic plants. The strategy is to connect either the transgene or the selectable marker gene with transposable sequence in such a way that the two transgenes can be separated from each other in a controlled reaction after transformation and selection. Two types of vector systems have been developed for transgene elimination. In the first type, genes of interest (GOI) are inserted between the *Ds* inverted repeats (Yoder and Goldsbrough, 1994). In the second type, the selectable marker gene is flanked by the *Ds* repeats. In the presence of an active transposase, which can be introduced into the plant as an additional component of the T-DNA, or by a secondary transformation, or by crossing to a plant containing transposase, the *Ds* element containing the marker gene will transpose to a new genomic location. The *Ds* element, either flanking the GOI or the selectable marker gene will reinsert in about 90% of the events, and about 50% of these insertion events will be genetically unlinked to the primary site (Belzile et al., 1989). Then, sexual crossing can result in the transformant carrying GOI only (Afolabi, 2007).

An alternative strategy for exploiting the *Ac*/*Ds* system is to transpose the selectable marker genes and the transposase from the T-DNA leaving only the gene of interest in the inserted locus of the T-DNA. This research generated the *ipt*-type MAT (multi-auto-transformation) vector system which uses the isopentenyl transferase (*ipt*) gene as a selectable marker and is designed to remove the *ipt* gene after transformation by using the *Ac* transposable element (Ebinuma et al., 1997b). The *ipt* gene was interesting in this study as it was used as both a negative and positive selectable marker. In the first positive selection step, transformed tissue proliferated as adventitious shooty material that was abnormal in morphology and could not regenerate due to the overproduction of cytokinin. In the second negative selection step, after several weeks or months in culture, normal shoots appeared (due to the elimination of the *ipt* and transposase genes by transposition) and regenerated into transformed marker-free plants. This occurred at a frequency of about 5% (Ebinuma et al., 1997a). A similar approach was used to create Hygromycin β-phosphotransferase (*hpt)*-free rice plants that expressed the Bt endotoxin coded by the *cry* 1B gene. Sixty-eight independent rice transformants were regenerated and molecularly analyzed revealing excision and reinsertion of the *Ds-cry* 1B element in 37% and 25% respectively of the transformation events (Cotsaftis et al., 2002). Koprek et al. (2001) developed a system in barley based on the *Ac/Ds*, which requires only a few original transformation events and results in high numbers of single-copy transgenic plants, each carrying a stabilized transgene at different genomic locations.

Recently, two new transposition-mediated maker gene excision systems have been established. Yan and Rommens (2007) reported a system in which T-DNAs were used to deliver *Ds* elements into the nuclei of potato (*Solanum tuberosum*) cells, and a double–selection system was applied to enrich for plants that only contained a transposed *Ds* element. This system consists of a positive selection for the neomycin phosphotransferase (*nptII*) gene positioned within *Ds* followed by a negative selection against stable integration of the cytosine deaminased (*codA*) gene-containing T-DNA. As a result, sixteen of 29 transgenic plants were found to contain a transposed element while lacking any superfluous T-DNA sequences. The frequency of single-copy *Ds* transformation was 0.3%, which is only about 2.5-fold lower than the potato transformation frequency for backbone-free and single-copy T-DNAs. This transposition-based plant gene transformation method provides new means of genetically modifying plants without inserting bacterial DNA into plant genome. On the other hand, Charng et al. (2008) reported another new strategy using maize Ac/*Ds* transposition system to generate selectable marker-off transgenic rice plants with no need for out-crossing. In this system, the marker gene was accompanied by the inducible *Ac*-based transposon KCEH system, but one end of the transposon was located in the intron of the marker gene, the rice 5-enolpyruvylshikimate-3-phosphate synthase (*epsps*) was modified for glyphosate tolerance. The transposon contained *Ac* transposase, which fused with the promoter of the inducible gene for pathogenesis-related protein 1a (PR-1a). This construct was engineered into an expression vector pCAMBIA1300, harboring a hygromycin-resistant gene. The construct was first transformed into rice calli, and transformed plants were selected on hygromycin-containing medium. The stably transformed calli underwent determination for normal transcripts and tolerance to glyphosate. The results were applied to a rice transformation with the same construct, but using glyphosate as the selective agent. After obtaining stable transgenic plants, transposition could be induced by application of chemical inducer, salicylic acid, to eliminate the transposon, which leads to the marker gene truncation and expression termination. With a transposon system to terminate the marker gene, the GOI-containing transposon could be used to create more independent transgenic lines from one successful transformant.

Site-specific Recombinase-mediated Excision of Marker Genes in Transgenic Plants

Classification of Site-specific Recombinases

Site-specific recombination systems are common in prokaryotes and lower eukaryotes. Based on amino acid sequence homology and mechanistic relatedness most site-specific recombinases are grouped into one of two families: the tyrosine recombinase family or the serine recombinase family. Serine recombinase family is also sometimes known as resolvase/invertase family, while tyrosine recombinases are known as the integrase family, which reflects the types of reaction that most known members in each family have evolved to catalyse. Typical examples of tyrosine recombinases are the well known enzymes such as the Cre of bacteriophage P1 (Sauer, 1987), the FLP of *Saccharomyces cerevisiae* (Broach and Hicks, 1981) and R of *Zygosaccharomyces rouxii* (Araki et al., 1985). Well known serine recombinases include Tn3 resolvase from the Tn3 transposon and φC31 integrase from the φC31 phage (Nash 1996; Stark and Boocock, 1995). Although the individual members of the two recombinase families can perform reactions with same practical outcomes, the two families are unrelated to each other, having different protein structures and reaction mechanisms. The β recombinase, encoded by the gram-positive broad-host-range plasmid pSM19035 from *Streptococcus pyogenes* belongs to the same subgroup of serine recombinases as the Gin recombinase (Smith and Thorp, 2002),

Mechanism of the Site-specific DNA Recombination Reactions

Recombination between two DNA sites begins by the recognition and binding of these sites by the recombinase protein. This is followed by the synapsis i.e. bringing the sites together to form the synaptic complex. It is within this synaptic complex that the strand exchange takes place, as the DNA is cleaved and rejoined by controlled transesterification reactions. During strand exchange, the DNA cut at fixed points within crossover region of the site releases a deoxyribe hydroxyl group, while recombinase protein forms a transient covalent bond to a DNA backbone phosphate. This phosphodiester bond between hydroxyl group of the nucleophilic, serine or tyrosine residue conserves the energy that was expended in cleaving the DNA. Energy stored in this bond is subsequently used for the rejoining of the DNA to the corresponding deoxyribose hydroxyl group on the other site. The entire process therefore goes through without the need for external energy rich cofactors such as ATP. As stated previously, the recombination sites are slightly asymmetric, which allows the enzyme to tell apart the left and right ends of the site. When generating products left ends are always joined to the right ends of their partner sites and vice versa. This causes the recombination sites to be reconstituted in the recombination products. Joining of left ends to left or right to right is avoided due to the asymmetric "overlap" sequence between the staggered points of top and bottom strand exchange. Left-left or right-right half-site recombinants would contain mismatched base pairs (Stark and Boocock, 1995). Although the basic chemical reaction is the same for both tyrosine and serine recombinases there are marked differences. Tyrosine recombinases, such as Cre or FLP, cleave one DNA strand at the time at points that are staggered by 6-8 bp, linking 3' end of DNA to the hydroxyl group of the tyrosine nucleophil (Van Duyne, 2002). Strand exchange than proceeds via a crossed strand intermediate analogous to the Holliday junction (Holliday, 1964) in which only one pair of strands has been exchanged. Conversely, serine recombinases like Tn3 resolvase cut all four DNA strands simultaneously at points that are staggered by 2-bp (Stark et al., 1992). During cleavage protein-DNA bond is formed via transesterification reaction in which a phosphodiester bond is replaced by a phosphoserine bond between a 5' phosphate at the cleavage site and hydroxyl group of the conserved serine residue (S10) in resolvase (Reed and Grindley, 1981; Reed and Moser, 1984).

Site-specific Recombinases Used in Plants

A lot of effort has been made in order to understand the mechanisms of site-specific recombination in plants for the purpose not only to excise marker genes from transgenic plants but also to deliver DNA more efficiently into the plant genome (Ow, 2002). The three site-specific recombination systems that are currently mainly used in plants, Cre/*loxP*, FLP/*FRT*, and R/*RS*, have in common the property that a single polypeptide recombination between two sites of identical or nearly identical sequences. Each recombination site consists of a short (2 to 8 bp) and asymmetric spacer sequence where strand exchange takes place, flanked by an inverted repeat of some 12 to 13 bp, where recombinases bind. The asymmetry of the spacer sequence gives an orientation to the recombination site, and dictates the outcome of a recombination reaction. Recombination between directly or indirectly oriented sites *in cis* excises or inverts the intervening DNA. Recombination between sites *in trans* causes a reciprocal translocation of two linear DNA molecules, or integration if at least one of the two molecules is circular, since the product sites generated by recombination, the reaction is freely reversible. In practice, however, excision is essentially irreversible because the probability of an intramolecular interaction, where the two recombination sites are closely linked, is much higher than an intermolecular interaction between unlinked sites. The corollary is that the DNA molecule inserted into a genomic recombination site will readily excise out (Ow, 2002).

Site-specific Recombinase-mediated Excision of Marker Genes in Plants

Site-specific recombination systems remove the marker gene by the action of single enzymes (e.g., Cre) acting on specific target sequences. Each of the target sites containing a few oligonucleotides surrounded by a few inverted repeats to determine the orientation of the target site. The expression of the recombinase causes recombination between the target sites and results in the loss of the marker gene flanked by the sites. To make use of site-specific recombinases, their target sites must flank antibiotic resistance genes with direct repeats. These direct repeats define the borders of a gene elimination cassette, in the absence of site-specific recombinases, antibiotic marker genes are stable and can be used as selectable markers. Once transgenic plants have been isolated using antibiotic selection, expression of the appropriate site-specific recombinase results in excision of its cognate elimination cassette. Trait genes that are located outside elimination cassettes are retained. Up to date, these site-specific recombination systems have been applied in plant transformation for excision of marker genes from transgenic plants in which single enzyme (e.g. Cre, FLP, R, Gin and β) acts on specific target sequences (*lox*, *FRT*, *RS*, gix, six, respectively) (Ow, 2002).

Cre–lox

The Cre–*lox* system from bacteriophage P1 was the first of the recombination systems shown to be effective in the generation of marker-free plants (Dale and Ow, 1991). Since then, a number of strategies have been used with Cre/*lox* to remove marker gene from transgenic plants. In all strategies the marker gene is flanked by directly repeated lox sites, and excision occurs when Cre activity is present. The strategy differs in how Cre function is delivered. Strategy 1) transgenic plants that carry a marker gene flanked by lox sites were retransformed with a cre expressing plasmid. The cre gene can then be segregated away in the next generation (Odell et al., 1990, Dale and Ow, 1991; Russell et al., 1992). Strategy 2) the lox plants can be crossed with plants that express Cre activity, in which case marker gene excision occurred in the F_1 progeny, followed by loss of the cre gene by genetic segregation in the F_2 generation (Odell et al., 1990; Bayley et al., 1992; Russell et al., 1992; Hoa et al., 2002; Arumugam et al., 2007). Strategy 3, the cre gene can be included on the DNA segment that is flanked by lox sites and be regulated such that expression is activated (induced) after the marker gene is no long needed. It has been demonstrated by using chemically inducible promoter in Arabidopsis (Sugita et al, 2000; Zuo et al., 2001), in rice (Sreekala et al., 2005); and using heat shock-inducible promoter in Arabidopsis (Hoff et al., 2002), in tobacco (Wang et al., 2005; Liu et al., 2005) and potato (Cuellar et al, 2006). Cre recombinase can also be introduced by virus infection into transgenic plants harboring a transgenic construct [lox-marker-lox] to initiate marker gene excision (Jia et al., 2006; Kopertekh et al., 2004).

A significant refinement of the site specific recombination strategies was developed using the β-estradiol-inducible promoter system in which an artificial transcription factor, XVE was constructed for use in plants with its target promoter (Zuo et al., 2001). In this system, the gene of interest was separated from its promoter by a fragment containing the genes coding for the XVE transcription factor, the *npt*II selectable marker and the Cre recombinase (under the control of the inducible promoter) surrounded by *lox* sites. Transformation of *Arabidopsis* was achieved by selection for kanamycin resistance. Subsequent induction with β-estradiol resulted in the excision of the complete induction system along with the Cre recombinase and selectable marker genes. The final product was the reconstituted gene of interest, in this case GFP. In *Arabidopsis*, excision occurred in all of the plants with high efficiency in the germline cells (29–66%) using a single transformation (Zuo et al., 2001).

Zhang et al. (2003) adopted two strategies, crossing and autoexcision, for demonstrating the use of Cre/*lox* system in maize. In the crossing strategy, plants expressing the Cre recombination are crossed with plants bearing a transgene construct in which the selectable marker gene is flanked by directly repeated lox sites. Complete somatic and germline marker gene excision occurred in the F1 plants from most crosses with multiple independent Cre and lox lines. In the autoexcision strategy, the cre gene under the control of a heat shock-inducing promoter is excised along with the *nptII* marker gene. There are several advantages of the auto excision strategy over the crossing strategy. First, it can be used for vegetatively propagated crops, such as potato, or crops that are tedious to cross, such as soybean. Secondly, it does not require outcrossing to bring in the cre gene, nor a second generation to eliminate the cre gene, saving at least one plant generation. Additionally, it allows one to control the timing for marker removal. The autoexcision strategy was recently demonstrated in tobacco with a chemically inducible promoter and the R/*Rs* site-specific recombination system (Sugita et al. 2000) and in Arabidopsis using an estrogen receptor-based fusion transacitator XVE that is B-estradiol-inducible and Cre/*lox* (Zuo et al., 2001).

A Cre/*loxP*-mediated self-activating gene excision system has been reported (Li et al., 2007). In this system, a selectable marker gene *hpt* and the cre recombinase gene under control of the *app1*, a developmentally regulated embryo-specific gene promoter, were flanked by two *lox*P recombination sites. The entire sequence was then placed between a constitutive promoter and a coding region for either β-glucuronidase (Gus) or glyphosate acetyltransferease (Gat). Gene excision removes the entire sequence between the two *lox*P sites and brings the coding region to the constitutive promoter for expression. Using this system marker gene excision occurred in over 30% of the stable soybean transgenic events.

Mlynarova et al. (2006) and Luo et al. (2007) showed that a promoter that is specifically functional during microsporogenesis, in pollen or in seed, can efficiently drive the site-specific recombinase in removing selective markers. Then, Verweire et al. (2007) reported a new system, called genetically programmed auto-excision of marker gene from transgenic Arabidopsis plants without any extra handling and in the same time frame. This genetic programming is established by introducing a germline-specific promoter auto-excision vector (GSA) in which a germline-sepcific promoter is used to control the Cre/*lox* recombination system. Two modules of this genetic program were performed. In the first module, the promoter, placed upstream of the cre gene, confers CRE functionality in both the male and the female germline or in the common germline (e.g. floral meristem cells). In the second module, a promoter conferring single germline-specific CRE functionality was introduced upstream of the cre gene. Promoter sequences used in this work are derived from the APETALA1 and SOLO DANCERS genes from Arabidopsis Columbia-0 conferring common germline and single germline functionality, respectively. Those results clearly indicate that germline-specific auto-excision is an efficient, flexible, and versatile system to remove selectable markers from transgenic plants.

The Cre/*lox* system has also been used to excise marker genes from transgenic plastids. Hajdukiewicz et al. (2001) tested the Cre/lox recombination system for antibiotic marker gene removal from plastids. And Corneille et al. (2001) adapted Cre/lox recombination system for the elimination of marker genes from tobacco plastid genome. Three pathways of Cre-mediated recombination occur in plastids. The first and most prevalent pathway results in the predicted cre/lox-mediated excision of the antibiotic resistance marker used for selection of transplastomic plants. The excision event activated GFP expression, resulting visible marker to monitor and timing of excision, transgene excison was complete in nearly all the lines examined. The second pathway of Cre-mediated recombination in plastids arose from homologous recombination between a 117 bp directly repeated, non-tandem transgenic and endogenous Prrn sequences. The third general recombination pathway revealed a “hot spot” of recombination composed of multiple direct repeats of a 5-bp sequence motif, which recombined with *lox* independent of sequence homology which had not been observed previously.

FLP–FRT

The FLP–*FRT* system derived from the 2-μm plasmid of *Saccharomyces cerevisiae* has also been widely tested in plants. The first plant study on the functionality of the FLP-*FRT* system was published by Lyznik et al. (1993, 1996) in maize and rice protoplasts. In tobacco and *Arabidopsis*, plants transformed with the FLP recombinase were crossed with plants transformed with T-DNA in which the GUS coding region is separated from the 35S promoter by an *hpt* gene bracketed by *FRT* sites. This resulted in excision of the *hpt* gene and activation of the GUS gene in all cases (Kilby et al., 1995). Additional studies on the use of the FLP-*FRT* system in maize protoplasts followed (Lyznik et al., 1996). FLP activity has also been shown in tobacco and *Arabidopsis* (Bar et al., 1996; Davies et al., 1999; Gidoni et al, .2001; Kilby et al., 1995; Lloyd and Davis, 1994; Sonti et al., 1995). In these cases, the *FLP* gene was under the control of a constitutive cauliflower mosaic virus 35S promoter (CaMV 35S) or inducible soybean heat-shock promoter. Studies on the efficiency of FLP-mediated excision and germinal transfer of the recombined locus in *Arabidopsis* and tobacco have indicated that recombination is poor in these systems (Davies et al., 1999; Gidoni et al., 2001; Kilby et al., 1995), but it is not clear whether this low efficiency of FLP-mediated recombination between the target *FRT* sites is due to the suppression of the *FLP* locus or inherent to FLP-*FRT* system. Radhakrishnan and Srivastava (2005) introduced the gene for FLP recombinase, under the control of the maize ubiquitin-1 promoter, into the rice genome. FLP activity was monitored in callus and regenerated plants by an assay based on the deletion of the *FRT*-flanked DNA fragment, leading to the activation of the β-glucuronidase gene. FLP activity was detected both in the callus and leaves of some of the transgenic lines.

Luo et al. (2000) described a system, called the hybrid/recombinase system, for making Arabidopsis hybrids. In transgenic male-sterile plants, the expression of antisense bcp1 gene, driven by its own promoter or by a constitutive 35S promoter, causes male sterility. Cross-pollination with pollen from a FLP-expressing plant brings FLP recombinase into the male-sterile plants, and the function of FLP results in the removal of the antisense *bcp1* gene, completely restoring plant fertility in the hybrid progeny.

Hu et al. (2008) evaluated the feasibility of the FLP/*FRT* site-specific recombination system in rice. In the experiment, the FLP recombinase and *FRT* target sites were introduced separately in rice plants and brought together in hybrids by the cross-pollination of parental transgenic plants. The recognition of chromosomally integrated *FRT* sites and site-specific DNA recombination by FLP recombinase was visualized by excisional activation of the gusA gene. The recombined locus was shown to be stably inherited by the progeny, which confirm that the FLP/*FRT* recombination system functions effectively in the cereal crops.

R–RS

The R–*RS* system from the pSR1 plasmid of *Zygosaccharomyces rouxii* was first reported functioning efficiently in tobacco and Arabidopsis (Onouchi et al., 1991 and 1995). It then was used in the multi-auto-transformation (MAT) vectors system, that combines positive selection, using the isopentenyl transferase (*ipt*) gene, with a site-specific recombination and DNA removal system (R/*Rs*) that generates marker-free plants (Ebinuma et al. 2001). Tobacco plants were transformed with T-DNA vectors in which the *ipt* selectable marker gene and the gene coding for the R recombinase were surrounded by *RS* sites. The *ipt* gene provided the initial selection for morphological abnormalities (i.e. the shooty phenotype). The *A. rhizogenes rol* genes which confer the rooty phenotype have also been used (Cui et al., 2001). Co-expression of the *R* recombinase, under the control of the 35S promotor, eventually excised the *ipt* and *R* genes resulting in the development of normal marker-free shoots at very high frequencies (39–70%) from tobacco (Sugita et al., 1999), citrus (Ballester et al., 2007) and strawberry (Schaart et al., 2004). In this system, neither antibiotic nor herbicide resistance genes are necessary as a selective marker. This system was modified into a single-step transformation system for generating marker-free transgenic rice and needs no selective agent and no sexual crossing for marker free transgenic plants (Endo et al., 2002).

R/*Rs* recombination system has been demonstrated active in potato by Kondrak et al (2006). In their study, a binary vector PROGMO was constructed using a constitutively expressed R recombinase and a codA-nptII bi-functional, positive/negative selectable marker gene. After integration of the complete binary vector the *Rs* sites are orientated and located as such that the active recombinase enzyme will recombine and delete the recombinase gene, the markers and backbone sequence out of the plant genome resulting in marker-and backbone-free transgenic potato plants.

***β* -six**

The β recombinase from the pSM19035 plasmid of *Streptococcus pyogenes* belongs to the same subgroup of serine recombinase as the Gin recombinase although the two proteins are not very close relatives (Smith and Thorp, 2002). The functionality of serine recombinase in plant genome has not been tested until Gronlund et al. (2007) reported their results in tobacco and *Arabidopsis*. In their experiment, to mediate intra-molecular site-specific recombination, two β recombinase dimmers bind cooperatively to each asymmetric 90-bp recombination site which terms *six* sites. The ability of the β /*six* recombination system to catalyze site-specific deletion of target DNA sequences in tobacco and Arabidopsis plants was examined in the plants expressing the β recombinase under control of the constitutive CaMV 35S promoter. β recombinase stably expressed in tobacco plants can catalyze deletion of a spacer region that is flanked by directly oriented *six* sites and has been placed between promoter and a GUS reporter gene. Similarly, β recombinase in stably double-transformed Arabidopsis plants deletes specifically the spacer region of a reporter construct that has been incorporated into the genome.

Intrachromosomal Recombination

Superfluous T-DNA elimination can be also achieved with the approach of intrachromosomal recombination, which relies on the occurrence of base pairing between identical sequences in close proximity during DNA replication. Homologous recombination can cause the excision of a DNA fragment located between the two repeated DNA sequences of the homologous recombination substrate. Using 352-bp bacteriophage λ attachment sequence (attP), Zubko et al. (2000) made a binary vector containing a gene of interest, transformation booster sequence (Galliano et al., 1995) and an attP cassette. In the cassette, reporter genes *gfp* and *nptII* as well as an auxin-produced gene *tms2* were flanked by two attP regions. Intrachromosomal recombination results in the deletion of whole fragment between the attP regions even though the bacterial proteins that react against attP sequence were not present. Enhanced auxin biosynthesis due to the activity of *tms2* gene prior to the combination caused the callus production of the transformants carrying the cassette, but not rooting. On the contrary, the transformants without the cassette could be easily identified with the ability to produce roots. PCR analysis on all the rooted plants confirmed that the attP cassette was deleted in all the rooted transformants. Currently, intrachromosomal recombination is not yet efficient for routine production of marker-free nuclear transgenic plants due to its very low frequency. However, it has been successfully employed for construction of marker-free plastid transgenic plants (Dufourmantel et al., 2007). Furthermore, a lot of effort has been made to increase recombination frequency for practical applications including mark-free nuclear transgenic plant generation. Xiao and Peterson (2000) employed a maize *Ds* transposable element inserted between two overlapping GUS reporter gene segments and *Ac* transposase, resulting in an over 1000-fold increase in the incidence of recombination relative to spontaneous recombination events. Although the mechanisms underlying attP cassette excision and transposon enhanced homologous recombination are obscure, it would be interesting to test the generality of the approaches in other plant species. Additionally, recombination substrate size and substrate homology are two key factors that affect the recombination efficiency. In plants, it is clear that a positive correlation between the length of homologous overlapped sequences and the recombination efficiency exists (Baur et al., 1990; Putchta and Hohn, 1991). Li et al. (2004) used duplicated copies of different sizes of the intron segments (153, 418 and 589 bp) to construct the recombination substrates. They found that 153 bp of overlap substrate resulted in very low recombination events while longer substrates (418 and 589 bp) resulted in 44 and 183 times more events per plant. Additionally, Ilnytskyy et al (2004) also reported that in Arabidopsis, the frequency of intrachromosomal recombination increased 16-fold when the length of the substrate increased from 566 to 1213 bp. Moreover, induction of double-strand break (DSB) between repeats designed for intrachromosomal recombination events may enhance the recombination events. Siebert and Puchta (2002) showed that one-third of plant genomic DSB are repaired by predicted HR events, suggesting that elimination of superfluous foreign DNA could be achieved efficiently.

Recent studies on plant recombination showed that plant recombination plays an important role in the repair of DNA damage and genetic diversity and that is affected by various factors such as DNA sequence similarity and activation of transposons (Li et al., 2007). Homologous recombination has been employed for common practice of gene targeting in mice (Evans et al., 2001), but there are only a few successful reports in plants (Kempin et al., 1997; Hanin et al., 2001; Terada et al., 2007). Recently, zinc-finger nucleases were reported to target specific genome sequences for mutagenesis (Lloyd et al., 2005) and gene modification (Wright et al., 2005). Manipulations of plant endogenous recombination pathways and zinc-finger nucleases may result in efficient protocols for targeted gene replacement and marker free transgenic plant generation.

Co-Transformation

The marker gene can be eliminated after segregation of the foreign T-DNAs, including the gene of interest in the progeny of transgenic plants unless the T-DNAs are tightly linked together. Conventionally, the gene of interest, which confers the desired trait, and the marker gene are coupled together. They are usually passed on to the progeny as a single locus. In order to eliminate the marker from transgenic plants, the trait and marker genes have to be introduced into the genome independently. Then in the progeny of the transformants, segregation of the trait and marker genes occur, which makes possible to obtain the transgenic plant carrying trait gene only. To date, this approach is considered mature, as it has been used in many plant species (McKnight et al., 1987; De Neve et al., 1997; Ye and Qin, 2007).

In order to get the foreign T-DNAs, including trait-conferring and marker genes integrated into plant genome independently, several approaches for plant co-transformation have been reported. The different T-DNAs can either be present in one vector/one strain (Komari et al., 1996) or two vectors/ one strain (De Framond et al., 1986; Daley et al., 1998) or two vectors/ two strains (Mcknight et al., 1987; De Block and Debrouwer, 1991). All these approaches resulted in quite high co-transformation efficiency. Two separate bacterium strains were used to co-transform tobacco and/or tomato, resulting in about 50% co-transformation frequencies and large proportion of unlinked T-DNA insertions (Depicker et al., 1985; Mcknight et al., 1987). A single bacterium strain containing two separate T-DNAs resulted in even higher co-transformation frequencies (>50%) in tobacco and *Brassica napus* (Daley et al., 1998). A 'super binary' vector carrying more than one T-DNA region can result in even much higher co-transformation frequencies, for example, 66% in barley (Matthews et al., 2001), 93% in maize (Miller et al., 2002), 82-90% in rice (Breitler et al., 2004). Consequently, segregation frequencies for marker free transformants obtained by these approaches are also quite high.

Incorporation of a negative selection step can significantly enhance the screening process for the transgenic plants only carrying trait conferring gene. The use of a negative selection marker gene coupled with a positive selection marker gene in the same construct is a powerful method for obtaining marker-free transgenic plants. The plasmids are constructed to transfer the target gene and marker genes independently of one another and integrate them separately into the genome of a plant. The first round positive selection results in the transgenic plant carrying both marker and target genes. A subsequent negative selection kills the individuals carrying the marker gene, leaving the transgenic plants carrying only the gene of interest (Jeongmoo et al., 2004). A conditional marker gene was recently found (Erikson et al., 2004), which allows both positive and negative selection in plants. This marker will surely promote marker-free plant production through the co-transformation system.

However, it is clear that the co-transformation system has its limit in marker–free transgenic plant production. First, it is not suitable for the vegetative propagated plants such as fruit trees. Second, it is not suitable for the transgenic plants generated with gene bombardment because these transformants usually contain multiple copy number and tandem repeats of transgenes, which makes it difficult to obtain marker-free transgenic plants. Furthermore, it is laborious and time consuming because the transgenic plants after co-transformation have to be crossed and enough progeny have to be grown in order to obtain sufficient segregated plants without superfluous T-DNAs.

FUTURE PERSPECTIVES

For nearly two decades, research on recombinase-based applications has steadily advanced the surgical capabilities of plant genome rearrangement. In 2006, a first crop variety, corn LY038, developed by Renessen, a joint venture between Cargill and Monsanto, has been approved (Ow, 2007) for release under the trade name Mavera™, High Value Corn with Lysine. LY038 is derived from the biolistic transformation of maize to incorporate a *cordapA* ' gene that directs the seed-specific production of a lysine-insensitive dihydrodipicolinate synthase enzyme. The kanamycin resistance marker used for selection after transformation was subsequently removed by site-specific recombination (Ow, 2007). Although corn LY038 is the first marker-free transgenic plant released, it provides evidence that the use of recombinase for transgene excision or resolving complex transgenic loci will likely become common tools in the production of transgenic plants. Many companies that produce transgenic crops are now taking this approach in their research and marketing strategies, resulting in less regulatory and consumer-based concerns for their products. Concurrently with implementing this corporate strategy, there has been a recent explosion of technologies that, after further improvement, will allow the removal of DNA in plants almost as easily as it is inserted today.

It is noticed that site-specific recombination systems have been used in two different applications for transgenic crops: deletion of a marker gene from the transgenic plant, and site-specific integration of a gene into a pre-determined genomic location in a precise, single-copy pattern. These two applications have been developed independently, and each has great potential for transgenic research. The application of site-specific recombination to delete DNA – notably, the genes for antibiotic resistance – has progressed to using inducible expression systems. Chemical or heat induction systems keep the recombinase gene switched 'off' until the stage at which the DNA needs to be removed. This advance has extended the technology for marker gene removal to plant species in which the recombinase gene is not routinely introduced by a genetic cross (e.g. in many vegetatively propagated plants) and/or is inconvenient (e.g. in several tree species with long generation times). Therefore, Srivastava and Ow (2004) proposed a new strategy by joining site-specific integration and deletion into a 'combined step' method for introducing DNA into plants. This strategy makes use of two site-specific recombination systems: one for integrating the DNA, and a second for removing sequences that are no longer needed after DNA transfer. For example, (a) Use of Cre–*lox* to integrate a circular DNA, followed by use of the inducible FLP–*FRT* system to delete unneeded DNA from the integration locus. Delivery of circular DNA containing a *lox75* (left arm mutant) site into cells containing a *lox76* (right arm mutant) locus results in site-specific integration of the gene of interest (*GOI*) and formation of a double mutant *lox* site, which stabilizes the integration locus. After the site-specific integration lines are identified, the *FLP* gene is induced to remove unneeded DNA. (b) Use of R–*RS* system to integrate a linear T-DNA, delivered by *Agrobacterium*, into target *RS* sites, followed by use of the inducible Cre–*lox* system to remove unneeded DNA from the integration locus. Although this system remains to be tested, it will be a more expedient method in the future research on transgenic plants. There is no doubt that marker gene removal will eventually be routine as the techniques reviewed here are perfected.

Furthermore, the precise integration of DNA in planta can be achieved through site-specific recombination. One of the future research aspects will be to develop a recombinase-based plant transformation system, in particular, a gene stacking strategy that may be used by *Agrobacterium*-based delivery of the integrating DNA for stacking DNA, i.e. to append new DNA sequentially to a target site which permits the continual use of a previously characterized chromosome location. Stacking transgenic traits at a limited number of target sites is also preferable to scattering transgenes all over the genome (Ow 2005).The new transformation operating system can not only generate transgenic crops with good complex traits but also with precise structural fidelity and faithful expression. It will accelerate the process of research and development of elite cultivars, as well as help mitigate biosafety concerns of plant genetic modifications.

REFERENCES

Afolabi SA (2007) Status of clean gene (selection marker-free) technology, *African J Biotechnol* 6: 2910-2923.

Araki H, Jearnpipatkul A, Sakurai T, Ushio K, Muta T and Oshima Y (1985) Molecular and functional organization of yeast plasmid pSR1, *J Mol Biol* 182: 191-203.

Arumugam N, Gupta V, Jagannath A, Mukhopadyay A, Pradhan AK, Burma PK and Pental D (2007) A passage through in vitro culture leads to efficient production of marker-free transgenic plants in Brassica juncea using the Cre-loxP system, *Trangenic Res* 16:703-712.

Ballester A, Cervera M and Pena L (2007) Efficient production of transgenic citrus plants using isopentenyl transferase positive selection and removal of the marker gene by site-specific recombination, *Plant Cell Rep* 26:39-45.

Bar M, Leshem B, Gilboa N and Gidoni D (1996) Visual characterization of recombination at FRT-gusA loci in transgenic tobacco mediated by constitutive expression of the native FLP-recombinase, *Theor Appl Genet* 93: 407-413.

Barley CC, Morgan M, Dale EC and Ow DW (1992) Exchange of gene activity in transgenic plants catalyzed by the Cre-lox site-specific recombination system, *Plant Mol Biol* 18: 353-361.

Baur M, Potrykus I and Paszkowski, J (1990) Intermolecular homologous recombination in plants, *Mol Cell Biol* 10, 492– 500.

Davies GJ, Kilby NJ,. Riou-Khamlichi C and Murray JAH (1999) Somatic and germinal inheritance of an FLP-mediated deletion in transgenic tobacco, *J Exp Bot* 50: 1447-1456.

Bender J, Kuo J and Kleckner N (1991) Genetic evidence against intramolecular rejoining of the donor DNA molecule following IS10 transposition, *Genetics* 128:687-694.

Baszczynski CL. et al. (2003) Site-specific recombination systems and their uses for targeted gene manipulation in plant systems. In *Transgenic plants-current innovations and future trends,* edited by Stewart, C. C. Jr. Chapter 6, pp157-178.

Belzile F, Lassner MW, Tong Y, Khush R and Yoder JI (1989) Sexual transmission of transposed activator elements in transgenic tomatoes, *Genetics* 123: 181-189.

Breitler JC, Meynard D, Van Boxtel J, Royer M, Bonnot F, Cambillau L and Guiderdoni E. (2004) A novel two T-DNA binary vector allows efficient generation of marker-free transgenic plants in three elite cultivars of rice (Oryza sativa L.), *Transgenic Res* 13: 271-287.

Broach JR and Hicks JB (1980) Replication and recombination functions associated with the yeast plasmid, 2-μ circle, *Cell* 21: 501-508.

Charng YC, Li KT, Tai HK, Lin NS and Tu J (2008) An inducible transposon system to terminate the function of a selectable marker in transgenic plants, *Mol Breeding* 21:359-368.

Christou P and Twyman RM (2004) The potential of genetically enhanced plants to address food insecurity, *Nutr Res Rev* 17:23-42.

Corneille S, Lutz K, Svab Z and Maliga P (2001) Efficient elimination of selectable marker gene from the plastid genome by the CRE-lox site-specific recombination system, *Plant J* 27: 171-178.

Cotsaftis O, Sallaud C, Breitler JC, Meynard D, Greor R, Pereira A and Guiderdon (2002) Transposon-mediated generation of T-DNA and marker-free rice plants expressing a Bt endotoxin gene, *Mol Breeding* 10: 165-180.

Craig NL (1997) Target site selection in transposition. *Annu. Rev. Biochem.* 66: 437-474.

Cuellar W, Gaudin A, Solorzano D, Casas A, Nopo L, Chudakayandu P, Medrano G, Keruze J and Ghislain M (2006) Self-excision of the antibiotic resistance gene nptII using a heat inducible Cre-loxP system from transgenic potato, *Plant Mol Biol* 62: 71-82.

Cui M, Takayanagi K, Kamada H, Nishimura S and Handa T (2001) Efficient shoot regeneration from hariy roots of Anitirrhinum majus L. transformed by rol type MAT vector system, *Plant Cell Rep* 20: 60-66.

Dale EC and Ow DW (1990) Intra- and intermolecular site-specific recombination in plant cells mediated by bacterialphage P1 recombinase, *Gene* 91:79-85.

Dale EC and Ow DW (1991) Gene transfer with subsequent removal of the selection gene from the host genome, *Proc Natl Acad Sci* USA 88:10558-10562.

Daley M, Knauf VC, Summerfelt KR and Turner JC (1998) Co-transformation with one Agrobacterium tumefaciens strain containing two binary plasmids as a method for producing marker-free transgenic plants, *Plant Cell Re* 19: 489-496.

De Block M and Debrouwer D (1991) Two T-DNAs co-transformed into *Brasicca napus* by a double Agrobacterium infections are mainly integrated at the same locus, *Theor Appl Genet* 82: 257-263.

De Framond AJ, Back EW, Chilton WS, Kayes L and Chilton MD (1986) Two unlinked T-DNAs can transform the same tobacco plant cell and segregate in F1 generation, *Mol Gen Genet* 202:125-131.

De Neve M, De Buck S and Jacobs A (1997) T-DNA integration patterns in co-transformed plant cells suggests that T-DNA repeats originate from co-integration of separate T-DNAs, *Plant J* 11:15-29.

Depicker A, Herman L, Jacobs A, Schell J and Van Montagu M (1985) Frequencies of simultaneous transformation with different T-DNAs and their relevance to the Agrobacterium/plant cell interaction, *Mol Gen Genet* 201: 477-484.

Dufourmantel N, Dubald M, Matringe M, Canard H, Garcon F, Job C, Kay E, Wisniewski JP, Ferullo JM, Pelissier B, Sailland A and Tissot G. (2007) Generation and characterization of soybean and marker-free tobacco plastid transformants over-expressing a bacterial 4-hydroxyphenylpyruvate dioxygenase which provides strong herbicide tolerance, *Plant Biotechnol* J 5: 118-133.

Ebinuma H, Sugita K, Matsunga E and Yamakado M (1997a) Selection of marker-free transgenic plants using the isopentenyl transferase gene, *Proc Natl Acad Sci* USA 94: 2117-2121.

Ebinuma H, Sugita K, Matsunaga E ,Endo S, Yamada K and Komamine A (1997b) Principle of MAT vector, *Plant Biotech* 14: 133-139.

Ebinuma H, Sugita K, Matsunaga E, Endo S, Yamada K and Komamine A (2001) Systems for the removal of a selection marker and their combination with a positive marker, *Plant Cell Rep* 20: 383-392.

Endo S, Sugita K, Sakai M, Tanaka H and Ebinuma H (2002) Single step transformation for generating marker-free transgenic rice using the ipt-type MAT vector system, *Plant* J 30: 115-122.

Erikson O, Hertzberg M and Nasholm T (2004) A conditional marker gene allowing both positive and negative selection in plants, *Nat. Biotechnol.* 22: 455–458.

Evans MJ, Smithies O and Capecchi MR (2001) Mouse gene targeting, *Nat Med* 7: 1081–1090.

Fedoroff NV (1989) Maize transposable elements, p.375-412. in: Mobile DNA Berg.DE and Howe MM (Eds). *American Society for Microbiology*. Washington DC.

Ferry N, Edwards MG, Gatehouse J, Capell T, Christou P and Gatehouse AMR (2006) Transgenic plants for insect pest control, A forward looking scientific perspective, *Transgenic Res* 15:13-19.

Galliano H, Müller AE, Lucht JM, and Meyer P (1995) The transformation booster sequence from Petunia hybrida is a retrotransposon derivative that binds to the nuclear scaffold, *Mol Gen Genet* 247: 614–622.

Gidoni D, Bar M, Leshem B, Gilboa N, Mett A and Feiler J (2001) Embryonal recombination and germline inheritance of recombined FRT loci mediated by constitutively expressed FLP in tobacco, *Euphytica* 121: 145-156.

Gorbunova V and Levy AA (2000) Analysis of extrachromosomal Ac/Ds transposable elements, *Genetics* 155:349-359.

Gronlund JT, Stemmer C, Lichota J, Merkle T and Grasser KD (2007) Functionality of the B/six site-specific recombination system in Tobacco and Arabidopsis: A novel tool for genetic engineering of plant genomes, *Plant Mol Biol* 63:545-556.

Hagemann AT and Craig NL (1993) Tn7 transposition creates a hotspot for homologous recombination at the transposon donor site, *Genetics* 133:9-16.

Hajdukiewicz PT, Gilbertson L and Staub JM (2001) Multiple pathways for Cre-lox-mediated recombination in plastids, *Plant J* 27:161-170.

Hallet B and sherratt DJ (1997) Transposition and site-specific recombination: adapting DNA cut-and-paste mechanisms to a variety of genetic rearrangements, *FEMS Microbiol Rev* 21:157-178.

Hanin M, Volrath S, Bogucki A, Briker M, Ward E and Paszkowski J (2001) Gene targeting in *Arabidopsis*, *Plant J* 28: 671–677.

Hare DP and Chua NH (2002) Excision of selectable marker genes from transgenic plants, *Nat Biotechnol* 20: 575-580.

Hoa TTC, Bong BB, Hug E and Hodges TK (2002) Cre/lox site-specific recombination controls the excision of a transgene from the rice genome, T*heor Appl Genet* 104:518-525.

Hoff T, Schnorr KM and Mundy J (2001) A recombinase-mediated transcriptional induction system in transgenic plants, *Plant Mol Biol* 45: 41-49.

Hohn B, Levy AA and Puchta H (2001) Elimination of selection markers from transgenic plants, *Curr Opin Biotechnol* 12:139-143.

Holliday R (1964). A mechanism for gene conversion in fungi. Genet. Res. 5, 282-304.

Grainge, I and Jayaram, M. (1999) The integrase family of recombinase: organisation and function of the active site, *Mol Microbiol* 33: 449-456.

Hu Q, Kononowicz-Hodges H, Nelson-Vasilchik K, Viola D, Zeng PY, Liu HB, Kausch AP, Chandlee JM, Hodges TK and Luo H (2008) FLP recombinase-mediated site-specific recombination in rice, *Plant Biotechnol J* 6: 176–188.

Ilnytskyy Y, Boyko A and Kovalchuk I (2004) Luciferase-based transgenic recombination assay is more sensitive than β-glucuronidase-based. *Mutat Res* 559:189– 197.

Jeongmoo P, Young L, Bong K and Won C (2004) Co-transformation using a negative selectable marker gene for the production of selectable marker gene-free transgenic plants, *Theor Appl Genet* 109: 1562–1567.

Jia H, Yong PY, Chen X and Fang R (2006) Removal of the selectable marker gene from transgenic tobacco plants by expression of Cre recombinase from tobacco mosaic virus vector through agroinfection, T*ransgenic Res* 15:375-384.

Jones JDG Carland F, Lim E, Ralston E and Dooner HK (1990) Preferential transposition of the maize element Activator to linked chromosomal locations in tobacco, *Plant Cell 2:* 701-707.

Kempin SA, Liljegren SJ, Block LM, Rounsley SD, Yanofsky MF and Lam E (1997) Targeted disruption in Arabidopsis, *Nature* 389: 802-803.

Kilby NJ, Davies GJ, Snaith MR and Murray JAH (1995) FLP recombinase in transgenic plants: constitutive activity in stably transformed tobacco and generation of marked cell clones in Arabidopsis, *Plant J* 8: 637-652.

Komari T, Hiei Y, Saito Y, Murai N and Kumashiro T (1996) Vectors carrying two separate T-DNAs for co-transformation of higher plants mediated by *Agrobacterium tumefaciens* and segregation of transformants free from selection markers. *Plant J* 10: 165-174.

Kondrak M, van de Meer IM and Banfalvi Z (2006) Generation of marker- and backbone-free transgenic potatoes by site-specific recombination and a bi-functional marker gene in a non-regular one-border Agrobacterium transformation vector, *Transgenic Res* 15: 729-737.

Kopertekh L, Juttner G and Schiemann J (2005) PVX-Cre-mediated marker gene elimination from transgenic plants, *Plant Mol Biol* 55: 491-500.

Koprek T, Rangel S, McElroy D, Louwerse JD, Williams-Carrier RE, Peggy G and Lemaux P (2001) Transposon-Mediated Single-Copy Gene Delivery Leads to Increased Transgene Expression Stability in Barley. *Plant Physiol* 125:1534-1562.

Kuiper HA, Kelter GA, Noteborn HPJM and Kok EJ (2001) Assessment of food safety issues related to genetically modified foods. *Plant J* 27:503-528.

Kunze R and Starlinger P (1989) The putative transposase of transposable element Ac from Zea mays L. interacts with subterminal sequences of Ac, *EMBO J* 8:3177-3185.

Li J, Hsia AP and Schnable PS (2007) Recent advances in plant recombination. *Curr Opin Plant Biol* 10:131-135.

Li L, Santerre-Ayotte S, Boivin EB, Jean M and Belzile F (2004) A novel reporter for intrachromosomal homologous recombination in Arabidopsis thaliana, *Plant J* 40:1007–1015.

Lioyd AM and Davis RW (1994) Functional expression of the yeast FLP/FRT site-specific recombination system in *Nicotiana tobacum*, *Mol Gen Genet* 242: 653-657.

Lloyd L, Plaisier CL, Carroll D and Drews GN (2005) Targeted mutagenesis using zinc-finger nucleases in *Arabidopsis*, *Proc Natl Acad Sci* USA 102:2232–2237.

Li ZS, Xing AQ, Moon BP, Burgoyne SA, Guida AD, Liang HL, Lee C, Caster CS, Barton JE, Klein TM and Falco SC (2007) A Cre/loxP-mediated self-activating gene excision system to produce marker gene free transgenic soybean plants, *Plant Mol Biol* 65:329-341.

Liu HK, Yang C and Wei ZH (2005) Heat-shock-regulated site-specific excision of extraneous DNA in transgenic plants, *Plant Sci* 168:997-1003.

Luo H, Lyznik LA, Gidoni D and Hodges TK (2000) FLP-mediated recombination for use in hybrid plant production, *Plant J* 23:423-430.

Luo KM, Duan H, Zhao DG, Zheng XL, Deng W, ChenYQ Stewart Jr CN, McAvoy R, Jiang XN, Wu YH, He AG, Pei Y and Li Y (2007) 'GM-gene-deletor': fused loxP-FRT recognition sequences dramatically improve the efficiency of FLP or CRE recombinases on transgene excision from pollen and seed of tobacco plants, *Plant Biotech J* 5: 263-274.

Lyznik LA, Mitchell JC, Hirayama L and Hodges TK (1993) Activity of yeast FLP recombinased in maize and rice protoplasts, *Nucelic Acids Res* 21: 969-975.

Lyznik LA, Rao KV and Hodges TK (1996) FLP-mediated recombination of FRT sites in the maize genome, *Nucleic Acids Res* 24: 3784-3789.

Matthews PR, Waterhouse PM, Thornton S, Fieg SJ, Gubler F and Jacobsen JV (2001) Marker gene elimination from transgenic barley, using co-transformation with adjacent 'twin T-DNAs' on a standard Agrobacterium transformation vector, *Mol Breeding* 7: 195–202.

McKnight TD, Lillies MT and Simpson RB (1987). Segregation of genes transferred to one plant cell from two Agrobacterium strains, *Plant Mol Biol* 8:439-445.

Miller M, Tagliani L, Wang N, Berka B, Bidney D and Zhao ZY (2002) High efficiency transgene segregation in co-transformed maize plants using an Agrobacterium tumifaciens 2 T-DNA binary system, *Trans Res* 11: 381–396.

Miki B and McHugh S (2004) Selectable marker genes in transgenic plants: applications, alternatives and biosaftey, *J Biotechnol* 107:193-232.

Mlynarova L, Conner AJ and Nap JP (2006) Directed micro-specific recombination of transgenic alleles to prevent pollen-mediated transmission of transgenes, *Plant Biotechnol J* 4: 445-452.

Nash HA (1996) Site-specific recombination: intergration, excision, resolution, and inversion of defined DNA segments. In *Escherichia Coli* and *Salmonella Typhimurium*: Cellular and Molecular Biology. Neidhart, F. C (ed), Washington, DC: *American Society for Microbiology,* Vol.2, pp.2263-2376.

Odell J, Caimi P, Sauer B and Russell S (1990) Site-directed recombination in the genome of transgenic tobacco, *Mol Gen Genet* 223: 369-378.

Onouchi H, Yokoi K, Machida C, Matsuzaki H, Oshima Y, Marsuoka K, Nakamura K and Machida Y (1991) Operation of an efficient site-specific recombination system of Zygosaccharomyces rouxii in tobacco cells, *Nucelic Acids Res* 19:. 6373-6378.

Onouchi H, Nishihama R, Kudo M Machida Y and Machida C (1995) Visualization of site-specific recombination catalyzed by a recombinase from Zygosaccharomyces rouxii in Arabidopsis thaliana, *Mol Gene Genet* 247: 653-660.

Ow DW (2002) Recombinase-directed plant transformation for the post-genomic era, *Plant Mol Biol* 48:183-200.

Ow DW (2005) Transgene management via multiple site-specific recombination systems, In *Vitro Cell Dev Biol Plant* 41:213-219.

Ow DW (2007) GM maize from site-specific recombination technology, what next? C*ur rent opinion in Biotechnology* 18:115-120.

Poppy GM (2004) Geneflow from GM plants – towards a more quantitative risk assessment. *Trends Biotechnol.* 22, 436-438.

Puchta H and Hohn B (1991) A transient assay in plant cells reveals a positive correlation between extrachromosomal recombination rates and length of homologous overlap, *Nucleic Acids Res* **19**, 2693– 2700.

Puchta H (2003) Marker-free transgenic plants, *Plant Cell Tiss. Organ Cult.* 74:123–134.

Radhakrishnan P and Srivastava V (2005) Utility of the FLP-FRT recombination system for genetic manipulation of rice, *Plant Cell Rep* 23:721-726.

Ramessar K, Perermarti A, Gomez-Galera S, Naqvi S, Moralejo M, Munoz P, Capell T and Christou P (2007) Biosafety and risk assessment framework for selectable marker genes in transgenic crop plants: a case of the science not supporting the politics, *Transgenic Res* 16:261-280.

Reed RR and Grindley ND (1981) Transposon-mediated site-specific recombination in vitro: DNA cleavage and protein-DNA linkage at the recombination site, *Cell* 25: 721-728.

Reed RR and Moser CD (1984) Resolvase-mediated recombination intermediates contain a serine residue covalently linked to DNA, *Cold. Spring. Harb. Symp. Quant. Biol.* 49, 245-249.

Russell SH, Hoopes JL and Odell JT (1992) Directed excision of a transgene from the plant genome, *Mol Gen Genet* 234: 49-59.

Sadowski PD (1993) Site-specific genetic recombination: hops, flips, and flops, *FASEB J* 7: 760-767.

Sauer B (1987) Functional expression of the *cre-lox* site-specific recombination system in the yeast *Saccharomyces cerevisiae*, *Mol. Cell Biol.* 7: 2087-2096.

Schaart JG, Krens FA, Pelgrom KTB, Mendes O and Rouwendal GJA (2004) Effective production of marker-free transgenic strawberry plants using inducible site-specific recombination and a bifunctional selectable marker gene, *Plant Mol Biol* 55: 491-500.

Screekala C, Wu L, Gu K, Wang D, Tian D and Yin Z (2005) Excision of a selectable marker in transgenic rice (*Oryza sativa* L.) using a chemically regulated Cre/loxP systems, *Plant Cell Rep* 24: 86-94.

Shapiro JA (1993) Natural genetic engineering of the bacterial genome. *Curr Opin Genet Dev* 3:845-848.

Sherratt DJ, Ariszewska LK, Blakely G, Colloms S, Grant K, Leslie N and McCulloch R (1995) Site-specific recombination and circular chromosome segregation, *Phil Trans R Soc Lond B Biol Sci* 347:37-42.

Siebert R and Puchta H (2002) Efficient repair of genomic double-strand breaks by homologous recombination between directly repeated sequences in the plant genome, *Plant Cell* 14:1121-1131.

Smith MCM and Thorp HK (2002) Diversity in the serine recombinases, *Mol Microbiol* 44:299-307.

Sonti RV, Tissier AF, Wong D, Viret JF and Signer ER (1995) Activity of yeast FLP recombinase in Arabidopsis, *Plant Mol Biol* 28: 1127-1132.

Srivastava V and Ow DW (2004) Marker-free site-specific gene integration in plants, *Trends Biotechnol* 22: 627-629.

Stark WM, Boocock MR and Sherratt DJ (1992) Catalysis by site-specific recombinases, *Trends Genet* 8: 432-439.

Stark WM and Boocock MR (1995) Topological selectivity in site-specific recombination. In *Mobile Genetic Elements,* Sherratt, D.J. (ed.), pp 101-129, Oxford University Press.

Sugita K, Matsunaga E and Ebinuma H (1999) Effective selection system for generating marker-free transgenic plants independent of sexual crossing, *Plant Cell Rep* 18: 941-947.

Sugita K, Kasahara T, Matsunaga E and Ebinuma H (2000) A transformation vector for the production of marker-free transgenic plants containing a single copy transgene in high frequncy, *Plant J* 22: 461-469.

Terada R, Johzuka-Hisatomi Y, Saitoh M, Asao H and Iida S (2007) Gene targeting by homologous recombination as a biotechnological tool for rice functional genomics, *Plant Physiol.* 144: 846-856.

Van Duyne GD (2002) A structural view of tyrosine recombinase site-specific recombination, In *Mobile DNA II* (N. L. Craig, R. Craigie, M. Gellert, and A.M. Lambowitz, eds.), pp. 93-117. ASM Press, Washington, DC.

Verweire D, Verleyen K, De Buck S, Claeys M and Angenon G (2007) Marker-free transgenic plants through genetically programmed auto-excision, *Plant Physiol* 145:1220-1231.

Wang Y, Chen B, Hu Y, Li J and Lin Z (2005) Inducible excision of selectable marker gene from transgenic plants by Cre-lox site-specific recombination system, *Transgenic Res* 14: 605-614.

Wright DA, Townsend JA, Winfrey Jr RJ, Irwin PA, Rajagopal J, Lonosky PM, Hall BD, Jondle MD and Voytas DF (2005) High-frequency homologous recombination in plants mediated by zinc-finger nucleases, *Plant J* 44: 693–705.

Xiao YL and Peterson T (2000) Intrachromosomal homologous recombination in Arabidopsis induced by a maize transposon, *Mol Gen Genet* 263: 22-9.

Yan H and Rommens CM (2007) Transposition-based plant transformation, *Plant Physiol* 143: 570-578.

Ye XG and Qin H. (2007) Obtaining marker-free transgenic soybean plants with optimal frequency by constructing three T-DNAs binary vector, *Sheng Wu Gong Cheng Xue Bao* 23: 138-144.

Yoder JI and Godsbrough AP (1994) Transformation systems for generating marker-free transgenic plants, *Bio/Technol* 12:263-267.

Zhang W, Subbarao S, Addae P, Shen A, Armstrong C, Peschke V and Gielbertson L (2003) Cre/lox-mediated marker gene excision in transgenic maize (Zea mays L.) plants, *Theor Appl Genet* 107:157-168.

Zubko E, Scutt C and Meyer P (2000) Intrachromosomal recombination between attP regions as a tool to remove selectable marker genes from tobacco transgenes, *Nat Biotechnol* 18: 442-445.

Zuo J, Niu QW, Moller SG and Chua NH (2001) Chemical-regulated site-specific DNA excision in transgenic plants, *Nat Biotechnol* 19:157-161.

In: Transgenic Plants: New Research
Editor: Oliver T. Chan

ISBN 978-1-60692-017-6
© 2009 Nova Science Publishers, Inc.

Chapter 6

DETECTION OF TRANSGENE COPY NUMBER IN DURUM WHEAT TRANSGENIC LINES USING REAL-TIME PCR

Agata Gadaleta[1*], Angelica Giancaspro[1*], Maria Francesca Cardone[2], Maria C. Guastadisegni[2], Mario Ventura[2], Antonio Blanco[1]

[1]Department of Environmental and Agro-Forestry Biology and Chemistry, sect. Genetics and Plant Breeding, University of Bari, Via Amendola 165/A – 70126 Bari, Italy
[2]Department of Genetics and Microbiology, University of Bari, Via Amendola 165/A – 70126 Bari, Italy

ABSTRACT

Genetic transformation has played a key role in gaining and applying knowledge of the roles of HMW-GS in wheat end-use properties. Reliable and stable expression of transgenes as well as the characterization and field adaptation of transgenic lines are prerequisites for the successful application of gene technology.

Loci that appear to be stably expressed initially can become progressively silenced over generations. The stability and the behaviour of transgenes are influenced by several factors, such as chromosomal location, transgene copy number and their interaction with the host genotype.

Traditionally such factors are characterized using Southern analysis which can be time consuming and laborious. Recent results obtained in various crops indicate that Real-Time PCR could be a powerful tool for the detection and characterization of transgene locus structures. The determination of transgenic locus number through Real-Time PCR overcomes the problems linked to phenotypic segregation analysis and can

* These authors contributed equally to this work. Corresponding author: Dr. Agata Gadaleta. Department of Environmental and Agro-Forestry Biology and Chemistry, sect. Genetics and Plant Breeding, University of

analyze hundreds of samples in a day, making it an efficient method for estimating copy number integrated in a transgenic line.

This study was conducted to determine transgene copy number in transgenic lines and to investigate potential differences in sensitivity, resolution and variability between two different Real-Time chemistries (SYBR Green I dye and TaqMan probes). We have applied Real-Time PCR to a set of four transgenic durum wheat lines previously obtained. A total of four experiments (two experiments for each gene) were conducted and standard curves were obtained from serial dilutions of the plasmids containing the genes of interest. The correlation coefficients of the standard curves were rather good, being in the range between 0.95 and 0.97. By using TaqMan quantitative Real-time PCR we were able to achieve estimates of 1 to 42 copies of transgenes per haploid genome in T_4 homozygous transformants. Conversely, SYBR Green dye method revealed unable to accurately quantify transgene copy number as it failed in detecting the inserted genes when integrated in few copies.

In our study we assessed Real-Time PCR as a fast, sensitive and reliable method for the detection of transgene copies in durum wheat, which can be a valid alternative to Southern analysis.

Keywords: Durum wheat, transformation, glutenin subunits, Real-Time PCR

INTRODUCTION

Despite the continuous improvements in plant genetic transformation procedures, common delivery methods such as *Agrobacterium* and particle gun usually still lead to integration of multiple transgene copies at the same or different chromosomal locations. As transgene copy number can influence the expression level and the genetic stability of both exogenous and endogenous genes, the estimation of gene copy number has become an important area of genetically modified (GM) crop research. While multiple copies of a transgene may be useful for over-expression experiments, single-copy transformation events are preferred for most applications because they are stable over generations of subsequent breeding (Meyer, 1998). Thus, when new transgenic plants are obtained one essential step is their molecular characterization. This allows researchers to select in early stages of plant development the primary transformants carrying only one or two integration events, which are likely to yield high-level expression of the exogenous gene and can be more successfully destined to field cultivation.

The method traditionally employed to determine transgene copy number is Southern blotting. While reliable, this classic molecular biology technique is laborious, time consuming, requires large amounts of starting plant material and may involve the use of hazardous radioisotopes (De Preter et al., 2002). Moreover, Southern analysis may fail detecting the exact number of transgene copies when these integrate in a rearranged pattern in the host genome, loosing some relevant restriction sites (Mason et al., 2002).

Recently a new detection method has been developed to estimate transgene copy number in transformed plants based on quantitative Real-Time PCR (Ingham et al., 2001). At now

Bari, Via Amendola 165/A-70126 Bari (Italy). Ph. +390805442526 Fax +390805442200 e-mail: agata.gadaleta@agr.uniba.it this new approach has successfully been applied to several species such as bread wheat (Li et al., 2004), maize (Bubner and Baldwin, 2004), rice (Yang et al., 2005), soybean and peanut (Schmidt and Parrot, 2001), *Nicotiana attenuata* (Bubner et al., 2004) and tomato (Mason et al., 2002).

Real-Time PCR can be employed for quantifying both DNA and RNA abundance. It is used in a huge number of research applications such as monitoring of viral load in infected plants (Mackay et al., 2002; Mackay, 2004) or detection of genes involved in toxinogenesis (Mayer et al., 2003; Shnerr et al., 2001). Today, following the compulsory requirements on food quality and safety, Real-Time PCR is being employed also in testing food adulteration or contamination with transgenic DNA (Vaitilingom et al., 1999). Interestingly this quantitative PCR assay is recently becoming a choice for a growing number of molecular analyses including gene expression quantification, expression profiling, SNPs analysis, validation of microarray data and detection of transgene copy number.

Real-Time PCR allows two methods for the quantification of a target nucleic acid sequence: absolute and relative quantification. In the absolute quantification the exact level of a sequence in a sample is directly inferred from a standard curve prepared from a dilution series of control template of known concentrations. On the other hand, relative quantification is mainly used in gene expression studies to determine the up- or down-regulation of a gene of interest relative to a calibrator.

Quantitative Real-Time PCR measures the amplified products as they accumulate at each cycle of the PCR reaction by registering the fluorescence intensity emitted by a fluorescent reporter dye bound to the amplified target. The first cycle at which the instrument can distinguish the amplification generated signal as being above the background noise is referred to as the threshold cycle (Ct) and is inversely proportional to the starting copy number of the template. As fluorescence output is linear to sample concentration over a very broad range, the linear correlation between PCR product and fluorescence intensity can be used to calculate the amount of starting DNA template in the sample. The absolute quantity of the target is inferred from a calibration curve obtained from a dilution series of a control template, covering the entire range of concentrations that are supposed to be measured in the assay. The standard curve is generated by plotting the log of the initial template copy number against the Ct generated for each dilution. The standard control is usually represented by a plasmid containing the cloned gene of interest.

The reporter molecules used in Real-Time reactions can be not-specific DNA-binding fluorophores (e.g. SYBR Green I), fluorophore-labelled primers (e.g. LUX) or sequence-specific probes (e.g. TaqMan). While DNA-binding dyes are low cost and easy to use, they show the disadvantage to bind to any double-stranded DNA leading to the detection of both target sequence and not-specific amplification products. On the contrary, oligonucleotide probes specifically bind to the target sequence allowing an higher level of detection specificity, and enabling the quantification of several targets in the same sample within a single PCR reaction.

In this paper we report for the first time the use of quantitative Real-Time PCR to determine transgene copy number in durum wheat (*Triticum turgidum* L. var. *durum*) lines transformed with *Glu-D1-1d* and *Glu-D1-2b* genes of bread wheat encoding the HMW glutenin subunits 1Dx5 and 1Dy10; two different chemistries were compared for transgene quantification: SYBR Green dye and TaqMan probes.

Materials and Methods

Transgenic Plants and DNA Preparation

Four independent transgenic lines of durum wheat *(Triticum turgidum L.* var. *durum)*, previously obtained in direct transformation experiments by biolistic method described by Gadaleta et al. (2008), were characterized by absolute Real-Time PCR. Lines DV1-4c, DC2-10-4, DC2-65s and DS2-127 were derived from cultivars Varano (DV1-4c), Creso (DC2-10-4, DC2-65s) and Svevo (DS2-127) co-transformed with the two bread wheat (*Triticum aestivum L.*) genes *Glu-D1-1d* and *Glu-D1-2b* encoding the HMW glutenin subunits 1Dx5 and 1Dy10. As negative controls for Real-Time experiments not transformed lines of the same cultivars were used. Genomic DNA from control and transformed wheat lines was extracted starting from 5 grams of frozen leaves using the Phenol-Chloroform method. DNA concentration was determined using a UV-Visible spectrophotometer detecting adsorbance at a wavelength of 260 nm. Purity of extracted DNA was assessed reading the 260 nm / 280 nm ratio, with a value of approximately 1.8 indicating a good quality. DNA was also checked on 0.8% agarose gel stained with ethidium bromide.

In order to have standard templates to build the calibration curves for the Real-Time experiments, plasmids pKs+Dx5 and pKs-Dy10 were cloned into *E. coli* competent cells by *heat-shock* method. Selection of transformed cells was performed on 1X LB medium containing 50 mg/l Ampicillin. Plasmid DNA was finally extracted from *E. coli* by using the *QIAGEN Plasmid MAXI Kit* following the manufacturer instructions. Each plasmid was finally checked for the inserted gene by PCR amplification with specific primer pairs, and by endonuclease digestion followed by electrophoretic separation on 1.8% agarose gel.

As durum wheat is an allo-tetraploid species, in our study we referred the transgene copy number to each haploid genome. With this aim we firstly calculated the exact DNA concentration of both transgenic and not transformed (control) lines using a fluorimetric measurement, and then we converted this value into the number of copies of genome/µl as follows:

$$\text{n° copies genome/µl} = \frac{\text{DNA concentration [g/l]}}{\text{Molecular weight [g/mol]}} \times \text{Avogadro n.} \times 10^{-6}$$

The same quantification was carried on plasmids pKs+Dx5 and pKs-Dy10 in order to prepare dilution series of known concentration to obtain the calibration curves for transgene absolute quantification.

Outline of Real-Time PCR Method

In order to detect *1Dx5* and *1Dy10* inserted copy number in the four transformed lines of durum wheat we used Absolute Quantification (AQ). The exact amount of each target gene was calculated by means of a calibration curve obtained from a dilution series of a standard template represented by the plasmid used for the transformation experiments. Vector pKs+Dx5 was used for the quantification of *1Dx5*, while pKs-Dy10 was employed to detect

1Dy10. The results of AQ experiments are reported in the same measure as the standard curve.

As *1Dx5* and *1Dy10* bread wheat genes show sequence similarity to other endogenous glutenin subunits, two different systems (*right* and *left*) were designed for transgenes detection, each consisting in one primer complementary to the vector sequence and the other one annealing to the gene of interest. The use of primer pairs encompassing the vector sequence ensured the amplification only of the transgenic insert and not of the endogenous genes.

In this work Real-Time PCR was employed to asses transgene copy number by comparing two different chemistries: SYBR Green dye and TaqMan probes. In SYBR Green assays the fluorigenic SYBR Green I dye allows the detection of amplification products as they accumulate during PCR by binding to double stranded DNA. The limit of this chemistry is that the dye binds non-specifically to all double-stranded DNA, giving a signal due to the amplification both of specific target and not-specific sequences. Nevertheless, any not-specific fragment can be detected after PCR amplification by performing the dissociation curve reporting fluorescence intensity versus temperature during the melting process. The presence of more than one population of PCR products is reflected as multiple thermal transitions in the fluorescence intensity.

TaqMan assay provides a higher level of detection specificity as it uses a specific oligonucleotide probe internal to the target sequence, which carries a fluorescent reporter at the 5'end and a quencher at the 3'. As long as the two molecules stay in proximity the fluorescence by the reporter is quenced; during the extension phase of PCR the 5'-3' exonuclease activity of *Taq* polymerase cleaves the probe releasing free reporter dye into solution and allowing fluorescence to be detected in each successive round of amplification.

Primers and Probes

TaqMan assays were performed using specific primer pairs and specific probes annealing to the target gene sequence slightly downstream of one of the primers. For *1Dx5-left* we designed the forward primer complementary to the vector sequence, and the reverse primer and the probe complementary to the insert sequence. On the contrary, for *1Dx5-right* and *1Dy10-right* the forward primer and probe were complementary to the insert sequence, while reverse primer was complementary to the vector sequence. Primers and probes were designed by using Primer Express software (Applied Biosystem Co., Foster City, CA, USA) and are described in Table 1.

For TaqMan chemistry Minor Groove Binder (MGB) DNA probes were used (Applyed Biosystems). MGB probe tipically consists of a 14 or 15 bp long oligonucleotide carrying the fluorescent dye FAM (6-carboxifluorescein, excitation wavelength = 494 nm, emission wavelength = 521 nm) at its 5'end, and the TAMRA quencher and MGB at its 3'end. MGB part of the probe binds to the minor groove of the double stranded DNA helix (consisting of the oligo part of the MGB probe and the complementary target sequence to which it is hybridized) irrespective of the nucleotide sequence, thus enhancing the stability of the probe-amplicon binding.

Table 1. Primers, probes and other components employed for PCR amplifications in Real-Time TaqMan assays. *F*: Forward primer; *R*: Reverse primer

Target gene	*Primer sequence (5' to 3')*	*MGB probe sequence (5' to 3')*	*Amplicon size (bp)*
1Dy10-right	*F:* TCATTTGGAGTTGTGCAGAATAGC *R:* GCTGCAGGAATTCTGGACTGA	ATCTCTGTTGTAGCCTTT	65
1Dx5-left	*F:* CGATAAGCTTGATATCGAATTCTAAAAG *R:* AGTCAAAATCGGATGATCCAATTT	CAATTGAGGTCTCCAATT T	73
1Dx5-right	*F:* AAATTATGATCGCACATTTGTAACG *R:* CCCGGGCTGCAGGAAT	CGCTGATTTGGTACTTG	60
Reporter	FAM (495 nm)		
Quencer	TAMRA		
Passive reference	ROX		

In SYBR Green assays PCR amplifications were run in presence of specific primer pairs for the target genes (Tab. 2) and a non-specific DNA binding dye.

SYBR Green I is an asymmetrical cyanine dye commonly employed as a nucleic acid stain in molecular biology. This molecule binds to double-stranded DNA but not to single-strand DNA, resulting in a DNA-dye-complex which absorbs blue light (λ_{max} = 488 nm) and emits green light (λ_{max} = 522 nm). SYBR Green is incorporated into the amplicon during the amplification reaction, so that when it binds to the minor groove of dsDNA the intensity of fluorescence emission increases allowing the detection of accumulating products. In order to check for the presence of undesired products such as primer dimers or not-specific fragments, dissociation curves were performed following PCR amplification for all the SYBR Green reactions.

Table 2. Primers employed for PCR amplifications in Real-Time SYBR Green I assays. *F*: Forward primer; *R*: Reverse primer

Target gene	*Primer sequence (5' to 3')*	*Amplicon size (bp)*
1Dy10-left	*F*: AGCTTGATATCGAATTCACTAGAAT *R*: CGAAATAGCCGGGTCCTC	67

Optimization of Primers and Probes Performance

As shorter amplicons work more efficiently, primers were designed that amplified short DNA segments in the 50 to 150 bp range. As starting point a standard PCR reaction and a PCR with a gradient of different annealing temperatures (ranging between 56 °C and 64 °C) were performed for each designed primer pair, in order to check for their specificity and determine the best temperature for their annealing. Amplification products were checked for the expected molecular size on 1.5% agarose gel stained with ethidium bromide (Fig. 1). The gradient PCR was carried out in the Mastercycler Gradient cycler (Eppendorf, Germany). All primers were used at an optimal annealing temperature of 60 °C.

For both TaqMan and SYBR Green systems, primer concentrations were optimised in preliminary experiments of Real-Time PCR. Reactions were run with nine different combinations of primer concentration (50, 300 and 900 nM for each primer), choosing those giving a high endpoint fluorescence and a low Ct value (Tab. 3). For the MGB probe an optimal concentration of 250 nM was used. In SYBR Green experiments all primer pairs were employed at the final concentration of 300 nM.

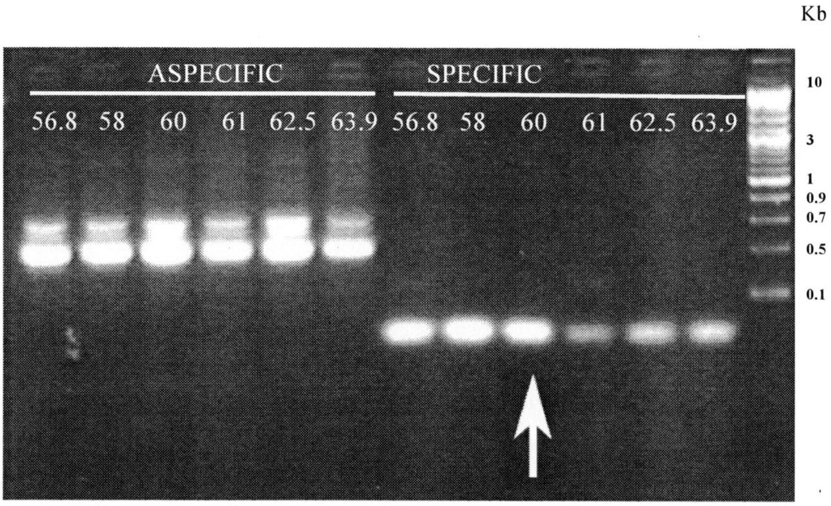

Figure 1. Electrophoretic pattern deriving from PCR amplification of 1Dy10-right gene with a gradient of different annealing temperatures (ranging between 56°C and 64°C). On the left are shown the fragments derived from the amplification with a non-specific primer pair. On the right is reported the amplification of the same gene with specific primers giving the expected product of 65 bp. The white arrow indicates the best annealing temperature of 60°C chosen for the Real-Time PCR reactions.

Table 3. Optimized primer concentrations employed in TaqMan experiments

Target gene	*Final For/Rev primer concentration (nM)*
1Dy10-right	300/300
1Dx5-left	900/900
1Dx5-right	300/300

While specificity of TaqMan amplification is ensured by using the probe, in SYBR Green assays primers were checked for giving the expected product by performing the melting curve following the amplification reaction. The presence of not-specific products was detected by multiple peaks, while specific products were represented by a single thermal transition (Fig. 2a, b).

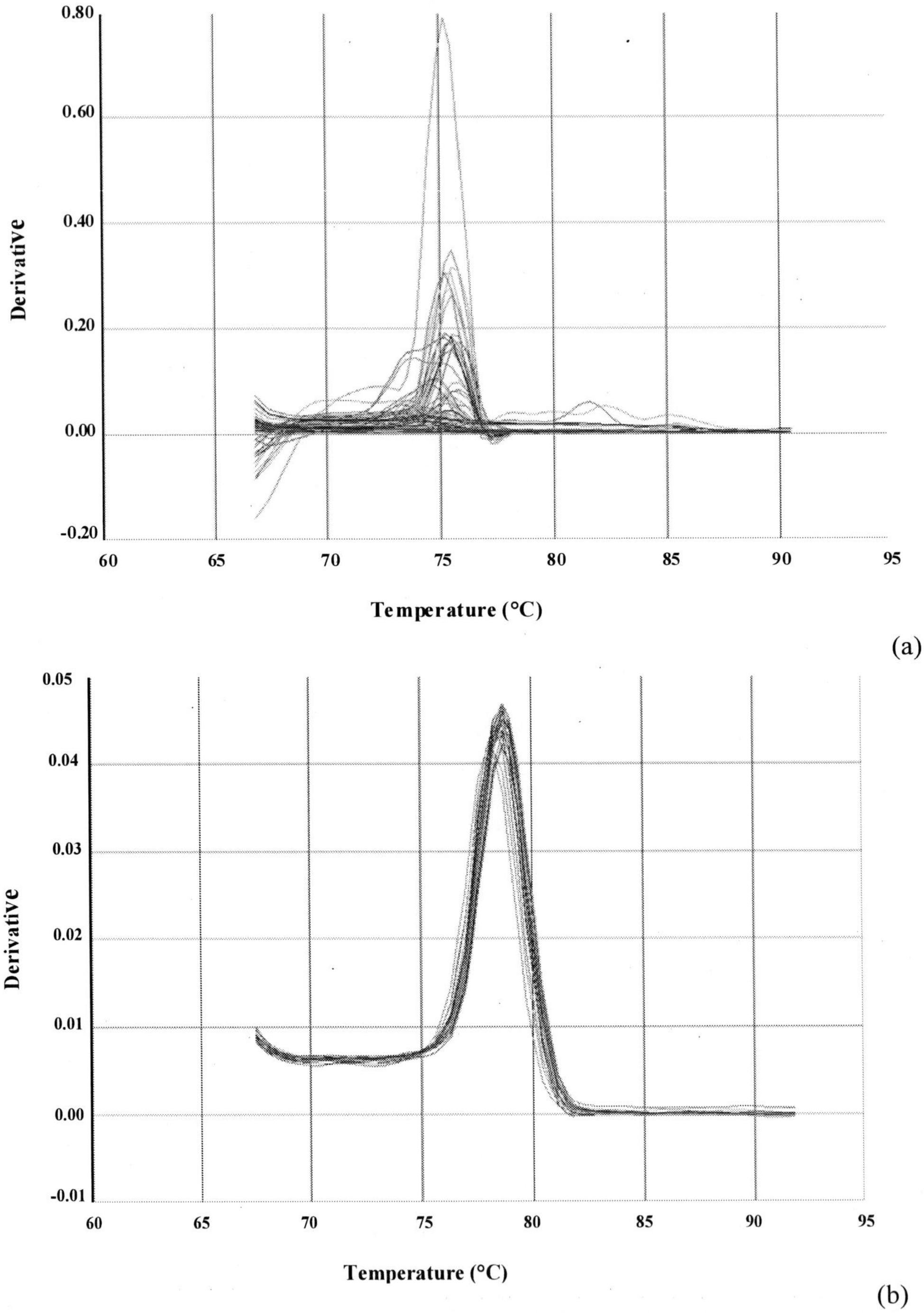

Figure 2. Melting curve of Dy10-right gene amplification products reporting the first derivative of raw fluorescence plotted against increasing temperature. (a) The higher peak accounts for the amplification of specific fragment; the lower second peak (black arrow) indicates the presence of small amount of a non-specific product like primer-dimer. (b) Dissociation curve registered after amplification of the same gene with a specific primer pair. The single melt peak indicates that a single specific product is being amplified in the samples.

Calibration Curves

As Real-Time absolute quantification of DNA is based on the linear correlation between starting amount of the template and fluorescence intensity at the Ct, preliminary SYBR Green experiments were carried out to test the linearity between Ct and log $_{\text{copy number}}$ both in *1Dx5* and *1Dy10 left* and *right* systems. This allowed to evaluate if the amount of the template and its Ct varied linearly during Real-Time PCR, and build calibration curves to use for transgenes quantification. Starting from known concentrations of standard templates (plasmid pKs+Dx5 for *1Dx5* and pKs-Dy10 for *1Dy10*), accurate serial dilutions were prepared and run in triplicate. In order to accurately quantify all the samples, the least and most diluted points of the curves were opportunely adjusted so that samples Ct values fell within the linear section of the curve. For each gene the standard curve was run on the same plate as the samples in order to ensure good reproducibility of Ct values, as these may vary if run at different times and plates.

Calibration curves were obtained for each primer-probe system by plotting the log of starting standard DNA amounts versus the corresponding Ct. Calibration curves were optimized by adjusting the threshold and the baseline in order to obtain an R square value (R^2 or Pearson Correlation Coefficient) very close to 1 and a good slope (around -3.3). The threshold is the fluorescence value slightly above the background fluorescence measured before exponential growth starts. Thresholds were either calculated according to an algorithm of the Real-Time PCR software or set manually. We used a six-point calibration curve with five-fold dilution for the two systems of *1Dx5* and three-fold dilution for *1Dy10*.

Real-Time PCR Reactions

TaqMan

TaqMan reactions were carried out in the *7300 Real-Time PCR System* (Applied Biosystem Co., Foster City, CA, USA) using 96-well reaction plates. Reaction mixtures consisted of 1X *Platinum Quantitative PCR SuperMix UDG* (Life Technologies-Invitrogen, containing AmpliTaq Gold DNA Polymerase, dNTPs, dUTP and the enzyme uracyl-N-glycosilase UNG), 500 nM of ROX passive reference, 250 nM of specific probe, specific primers at the optimal concentrations (see Tab. 3), 5 ng of sample DNA and water to a 15 µl final volume. For each transgenic line a single DNA extract was used in all the experiments to ensure a good reproducibility and to allow a correct quantification. The cycling parameters used were the following: one cycle at 50°C for 2 min for activation of UNG, one cycle at 95°C for 10 min for DNA polymerase activation, followed by 40 cycles of 95°C for 15 sec (denaturation) and 60°C for 1 min (combined annealing and extension). The use of 60°C (7-10°C below the probe Tm) allowed to avoid the displacement of the probe from the template if the temperature in the reaction is too high during the *Taq* polymerase extension (such as at the standard extension temperature of 72°C). In addition to the transgenic lines, both positive and negative controls were amplified in each experiment. The positive controls were the same transgenic plasmids used to obtain the standard curves. As the negative controls we used the No-Template-Control (NTC), that is a reaction mix including all the reaction components except for the DNA, and the DNA of cultivars Svevo, Varano and Creso. All the standard points were run in triplicate, while all the samples were run in six copies. Each experiment was repeated twice.

SYBR Green I

PCR reactions with SYBR Green dye were performed in the same device as TaqMan chemistry. Reaction mixtures consisted of 1X *SYBR Green PCR Master Mix* (containing SYBR Green I Dye, AmpliTaq Gold DNA Polymerase, dNTPs, dUTP and optimized buffer components), 500 nM of ROX passive reference, 300 nM of primers and 5 ng of sample DNA. Water was added to a 15 µl final volume. The cycling thermal parameters were the same as reported in TaqMan assays, with an additional dissociation step performed as follows: one cycle at 95°C for 15 sec, one cycle at 60°C for 30 sec and one cycle at 95°C for 15 sec. This final step allowed to register the melting curve enabling the detection of any not-specific amplification product including primer-dimers. Also in SYBR Green assays positive and negative controls were amplified along with the transgenic samples.

Calculation of Transgene Copy Number and Statistical Analysis

Transgene absolute quantity was determined in each transformed line by comparing the Ct of the unknown samples against the calibration curve obtained from serial dilutions of standard templates represented by plasmid pKs+Dx5 for *1Dx5* gene and pKs-Dy10 for *1Dy10*. The software used in Real-Time experiments was the *Applied Biosystem 7300/7500 Sequence Detection System* (*SDS Software*, Applied Biosystem Co., Foster City, CA, USA). This software determines the internal fluorescence emission of the passive reference (ROX) and that of the reporter dye in order to calculate the normalized reporter (R_n) as follows:

R_n = Reporter dye fluorescence emission / ROX fluorescence emission

The magnitude of the signal generated by the specified set of PCR conditions is called $\Delta R_n = R_n -$ baseline (the baseline is the initial cycle in which there is little change in fluorecence signal). For each sample the amplification results are reported in a ΔR_n versus Cycle view, that displays the log of dye fluorescence (ΔR_n) as a function of cycle number.

The precise number of integrated transgenic sequences for each wheat line was finally obtained from the ratio between the absolute quantity of each gene and the genome copy number.

RESULTS

In the last decade transformation experiments with high molecular weight glutenin subunits (HMW-GS) genes in wheat have been published by several research groups (reviewed and referenced in Jones, 2005 and Blechl et al., 2007) because they play a major role in determining the viscoelastic properties that underlie bread dough formation (reviewed and referenced in Shewry et al., 2003). In order to study their effects on dough properties, HMW-GS genes have also been used to transform *Tritordeum* (Rooke et al., 1999a), maize (Sangtong et al., 2002), and rye (Alpeter et al., 2004). The HMW-GS genes used were *1Dx5*, *1Ax1* or a hybrid between *1Dy10* and *1Dx5* genes (Blechl and Anderson, 1996). Stable integration and expression of *1Dy10* gene has been only reported by Altpeter et al. (2004) in rye and by Blechl et al. (2007) in bread wheat. In a previous study, transformation of four commercial durum wheat cultivars with HMW-GS genes *1Dx5* and *1Dy10* was carried out by Gadaleta et al. (2008) in order to study the expression of the two genes in different genotypes. Four transgenic lines were obtained. Two transgenic events expressed the 1Dx5 and 1Dy10 subunits; one transgenic event expressed the 1Dx5 subunit, and one transgenic event showed both co-suppression of transgenes *1Dx5* and *1Dy10* and the native genes encoding the endogenous HWM glutenin subunits. Reliable and stable expression of transgenes as well as molecular characterization and field adaptation of transgenic lines are prerequisites for the successful application of transgene technology. Many papers have reported significant variability in the behaviour of the same transgene in different lines (e.g. Iglesias et al., 1997). Loci that appear to be stably expressed initially can become progressively silenced over several generations (Srivastava et al., 1996). The stability and the behaviour of transgenes are influenced by several factors, such as chromosomal location, transgene copy number and arrangement, and interaction with the host genotype. The establishment of an efficient and fast protocol for the detection of the inserted transgene copy number has been required.

In this work the four durum wheat transgenic lines obtained by Gadaleta et al. (2008) above described, were characterized for the integrated copy number of the two transgenes *Glu-D1-1d* and *Glu-D1-2b* by Real-Time PCR. A total of four experiments (two experiments for each gene) were conducted. Detection of *1Dx5* gene was performed by employing the TaqMan chemistry for both *right* and *left* system, while for the *Dy10* gene SYBR Green and TaqMan chemistries were compared: TaqMan was employed for *right* system and SYBR Green for *left*. In all experiments samples were run in six copies. The absolute quantity of the transgenes in each wheat line was inferred from the respective standard curves; the correlation coefficients (R^2) of all the curves were quite good as they varied in a range of 0.95-0.97, indicating a good linearity between fluorescence output and DNA concentration over a broad range of values (Figures 3, 4). The precise number of transgenic sequences integrated in each wheat line was obtained from the ratio between the absolute quantity of each gene and the number of copies of genomes.

For each transgene the results of the two experiments were combined and shown in Tables 4 and 5 together with their standard deviation. As the obtained data fell within the upper and lower limits of the calibration curves, they can be considered acceptable as they reflect the linear correlation between starting DNA amount and Ct. No transgene was detected in not transformed cultivars (Svevo, Varano, Creso) and NCT (No-Template-Control) employed as negative controls to ensure the specificity of PCR amplifications. The testing of negative controls was one of the most important components of QPCR assays as it provided a mechanism to check for external contamination or other factors resulting in not-specific increase in fluorescence signal.

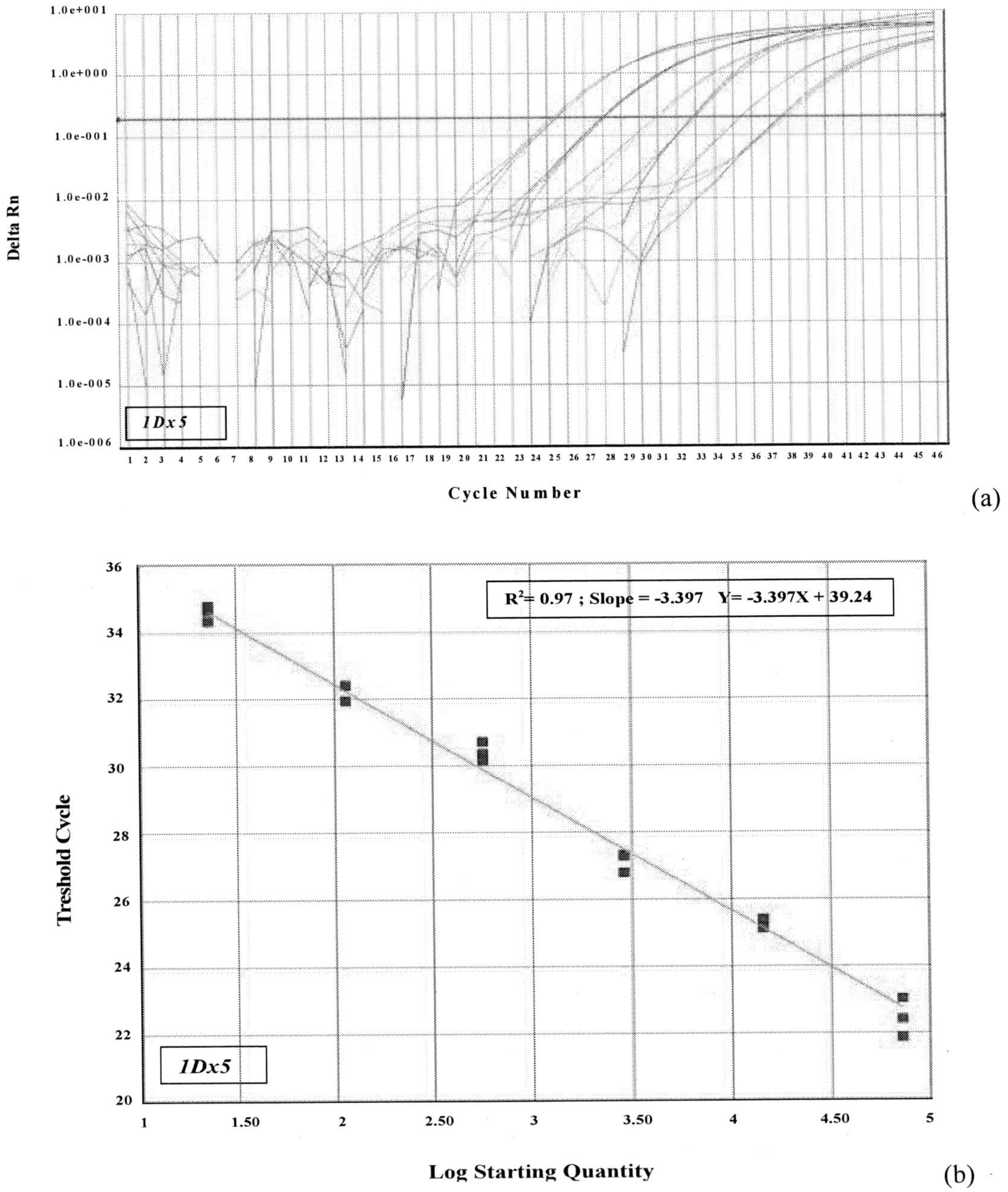

Figure 3. Real-Time PCR amplification and standard curve of *Glu-D1-1d* (*1Dx5-right*) transgene. (a) Logarithmic plot resulting from the amplification of six five-fold dilutions of standard DNA (plasmid pKS+Dx5). Each sample was run in triplicate. In green is reported the treshold. (b) Standard curve reporting the calculated Ct values plotted against the log of each starting quantity. Correlation coefficient and slope values are indicated.

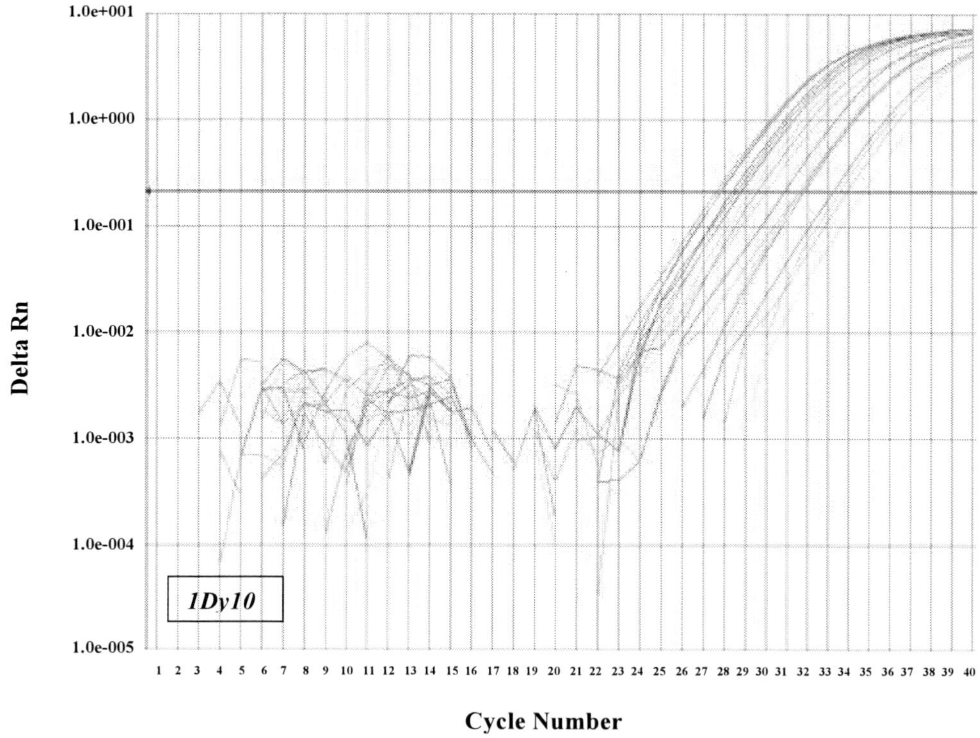

(a)

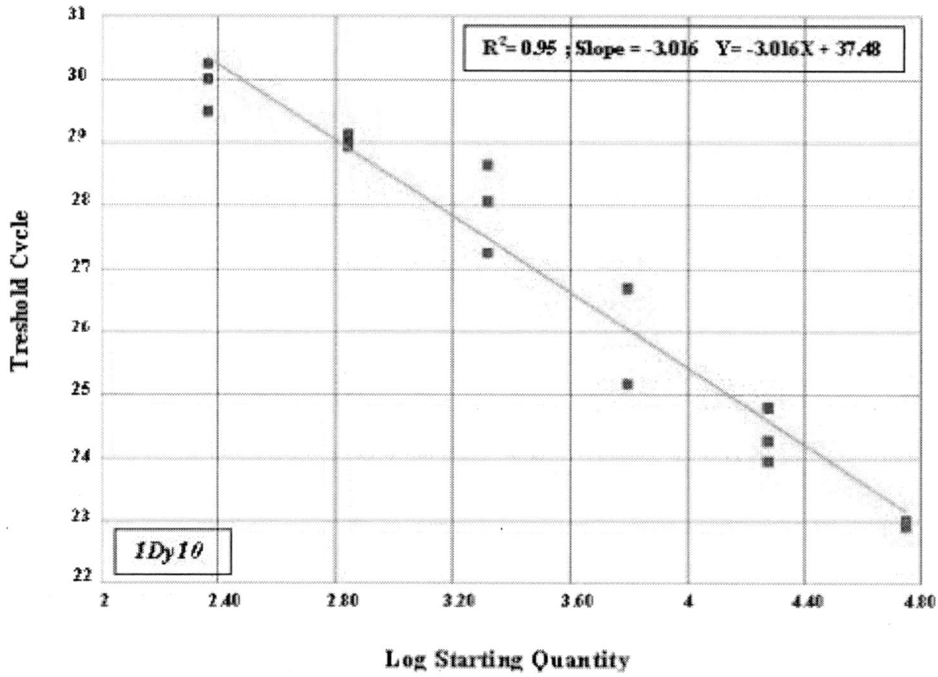

(b)

Figure 4. Real-Time PCR amplification and standard curve of *Glu-D1-1b* (*1Dy10-right*) transgene. (a) Logarithmic plot resulting from the amplification of six three-fold dilutions of standard DNA (plasmid pKS-Dy10). Each sample was run in triplicate. (b) Standard curve reporting the calculated Ct values plotted against the log of each starting quantity. Correlation coefficient and slope values are indicated.

Tables 4 and 5 report the copy number per haploid genome detected in the four transgenic lines for each transgene. In Table 4 results for *1Dx5* only refer to the *right* system as for this gene data from the two systems were completely comparable. Table 5 reports the comparison of *1Dy10* copy number as determined by TaqMan assay on *right* system and SYBR Green on *left*. In this case SYBR Green showed decreased sensitivity respect to TaqMan method as it failed in detecting very low copy number of inserted sequences. In fact, in line DV1-4c a single insertion was identify by TaqMan, but no transgene was detected by SYBR Green. For this reason, for *1Dy10* gene we considered acceptable the results coming from TaqMan experiments only.

Table 4. Copy number of inserted *1Dx5* gene in wheat transgenic lines estimated by Real-Time TaqMan experiments on *right* system. Starting quantities, as calculated by the software, are expressed in arbitrary units (a.u.)

Line	*Transgene*	*Starting absolute quantity (a.u.)*	*Standard deviation (a.u.)*	*Estimated copy number per haploid genome*
DV1-4c	*1Dx5-right*	613.3	54.7	5.4
DC2-10-4	*1Dx5-right*	1358.9	85.4	13.8
DC2-65s	*1Dx5-right*	8524.9	723.2	42.4
DS-127	*1Dx5-right*	1072.8	143.7	8.6

Table 5. Comparison of *1Dy10* integrated copy number estimated by TaqMan assay on *right* system and SYBR Green on *left*. Starting quantities, as calculated by the software, are expressed in arbitrary units (a.u.)

	TaqMan probe				*SYBR green I dye*			
Line	*Transgene*	*Starting absolute quantity (a.u.)*	*Standard deviation (a.u.)*	*Estimated copy number per haploid genome*	*Transgene*	*Starting absolute quantity (a.u.)*	*Standard deviation (a.u.)*	*Estimated copy number per haploid genome*
DV1-4c	*1Dy10-right*	3790.7	371.8	1.0	*1Dy10-left*	-	-	-
DC2-10-4	*1Dy10-right*	8422.2	375.5	12.8	*1Dy10-left*	7035.9	57.9	15.7
DC2-65s	*1Dy10-right*	7206.3	843.5	35.8	*1Dy10-left*	667.5	61.6	3.3
DS-127	*1Dy10-right*	2470.4	212.3	20.0	*1Dy10-left*	20492.0	558.3	60.0

The estimated number of integrated transgenes detected in our experiments ranged from 1 of *1Dy10* in line DV1-4c to 42 of *1Dx5* in line DC2-65s. This latter showed the higher number of inserted sequences for both *1Dy10* and *1Dx5* genes, while the line harbouring the lower copy number was DV1-4c.

The four transgenic lines were characterized in a previous work (Gadaleta et al., 2008) for HMW-glutenin protein expression and for chromosomal location of plasmid pKS+Dx5 and pKs-Dy10 by FISH.

Of the four lines, DS2-127 of cv. Svevo and DC2-10-4 of cv. Creso showed the expression of both 1Dx5 and 1Dy10 subunits in endosperm extracts. Line DS2-127 had two transgenic loci detected by FISH on different chromosomes: one carrying a co-insertion of the two plasmids and an additional one containing only the construct with the gene for Dy10 subunit. In this line the total number of transgenes detected by Real-Time PCR were 8.6 for *1Dx5* and 20 for *1Dy10*. By comparing the results from FISH and Real-Time PCR we supposed that the 8.6 copies of *1Dx5* gene were inserted at a single locus, while the 20 copies of *1Dy10* integrated in two loci with a mean value of 10 transgenes for each locus.

Line DC2-10-4 carries the plasmids pKS+Dx5 and pKs-Dy10 inserted at a single locus, containing 14 copies of *1Dx5* gene and 13 of *1Dy10* as detected by Real-Time PCR.

Line DV1-4c of cv. Varano showed expression of the *1Dx5* transgene, but the 1Dy10 HMW-GS was not detected in the endosperm extract. Five copies of *1Dx5* gene were detected by Real-Time PCR; on the other hand, the presence of one single copy of *1Dy10* revealed both by FISH and Real-Time suggests that in this line *1Dy10* gene was either silenced or altered in some way that prevented translation.

The fourth line DC2-65s carries the two plasmids in each of two insertion sites located in the telomeric and sub-telomeric region of the same chromosome, and showed a co-suppression of all HMW glutenin subunits including the endogenous ones. Moreover, co-suppression was found to be genetically stable in the generations analysed since T_1 to T_4 (Gadaleta et al., 2008). In the present work we detected a very high number of transgenes inserted in DC2-65s genome by Real-Time PCR, i.e. 42.4 copies for *1Dx5* and 36 for *1Dy10*.

Genetic engineering of plants sometimes results in transgene silencing after integration into the host genome, which may be related to a defence mechanism against foreign DNA expression (Kumpatla et al., 1998). In fact when analyzing the effects of the immediate genomic environment on transgene expression, one must also consider the potentially broader impact of the genome organization of the host plant with respect to the chromosomal distribution of structural genes, abundance, diversity, arrangement of repetitive DNA and the constitution of sub-genomes in polyploids (Iglesias et al., 1997). Transgene silencing has been reported for diploid, autopolyploids and allopolyploids (Srivastava et al., 1996; Pawlowski and Somers, 1996). These species all have distinctive genomic attributes that might influence the frequency and types of silencing effects that are observed.

The phenomenon of transgene silencing has often been reported in hexaploid wheat, and it may affect both the exogenous DNA sequences and the homologous genes, depending on a mechanism called "Homology Dependent Gene Silencing" (HDGS, Matzke and Matzke, 1998; Meyer and Heidmann, 1994) .

Alpeter et al. (1996) transformed bread wheat cultivar Bobwhite with the gene encoding HMW-GS 1Ax1 and speculated that its integration resulted in silencing of the *1Ax2* gene in one of their transformants. Blechl and Anderson (1996) transformed Bobwhite with a chimeric gene composed of *1Dy10* and *1Dx5* and observed co-suppression in one of their lines (Blechl et al., 1998). More recently silencing of HMW-GS genes in wheat has been reported by Alvarez et al. (2000).

There is surprising evidence to suggest that the expression of endogenous genes can be inhibited by the introduction of homologous constructs capable of transcribing mRNA of the same sense as the transcripts of endogenous genes (Meyer, 1995). In our work line DC2-65s contained two transgene insertions on the same chromosome carrying a high total gene copy number. This suggests that the interaction of multiple genes with the same mRNA on the same chromosome could have made co-suppression of *1Dy10* transgene and the endogenous homologues more likely.

In this work Real-Time PCR not only revealed a precise and sensitive method to estimate gene copy number, but it also allowed to assess molecular data deriving from previous cytogenetic analyses. Even if Real-Time can determine transgene absolute quantity without estimating the copy number inserted at each transgenic locus, in all the four wheat lines analysed the results were surprisingly in line with those obtained from FISH. The present research shows that besides other conventional methods, Real-Time PCR can be considered a valid tool which may be helpful in the characterization of transgenic lines. In lines containing a single insertion it can even discriminate between hemizygous and homozygous plants (Bubner and Baldwin, 2004), allowing these latter to be selected before they produce seeds, thereby avoiding laborious segregation analyses.

Conclusion

Real-Time PCR has traditionally been used to quantify RNA or DNA, but few studies have used it to determine transgene copy number (Callaway et al., 2002; Song et al., 2002). Transgenic plants with a single transgene copy are desirable because multiple copies often lead to gene silencing (Kumpatla et al., 1997). Unfortunately, the presence of more than one transgene locus is frequently observed in plants transformed via microprojectile bombardment (Blechl and Anderson, 1997; Pedersen et al., 1997; Rooke et al., 2003). The high frequency of co-integration of biolistically introduced DNAs carried on separate plasmids may be related to ligation of introduced DNA and concatamer formation (Pawlowski and Somers, 1996; Kohli et al., 1998).

As a method to determine gene copy number, Real-Time PCR shows a lot of advantages. In contrast to conventional PCR, where only the amount of end product is determined (Freeman et al., 1999), Real-Time PCR allows to track the change of PCR product during the reaction, and thus the starting point of the exponential growth can be precisely determined. Moreover, compared with other techniques of DNA quantification, Real-Time PCR allows the detection of the precise amount of a given nucleic acid sequence in a rapid, specific and sensitive way. Combining amplification and detection into one homogeneous assay, it obviates the need for gel electrophoresis to determine the amount of the DNA target accumulated after a fixed number of cycles. Real-Time assay also allows an increased precision in DNA quantification: as the Ct is determined during the exponential phase of the PCR, when reaction components are not limiting and do not affect the amplification efficiency, Ct values are highly reproducible (Higuchi et al., 1993).

Real-Time PCR can be performed by using specific probes or non-specific SYBR Green I dye. In our work the results obtained with SYBR Green and TaqMan chemistry for *1Dy10* gene were sensibly different, suggesting that the two chemistries are both applicable to transgene copy number detection, but sometimes SYBR Green shows decreased sensitivity as it may fail in revealing very low copy number of inserted sequences. On the other hand, SYBR Green chemistry shows important advantages such as the simple set-up of the assay, low material costs and no probe designing. Moreover, even if SYBR Green can bind both to specific and non-specific amplification products, the specificity of the reaction can always be checked by performing the dissociation curve.

The estimation of precise transgene copy number by Real-Time PCR usually needs an endogenous control gene, but choosing an appropriate internal gene can be problematic. The *puroindoline-b* gene highly conserved in several crops as wheat, maize and sorghum (Gautier et al., 2000) has been used as internal gene in some works (Li et al., 2004). In our work we optimized the detection of transgene copy number by using as a control the same transgene cloned in a plasmid. In this way a correct concentration of plasmid copies can be made in order to obtain an accurate standard curve that is a key element for the quantitative assay.

In conclusion, Real-Time PCR is a valid instrument for the characterization of transgenic lines. Above all, it can be very useful in detecting transgene copy number for the selection of plants harbouring low or single copy of the exogenous gene which can be more successfully destined to field cultivation.

ACKNOWLEDGMENTS

This work was supported by grants from Università degli Studi di Bari, Italy, project: Ateneo 2007, 2008 and from Ministero dell'Istruzione, dell'Università e della Ricerca, Italy, project: COFIN PRIN 2007.

REFERENCES

Altpeter, F; Popelka, JC; Wieser, H. Stable expression of *1Dx5* and *1Dy10* high molecular weight glutenin subunit genes in transgenic rye drastically increases the polymeric glutenin fraction in rye flour. *Plant Molecular Biology*, 2004 54, 783-792.

Altpeter, F; Vasil, V; Srivastava, V; Vasil, I K. Integration and expression of the high-molecular-weight glutenin subunit *1Ax1* gene into wheat. *Nature Biotechnoogy*, 1996 14, 1155-1159.

Alvarez, ML; Guelman, S; Halford, NG; Lustig, S; Reggiardo, MI; Ryabushkina, N; Schewry, P; Stein, J; Vallejos, RH. Silencing of HMW glutenins in transgenic wheat expressing extra HMW subunits. *Theoretical and Applied Genetics*, 2000 100, 319-327.

Blechl, AE; Anderson, OD. Expression of a novel high-molecular-weight glutenin subunit gene in transgenic wheat. *Nature Biotechnology*, 1996 14, 875-879.

Blechl, AE; Le HQ; Anderson, OD. Engineering changes in wheat flour by genetic transformation. *Plant Physioloy Journal*, 1998 152, 703-707.

Blechl, A; Lin, J; Nguyen, S; Chan, R; Anderson, OD; Dupont, FM. Transgenic wheats with elevated levels of Dx5 and/or Dy10 high-molecular-weight glutenin subunits yield doughs with increased mixing strength and tolerance. *Journal of Cereal Science,* 2007 .

Bubner, B; Baldwin, IT. Use of real-time PCR for determining copy number and zygosity in transgenic plants. *Plant Cell Reports,* 2004 23, 263-271.

Bubner, B; Gase, K; Baldwin, IT. Twofold differences are the detection limit for determining transgene copy numbers in plants by real-time PCR. *BMC Biotechnology*, 2004 4, 14.

Callaway, AS; Abranches, R; Scroggs, J; Allen, GC; Thompson, WF. High throughput transgene copy number estimation by competitive PCR. *Plant Molecular Biology Reporter,* 2002 20, 265-277.

De Preter, K; Speleman, F; Combaret, V; Lunec, J; Laureys, G; Eusson, BH; Francotte, N; Pearson, AD; De Paepe, A; van Roy, N; vander-sompele, J. Quantification of MYC-N, DDX I and NAG gene copy number in neuroblastoma using real time quantitative PCR assay. *Modern Pathology,* 2002 15, 150-166.

Freeman, WM; Walker, SJ; Vrana, KE. *Quantitative RT-PCR: pitfalls and potential. Biotechniques,* 1999 26, 112-122, 124-115.

Gadaleta, A; Blechl, AE; Nguyen, S; Cardone, MF; Ventura, M; Quick, JS; Blanco, A. Stably expressed D-genome-derived HMW glutenin subunit genes transformed into different durum wheat genotypes change dough mixing properties. *Molecular Breeding*, 2008 22, 267-279.

Gautier, MF; Cosson, P; Guirao, A; Alary, R; Joudrier; P. Pouroindoline genes are highly conserved in diploid ancestor wheat and related species but absent in teraploid Triticum species. *Plant Science*, 2000 153, 81-91.

Higuchi, R; Fockler, C; Dollinger, G; Watson, R. Kinetic PCR: Real-time monitoring of DNA amplification reactions. *Nature Biotechnology*, 1993 11, 1026-1030.

Iglesias, VA; Moscone, EA; Papp, I; Neuhuber, F; Michalowski, S; Phelan, T; Spiker, S; Matzke, M; Matzke, AJM. Molecular and cytogenetic analyses of stably and unstably expressed transgene loci in tobacco. *Plant Cell,* 1997 9, 1251–1264.

Ingham, DJ; Beer, S; Money, S; Hansen, G. Quantitative realtime PCR assay for determining transgene copy number in transformed plants. *Biotechniques,* 2001 31, 132-140.

Jones, HD. Wheat transformation: current technology and applications to grain development and composition. *Journal of Cereal Science*, 2005 41, 137-147.

Kohli, A; Leech, M; Vain, P; Laurie, DA; Christou, P. Transgene organization in rice engineered through direct DNA transfer supports a two-phase integration mechanism media*ted by the establishment of integration hot-spots. Proceedings of Natural Academy of Science* USA, 1998 95, 7203-7208.

Kumpatla, SP; Teng, W; Buchholz, WG; Hall, TC. *Epigenetic t*ranscriptional silencing and 5-azacytidine-mediated reactivation of a complex transgene in rice. *Plant Physiology*, 1997 115, 361-373.

Li, Z; Hansen, JL; Liu, Y; Zemetra, RS; Berger, PH. Using Real-Time PCR to Determine Transgene Copy Number in Wheat. *Plant Molecular Biology Reporter*, 2004 22, 179-188.

Mackay, IM. Real-time PCR in the microbiology laboratory. *Clin. Microbiol. Infect*. 2004 10, 190-212.

Mackay, IM; Arden, KE; Nitshe, A. Real-time PCR in virology. *Nucleic Acid Research*, 2002 30, 1292-1305.

Mason, G; Provero, P; Vaira, AM; Accotto, GP. Estimating the number of integrations in transformed plants by quantitative real-time PCR. *BMC Biotechnology*, 2002 1-10.

Matzke, AJM; Matzke, MA. Position effect and epigenetic silencing of plant transgenes. *Current Opinion in Plant Biolology*, 1998 1, 142-148.

Meyer, P.; Heidmann, I. Epigenetic variants of a transgenic petunia line show hypermetilation in transgene DNA, an indication for specific recognition of foreign DNA in transgenic plants. *Molecular Gene Genetics*, 1994 243, 390-399.

Mayer, Z; Bagnara, A; Farber, P; Geisen, R. Quantification of the copy number of *nor-1*, a gene of the aflatoxin biosynthetic pathway by real-time PCR, and its correlation to the cfu of *Aspergillus flavus* in foods. *International Journal of Food Microbiology*, 2003 82, 143-151.

Meyer, P. Stabilities and instabilities in transgene expression. In: Lindsey K, editor. *Transgenic plant research*. Harwood Academic, Amsterdam; 1998; 263-275.

Meyer, P. Understanding and controlling transgene expression. *Trends in Biotechnology*, 1995 13, 332-337.

Pawlowski, WP; Somers, DA. Transgene inheritance in plants genetically engineered by microprojectile bombardment. *Molecular Biotechnology*, 1996 6, 17-30.

Pedersen, C; Zimny, J; Becker, D; Janhne-Gartner, A; Lorz, H. Localization of introduced genes on the chrmomosomes of transgenic barley, wheat and triticale by fluorescence in situ hybridization. *Theoretical and Applied Genetics*, 1997 94, 749–757.

Rooke, L; Steele, SH; Barcelo, P; Shewry, PR; Lazzeri, P. Transgene inheritance, segregation and expression in bread wheat. *Euphytica*, 2003 129, 301-309.

Sangtong, V; Moran, DL; Chikwamba, R; Wang, KW; Woodman-Clikeman, MJ; Long, M; Lee, M; Scott, P. Expression and inheritance of the wheat *Glu-1Dx5* gene in transgenic maize.*Theoretical and Applied Genetics*, 2002 105, 937-945.

Schmidt, MA; Parrot, WA. Quantitative detection of transgenes in soybean [Glycine max (L.) Merrill] and peanut (Arachis hypogaea L.) by real-time polymerase chain reaction. *Plant Cell Reports*, 2001 20, 422-428.

Schnerr, H; Niessen, L; Vogel, RF. Real-time detection of the *tri5* gene in *Fusarium* species by LightCycler^{TM}-PCR using SYBR-Green I for continuous fluorescence minitoring. *International Journal of Food Microbiology*, 2001 71-53-61.

Song, P; Cai, CQ; Skokut, M; Kosegi, BD; Petolino, JF. Quantitative real-time PCR as a screening tool for estimating transgene copy number in WHISKER-derived transgenic maize. *Plant Cell Reports*, 2002 20, 948-954.

Srivastava, V; Vasil, V; Vasil, IK. Molecular characterization of the fate of transgenes in transformed wheat (*Triticum aestivum* L.). *Theoretical and Applied Genetics*, 1996 92, 1031-1037.

Vaitilingom, M; Pijnenburg, H; Gendre, F; Brignon, P. Real-time quantitative PCR detection of genetically modified Maximizer maize and Roundup Ready soybean in some representative foods. *Journal of Agricoltural Food Chemistry*, 1999 47-5261-5266.

Yang, L; Ding, J; Zhang, C; Jia, J; Weng, H; Liu, W; Zhang, D. Estimating the copy number of transgenes in transformed rice by real-time quantitative PCR. *Plant Cell Reports*, 2005 23, 759-763.

In: Transgenic Plants: New Research
Editor: Oliver T. Chan

ISBN 978-1-60692-017-6
© 2009 Nova Science Publishers, Inc.

Chapter 7

ENGINEERING OF BROAD SPECTRUM RESISTANCE TO VIRAL INFECTIONS IN TRANSGENIC PLANTS

Ekaterina A. Trifonova
Institute of Cytology and Genetics, Siberian Branch of Russian Academy of Sciences, Novosibirsk, Russia

ABSTRACT

Viruses are particularly fast evolving pathogens, so they are capable of overcoming natural and engineered virus resistance. This is why engineering of virus resistance has become a testing area for new transgenic approaches. Nowadays pathogen derived resistance is the most common strategy to protect crop plants against viruses through transgenesis. Generally, pathogen derived resistance provides moderate protection against a narrow range of related viruses (protein-mediated resistance) or high levels of protection only against closely related strains of a virus (RNA-mediated resistance). So far this strategy has not offered protection against a broad spectrum of heterologous viruses. At the same time, several alternative strategies were developed that have conferred protection against multiple viruses. It was shown that transgenic plants expressing human ά- and β-interferon; interferon-induced 2'-5' oligoadenylate synthetase, pokeweed antiviral proteins (PAP) and PAP analogues, intracellular ds-RNA specific RNases and extracellular bovine pancreatic RNase were broadly resistant to multiple viruses. Recently some modifications of pathogen derived strategy such as transgenic production of artificial microRNA or chimeric peptide "aptamers" were made to provide broad-spectrum resistance. All these approaches offer different degrees of protection, specificity and biosafety that will be reviewed in this article.

Introduction

Viral diseases are responsible for substantial agronomic losses of crop plants worldwide. It's well known that viruses are particularly fast evolving pathogens. Therefore, the control of viral infections requires different approaches such as crop rotation, production of virus-free plants and seeds, and selection of virus-resistant varieties. However, the transfer of resistance traits from a wild plant species without losing crop qualities is a laborious and difficult process.

Transgenesis has suggested new methods to protect plants against viruses. Nowadays pathogen derived resistance is the most common strategy to create virus-resistant transgenic plants. This concept originated with Sanford and Johnson (1985) proposed that transgenic expression of genes derived from pathogen itself can effectively protect plants from the pathogen. This concept was proved by expressing TMV coat protein in transgenic tobacco plants that demonstrated virus-resistant phenotype (Powell-Abel et al., 1986). Classically, pathogen derived resistance provides moderate protection against a narrow range of related viruses in the case of protein-mediated resistance or high levels of protection only against closely related strains of a virus in the case of RNA-mediated resistance. Until now this strategy has not demonstrated protection against a broad spectrum of heterologous viruses. At the same time, several alternative strategies were developed to attempt a difficult task protecting of transgenic plants against multiple viruses.

Plant R Genes

Plant R genes control resistance of the hypersensitive response (HR) type that characterized by a rapid formation of necroses. During the reaction, the pathogen fails to spread and the plant forms a state of systemic acquired resistance (SAR). The products of resistance genes of such type are considered to be involved in resistance to taxonomically different pathogens including viruses, bacteria and fungi. Tobacco R gene known as N gene was used for transformation of tomato plants. The plants thus obtained demonstrated a rapid formation of local necroses after infection by tobamoviruses and further cessation of virus replication and transport (Whitham et al., 1996).

Further advance in the production of virus resistant transgenic plants that express resistance genes of R type is restrained by the fact that this type of resistance requires interaction of several gene products. The interaction is species-specific, and a transferred gene may fail to function in the absence of secondary genes involved in the process (Cooper, 1999). An added bonus to using plant R genes over pathogen derived genes is that plant genes is thought to be more biosafe and give much less public concerns.

The products of R genes have nucleotide-binding site and leucine-rich repeats (LRR). The LRRs provide recognition of pathogen-derived elicitors. The natural potato Rx protein confers resistance only against several strains of potato virus X (PVX). In vitro mutation analysis of Rx gene demonstrated that selected mutants were effective against an additional strain of PVX and against the distantly related poplar mosaic virus. These results indicate that evolution of disease resistance in wild population might be toward broader spectrum resistance and might be useful for creation transgenic plants protected against multiple viruses (Farnham & Baulcombe, 2006).

INTERFERONS AND 2'-5'-OLIGOADENYLATE SYSTEM

Among secondary genes activated in the process of R-related resistance, homologous to human β-interferon AVF gene was identified (Sela et al., 1987). This fact has drawn the attention to the interferon-induced 2'-5'-oligoadenylate antivirus system of mammals and led to the construction of transgenic plants that express genes of interferons and 2'-5'-oligoadenylate system. It has also been shown that an exogenous addition of human natural and recombinant interferon inhibits reproduction of TMV in the tobacco cells and protoplasts (Orchansky et al., 1982). On the other hand, it has been demonstrated in certain studies that human interferon has no visible effect on reproduction of TMV and AlMV (Antoniw et al., 1984). Such a contradiction might be explained by the fact that the effect of human interferon on TMV reproduction shows only in the narrow range of interferon concentrations (Kaplan et al., 1988). An antiviral effect of expression of human β-interferon gene in the transgenic plants was demonstrated in tobacco and alfalfa plants (Rivkin et al., 1993). The resistance was evaluated as a delay in the infection symptom development and that of the virus accumulation in the leaves of transgenic plants. Virus concentration in the inocula was 0.05 μg/mL for TMV and 5 μg/mL for AlMV. For the expression of α-interferon gene in transgenic plants, the development of resistance of the HR type has been established with formation of necroses and infection localization (Smirnov et al., 1991).

In mammals interferon induces 2'-5'-oligoadenylate synthetase which polymerizes ATP with the formation of 2'-5'-oligoadenylates that activate specific RNAse L that hydrolyzes virus RNA. This two-component system is called mammal 2'-5'-oligoadenylate system and, like interferon, it was transferred into plants with the aim of making them resistant to a broad range of virus infections. Transgenic potato plants were produced expressing the genes of the 2'-5'-oligoadenylate system and they were proven to be resistant against potato PVX, PVY и PVS viruses (Truve et al., 1993; 1994). Like in the case of β-interferon, the resistance expressed itself as a delay in the development of disease symptoms.

On the other hand, these human enzymes were coexpressed in transgenic tobacco plants. Infection of leaves of the coexpressing transgenic plants by tobacco mosaic virus (TMV), alfalfa mosaic virus (AlMV), or tobacco etch virus (TEV) resulted in necrotic lesions (HR-like type resistance). In contrast, leaves of the control plants produced typical systemic infections (Mitra et al., 1996).

Additionally, transgenic tobacco plants that express the genes of 2'-5'-oligoadenylate system have been shown to be resistant against the cucumber mosaic virus, CMV (Honda et al., 2003; Ishida et al., 2002) and potato virus Y (Ishida et al., 2002). Molecular components of the obtained resistance were identified, expression of four PR proteins and five HR-related proteins was observed in the transgenic plants. The authors have come to the conclusion that transgenic expression of the genes of 2'-5'-oligoadenylate system leads to a formation of non-specific resistance of the HR-like type with further SAR development (Honda et al., 2003).

Ribosome Inactivating Proteins

The most investigated ribosome inactivating protein is the pokeweed antiviral protein (PAP). Transgenic tobacco and potato plants expressing PAP possessed various degrees of resistance against a broad range of viruses depending on the level of PAP accumulation (Lodge et al., 1993). Because of its toxicity (PAP removes catalytically a certain adenine residue from the large rRNA 60S subunit of the eukaryotic ribosome), use of the PAP gene as a transgene decreases transformation efficiency greatly. PAP II, a protein from the same plant homologous to PAP, but far less toxic, has been isolated. Two strains of transgenic tobacco expressing the gene PAP II have been produced and demonstrated resistance against TMV, PVX and *Rhizoctonia solani*. The resistance levels correlated with the degree of PAP II accumulation. Besides, these transgenic plants expressed constitutively PR1-protein in the absence of salicylic acid accumulation (Wang et al., 1998).

The type-2 ribosome-inactivating proteins (SNA-I, SNA-V, SNLRP) from elderberry (*Sambucus nigra* L.) and IRAb from iris (*Iris hollandica* L) are all devoid of rRNA N-glycosylase activity towards plant ribosomes. Transgenic tobacco plants were produced that express the gene encoding IRAb protein inactivating ribosomes from *Iris hollandica* L. The plants had phenotypic deviations but were fertile, the level of resistance also correlated with the level of accumulation of IRAb protein. The authors found no parallel expression of PR proteins and characterized the resistance obtained as one of entirely local nature (Vandenbussche et al., 2004a). Transgenic tobacco (*Nicotiana tabacum* L. cv Samsun NN) plants expressing the elderberry ribosome-inactivating proteins were generated and challenged with tobacco mosaic virus in order to analyze their antiviral properties. Although some transgenic plants were distinctly virus resistant, no clear correlation was observed between in planta virus resistance of transgenic tobacco lines expressing the different ribosome-inactivating proteins and the in vitro polynucleotide-adenosine glycosylase activity of the respective proteins towards tobacco mosaic virus genomic RNA. In addition, it is evident that the working mechanism proposed for pokeweed antiviral protein cannot be extrapolated to elderberry and iris ribosome-inactivating proteins because the expressions of SNA-V and IRAb are not accompanied by induction of pathogenesis-related proteins (Vandenbussche et al., 2004b).

Intracellular and Extracellular Ribonucleases

It has been shown that plants possess many various ribonuclease (RNase) activities (Green, 1994). It was reported that RNase activity was higher in diseased plants (Green, 1994; Lusso and Kuc, 1995; Galiana et al., 1997; Hugot et al., 2002; Sindelarova et al., 2002). Wounding also causes induction of RNase expression in *Arabidopsis thaliana* (LeBrasseur et al., 2002), *Zinnia elegans* (Ye and Droste., 1996), tobacco (Kurata et al., 2002), and tomato (Lers et al., 1998). Some plant extracellular RNases participate in defense against various pathogens. *Nicotiana tabacum* extracellular RNase NE was induced by elicitors and pathogens and provided resistance against TMV and *Phytophthora parasitica* (Galiana et al., 1997; Hugot et al., 2002).

It was suggested that introduction of RNase-encoding genes could be used to protect plants against various pathogens. Most plant viruses have single-stranded RNA genomes but virus reproduction includes formation of double-stranded RNA molecules. For specific impairment of replicative intermediates of plant viruses, transgenic tobacco plants were created that express yeast ribonuclease pac1 hydrolyzing specifically double-stranded RNA. The plants showed increased resistance against the tomato mosaic virus (ToMV) at concentration of 1 μg/mL, and against the cucumber mosaic virus (CMV) at concentration of 10 μg/mL. The resistance showed itself as a delay in symptoms development. Resistance against the potato virus Y upon inoculation with extracts from infected tobacco plants was also demonstrated as a delay in the appearance of the infection symptoms (Watanabe et al., 1995).

Sano et al. (1997) and Ishida et al. (2002) have shown that the transgenic potato plants expressing specific for double-stranded RNA ribonuclease are resistant to the potato spindle tuber viroid. Five potato strains (*Solanum tuberosum* L., cultivar Russet Burbank) that express ribonuclease pac1 from yeast were infected with potato spindle tuber viroid (PSTVd), but suppressed infection and PSTV accumulation. All the progeny of the potato tubers produced by the resistant plants was also viroid-free. Since the product of the pac1 gene hydrolyzes PSTVd in vitro, double-stranded regions of PSTVd molecule and/or replicative intermediates can serve as a target for ribonuclease pac1 in the transgenic plant (Sano et al., 1997). The pac1 transgenic chrysanthemum plants (*Dendranthema grandiflora* cv Reagan) were developed and demonstrated the resistance to chrysanthemum stunt viroid (Ishida et al., 2002).

With regard to well-known cytotoxic effect of ribonucleases, Langenberg and coworkers (1997) used a mutant bacterial RNAse III gene with a substitution in codon 117 that leads to the amino acid replacement of the glutamic acid to lysine. The product of this mutant gene was lacking hydrolase activity but capable of binding to the dsRNA molecules. The efficiency of both normal and mutant genes of RNAse III in the transgenic tobacco was attested by the increased resistance against viruses with divided genome (Langenberg et al., 1997). Later the authors used the mutant gene of RNAse III to produce transgenic wheat plants and proved the high resistance of the plants against the barley stripe mosaic virus (BSMV). The resistance was evidenced by the absence of infection symptoms and decreased accumulation of mature virions in the transgenic plants (Zhang et al., 2001).

Transgenic tobacco plants (*Nicotiana tabacum* SR1) expressing the extracellular pancreatic ribonuclease from *Bos taurus* and characterized by the increased level of ribonuclease activity in leaf extracts were challenged with tobacco mosaic virus. The transgenic plants exhibited a significantly higher level of protection against the virus infection than the control non-transformed plants (Trifonova et al., 2007). The protection was evidenced by the absence (or significant delay) of the appearance of typical mosaic symptoms and the retarded accumulation of infectious virus and viral antigen. These results demonstrate that modulation of extracellular nuclease expression can be efficiently used in promoting protection against viral diseases. In general, if the virus concentration in the inoculum was low or medium (0.01-0.1 μg/ml), transgenic plants were characterized by the absence or considerable delay of virus accumulation and appearance of disease symptoms, at higher TMV concentrations in inoculum (10 μg/ml), the differences between the control and transgenic plants were negligible in the later stages of infection (14-28 DAI). Expression of the heterologous secretory RNase is likely to enhance the intrinsic mechanisms of plant antiviral defense based on plant extracellular RNases (e.g., *Nicotiana tabacum* RNase NE (Galiana et al., 1997; Hugot et al., 2002) and *Arabidopsis thaliana* RNS1 (LeBrasseur et al., 2002)). Taken together, all these results demonstrate that the pathogen-targeted resistance strategy can be effectively employed in conferring resistance to viral diseases of crops.

Broad Spectrum Pathogen Derived Resistance

Dysfunctional Movement Proteins

Most viruses encode movement proteins (MPs) which promote intercellular virus movement in the infected tissues. It has been shown that some viruses can promote transport of heterologous viruses that are unable to perform this function (Atabekov and Taliansky, 1990). Hence one can inhibit spreading of many heterologous viruses that use common MPs, and transgenic plants with the impaired function of virus movement can appear resistant to a broad assortment of viruses.

In contrast to transgenic plants expressing CP genes, transgenic expression of functional MP genes did not raise plant virus-resistance. Moreover, it has been found that the transgenic expression of a wild type TMV MP increases virulence of the infecting virus instead of resistance against it (Cooper et al., 1995). The alternative to functionally active MPs are defective MPs which, acting as dominant negative mutants, interfere with wild type MPs. Malyshenko and coauthors have produced transgenic tobacco plants that express the gene of a temperature-sensitive MP of TMV (Malyshenko et al., 1993). When exposed to a high temperature, the transgenic protein lost functional activity, which led to the protection from TMV infection (3 μg/mL). The authors have shown that virus replication in the protoplasts is not inhibited by high temperature, therefore, it is the temperature sensitive dysfunctional MP that prevents spreading of the virus. The same study describes transgenic tobacco plants that express an MP from the brome mosaic virus (BMV) which cannot infect tobacco plants but can replicate in tobacco protoplasts meaning that the BMV MP might be nonfunctional in the tobacco. Transgenic plants that accumulate BMV MP have demonstrated a decrease in TMV accumulation and normal replication of the virus in plant protoplasts (Malyshenko et al., 1993).

Gafny et al. (1992) obtained dysfunctional TMV MP that was unable to support viral movement. The transgenic tobacco plants expressing this dMP were resistant to a wide range of virus pathogens both related to TVM and taxonomically far removed. The authors demonstrated that the virus replication and the number of infected sites in the plants thus obtained did not change, while the virus spreading was either blocked or delayed substantially. The fact that the genes of functional MPs expressed as transgenes did not protect plants from virus infection indicates that the mechanism of MP-mediated protection functions on the level of proteins, not RNA (Cooper et al., 1995).

Further study has revealed that transgenic expression of MPs in plants often leads to phenotypic deviations. It has been shown that transgenic tomato plants that express two MP genes (BV1 and BC1) of the bean dwarf mosaic virus (BDMV) have the phenotype of plants infected with the virus: decreased size, spotted, twisted and strained leaves. The same plants have shown a considerable delay in the development of tomato mottle virus (ToMoV) infection (Hou et al., 2000).

Transgenic potato plants that express a mutant allele of the gene MP17 from potato leaf roll virus (PLRV) proved to be resistant against a broad range of viruses including non-related to the transgene source PVX and PVY (Tacke et al., 1996). Further cloning of the MP17 gene in tobacco plants has shown that a high level of expression of this gene results not only in a rise of virus resistance, but also in a considerable accumulation of soluble sugars and starch in the leaves. The authors have suggested that the resistance to PVX and PVY can be caused by increased level of PR proteins resulting from an impairment of carbohydrate metabolism, not MP interference (Hofius et al., 2001).

Studies of transgenic tobacco plants constitutively expressing NSM – transport protein of tomato spotted wilt virus (TSWV) revealed that this protein induced formation of symptoms typical of severe virus infection. In such transgenic plants leaves were smaller and rosette-arranged because of delayed growth of the internodes. The accumulation of extra starch in the mesophyll forms the typical virus-induced chlorotic patches, so that leaves became gradually whiter and frailer. The authors showed that the above symptoms resulted from the blocking of plasmodesmata impaired by NSM protein via deposits of 1,3-β-D-glucans and callose. Collectively, this indicates that the symptoms of NSM plants are self-inflicted and due to a basal defense response that counteracts prolonged interference of the MP with plasmodesmata functioning. This type of defense may also play a role in the formation of symptoms during viral infection (Rinne et al., 2005). It has also been shown that some TMV MPs are suppressors of virus-induced silencing that prevent its systemic spreading (Voinnet et al., 2000; Ding et al., 2004)

The agricultural usage of MP-transgenic plants raises problems of the unpredictability of phenotypic changes produced by the transgenes and theoretical possibility of RNA-silencing suppression that would affect virus resistance negatively.

Peptide Aptamers

To contrive a new effective strategy to engineer virus-resistant plants, Rudolph and coworkers (Rudolph et al., 2003) investigated the molecular properties and functions of tomato spotted wilt virus (TSWV) proteins. The TSWV N protein was found to be involved in a number of molecular interactions, such as homopolymerization, packaging of the viral genomic RNA, and interaction with the TSWV movement protein. The authors isolated a peptide "microdomain" interacting with TSWV N protein and fused the peptide to a carrier protein. Transgenic *Nicotiana benthamiana* lines expressing the fusions were challenged with five different tospoviruses that N protein interacted with the peptide. The transgenic plants demonstrated strong resistance to tomato spotted wilt virus, tomato chlorotic spot virus, groundnut ring spot virus, and chrysanthemum stem necrosis virus. The special interest of the method presented in the work is that it relies on well characterized function of a viral protein. In some other cases virus resistance is obtained by expressing viral genes targeted against structures/functions completely unknown.

The peptide aptamers that bound to the AL1 protein of tomato golden mosaic virus (TGMV) and cabbage leaf curl virus (CLCV) were characterized (Lopez-Ochoa et al., 2006).

This finding might be used to engineer broad spectrum resistance to single-stranded DNA plant viruses.

Artificial miRNAs (amiRNAs) and Virus-Derived dsRNAs

Micro-RNAs (miRNAs) have been identified as important regulators of gene expression in both plants and animals. These are single-stranded RNAs approximately 22 nucleotides (nt) in length, generated from processing of longer pre-miRNA precursors (Bartel, 2004) by DCL1 in Arabidopsis (Xie et al., 2004). Plant miRNAs regulate the abundance of target mRNAs by guiding their cleavage in a sequence-specific manner. Niu et al. (2006) constructed modified *Arabidopsis thaliana* miR159 precursor targeting viral mRNA sequences encoding two gene silencing suppressors, P69 of turnip yellow mosaic virus (TYMV) and HC-Pro of turnip mosaic virus (TuMV). The transgenic plants that expressed both amiR-P69(159) and amiR-HC-Pro(159) from a dimeric pre-amiR-P69(159)/amiR-HC-Pro(159) transgene were resistant to both viruses. Moreover, the virus resistance trait was heritable through at least 3 generations. Finally, the authors concluded that broad spectrum resistance to several viruses can also be obtained by co-expression of appropriately designed multiple amiRNAs.

Artificial miRNA, targeting sequences encoding the silencing suppressor 2b of cucumber mosaic virus (CMV), were constructed by Qu and coworkers (2007). The transgenic tobacco plants expressing the construct were resistant to CMV infection. Moreover, the resistance level conferred by the transgenic expression of amiRNA was correlated to the amiRNA level.

Two constructs were produced to genetically transform cassava (*Manihot esculenta*) for resistance to cassava mosaic disease (CMD). These constructs contain a wild and mutant AC1 gene, which encodes the replication-associated protein from African cassava mosaic virus. The resulted transgenic plants were resistant to the homologous virus and provided strong cross-protection against two heterologous species of cassava-infecting geminiviruses. It was RNA-mediated resistance since siRNAs specific to the AC1 transgene were identified in the two most resistant transgenic plant lines prior to virus challenge (Chellappan et al., 2004).

To obtain broad tospovirus resistance, there were used N gene sequence fragments of the four major tomato-infecting tospoviruses (tomato spotted wilt virus (TSWV), groundnut ringspot virus (GRSV), tomato chlorotic spot virus (TCSV) and watermelon silver mottle virus (WSMoV)), in a single small chimaeric hairpin (hp) RNA construct (Bucher et al., 2006). Then this type of constructs is processed into siRNAs and induces RNA silencing, natural defense system in plants. The authors presented data that demonstrated virus resistance frequencies of over 80% of all transformed plants. It was shown that the plants became resistant to all four different tospoviruses. However, it was reported that virus and transgene-mediated RNA-silencing becomes attenuated at low temperatures which inhibit siRNA accumulation in plant cells. This temperature sensitivity breaks down siRNA-mediated virus resistance at 15°C (Szittya et al., 2003) while miRNA accumulation was permanent at low temperature (Niu et al., 2006).

Conclusion

The transgenic plants of the future should be resistant to multiple pathogens. In spite of all transgenic approach examined, nowadays there is no single strategy that creates strong resistance to a broad spectrum of heterologous viruses. A combination of several transgenic approaches might be required to solve this problem. For instance, transgenic expression of extracellular RNases can be efficiently used in combination with the expression of RNases of other types (e.g., intracellular PR-10-related ribonucleases (Park et al., 2004) or dsRNA-specific ribonucleases (Watanabe et al., 1995; Sano et al., 1997; Langenberg et al., 1997; Zhang et al., 2001)), as well as with a pathogen-derived resistance strategy to maximize protection efficiency and durability.

Acknowledgments

I thank Dr. Alex Kochetov for his helpful ideas and comments. This work was supported by the Siberian and Far East Branches of the Russian Academy of Sciences (complex integration projects 5.3 and 06-II-CO-06-024) and Russian Foundation for Basic Research grant 07-04-12085-ofr.

References

Antoniw J.F., White R.F., Carr J.P. 1984. An examination of the effect of human α-interferons on the infection and multiplication of tobacco mosaic virus in tobacco. *Phytopathol.* 109: 367-371.

Atabekov J.G., Taliansky M.E. 1990. Expression of a plant virus-coded transport function by different viral genomes. *Advances in Virus Research.* 38: 201-248.

Bartel D.P. 2004. MicroRNAs: genomics, biogenesis, mechanism, and function. *Cell.* 116: 281-97.

Bucher E., Lohuis D., van Poppel P.M., Geerts-Dimitriadou C., Goldbach R., Prins M. 2006. Multiple virus resistance at a high frequency using a single transgene construct. *J Gen Virol.* 87: 3697-701.

Chellappan P., Masona M.V., Vanitharani R., Taylor N.J., Fauquet C.M. 2004. Broad spectrum resistance to ssDNA viruses associated with transgene-indiced gene silencing in cassava. 2004. *Plant Mol. Biol.* 56: 601-611.

Cooper B. 1999. Genetic mechanisms for engineering host resistance to plant viruses. Handbook of biological control. *Academic press.* 557-574.

Cooper B., Lapidot M., Heick J.A., Dodds J.A., Beachy R.N. 1995. A defective movement protein of TMV in transgenic plants confers resistance to multiple viruses whereas the functional analog increases susceptibility. *Virology.* 206: 307-313.

Ding X.S., Liu J., Cheng N.H., Folimonov A., Hon Y.M., Bao Y., Katagi C., Carter S.A., Nelson R.S. 2004. The tobacco mosaic virus 126-kDa protein associated with virus replication and movement suppresses RNA silencing. *Mol. Plant Microbe Interact.* 17: 583-592.

Farnham G., Baulcombe D.C. 2006. Artificial evolution extends the spectrum of viruses that are targeted by a disease-resistance gene from potato. *Proc. Natl. Acad. Sci.* USA. 103: 18828-33.

Gafny R., Lapidot M., Berna A., Holt C.A., Deom C.M., Beachy R.N. 1992. Effects of terminal deletion mutations on function of the movement protein of tobacco mosaic virus. *Virology.* 187: 499-507.

Galiana E., Bonnet P., Conrod S., Keller H., Panabieres F., Ponchet M., Poupet A., Ricci P. 1997. RNase activity prevents the growth of a fungal pathogen in tobacco leaves and increases upon induction of systemic acquired resistance with elicitin. *Plant Physiol.* 115: 1557-1567.

Green P.J. 1994. The ribonucleases of higher plants. *Annu. Rev. Plant Physiol. Plant Mol. Biol.* 45: 421-445.

Hofius D., Herbers K., Melzer M., Omid A., Tacke E., Wolf S., Sonnewald U. 2001. Evidence for expression level-dependent modulation of carbohydrate status and viral resistance by the potato leafroll virus movement protein in transgenic tobacco plants. *Plant J.* 28: 529-543.

Honda A., Takahashi H., Toguri T., Ogawa T., Hase S., Ikegami M., Ehara Y. 2003. Activation of defense-related gene expression and systemic acquired resistance in cucumber mosaic virus-infected tobacco plants expressing the mammalian 2'-5'-oligoadenylate system. *Arch.Virol.* 143: 1017-1026.

Hou Y.M., Sanders R., Ursin V.M., Gilbertson R.L. 2000. Transgenic plants expressing geminivirus movement proteins: abnormal phenotypes and delayed infection by Tomato mottle virus in transgenic tomatoes expressing the Bean dwarf mosaic virus BN1 or BC1 proteins. *Mol. Plant Microbe Interact.* 13: 297-308.

Hugot K., Ponchet M., Marais A., Ricci P., Galiana E. 2002. A tobacco S-like RNase inhibits hyphal elongation of plant pathogens. *Mol. Plant Microbe Interact.* 15: 243-250.

Ishida I, Tukahara M, Yoshioka M, Ogawa T, Kakitani M, Toguri T. 2002. Production of anti-virus, viroid plants by genetic manipulations. *Pest Manag Sci.* 58: 1132-6.

Kaplan I.B., Taliansky M.E., Malyshenko S.I., Ogarkov V.I., Atabekov J.G. 1988. Effect of human interferon of reproduction of plant- and mycoviruses. *Arch. Phytopathol.* 24: 3-8.

Kurata N., Kariu T., Kawano S., Kimura M. 2002. Molecular cloning of cDNAs encoding ribonuclease-related proteins in Nicotiana glutinosa leaves, as induced in response to wounding or to TMV-infection. *Biosci Biotechnol Biochem.* 66: 391-7.

Langenberg W.G., Zhang L., Court D.C., Giunchedi L., Mitra A. 1997. Transgenic tobacco plants expressing the bacterial rnc gene resist virus infection. *Mol. Breeding.* 3: 391-399.

LeBrasseur N.D., MacIntosh G.C., Perez-Amador M.A., Saitoh M., Green P.G. 2002. Local and systemic wound-induction of RNase and nuclease activities in Arabidopsis: *RNS1 as a marker for a JA-independent systemic signaling pathway.* 29: 393-403.

Lers A., Khalchitski A., Lomaniec E., Burd S., Green P.J. 1998. Senescence-induced RNases in tomato. *Plant Mol. Biol.* 36: 439-449.

Lodge J.K., Kaniewski W.K., Tumer N.E. 1993. Broad spectrum virus resistance in transgenic plants expressing pokeweed antiviral protein. *Proc. Natl. Acad. Sci.* USA. 90: 7089-7093.

Lopez-Ochoa L., Ramirez-Prado J., Hanley-Bowdoin L. 2006. Peptide aptamers that bind to a geminivirus replication protein interfere with viral replication in plant cells. *J Virol.* 80: 5841-53.

Lusso M. and Kuc J. 1995. Evidence for transcriptional regulation of □-1,3-glucanase as it relates to induced systemic resistance of tobacco to blue mold. Molec. *Plant Microbe Interact.* 8:473-475.

Malyshenko S.I., Kondakova O.A., Nazarova Ju. V., Kaplan I.B., Taliansky M.E., Atabekov J.G. 1993. Reduction of tobacco mosaic virus accumulation in transgenic plants producing non-functional viral transport proteins. *J. Gen. Virol.* 74: 1149-1156.

Mitra A., Higgins D.W., Langenberg W.G., Nie H., Sengupta D.N., Silverman R.H.1996. A mammalian 2-5A system functions as an antiviral pathway in transgenic plants. *Proc. Natl. Acad. Sci.* U S A. 93: 6780-5.

Niu Q.W., Lin S.S., Reyes J.L., Chen K.C., Wu H.W., Yeh S.D., Chua N.H. 2006. Expression of artificial microRNAs in transgenic Arabidopsis thaliana confers virus resistance. *Nat Biotechnol.* 24: 1420-8.

Orchansky P., Rubinstein M., Sela I. 1982. Human interferons protect plants from virus infection. *Proc. Natl. Acad. Sci.* USA. 79: 2278-2280.

Park C.-J., Kim K.-J., Shin R., Park J.M., Shin Y.-C., Paek K.-H. 2004. Pathogenesis-related protein 10 isolated from hot pepper functions as a ribonuclease in an antiviral pathway. *The Plant J.* 37: 186-198.

Powell-Abel P., Nelson R.S., De B. et al. 1986. Delay of disease development in transgenic plants that express the tobacco mosaic virus coat protein gene. *Science.* 232: 738-743.

Qu J., Ye J., Fang R. 2007. Artificial microRNA-mediated virus resistance in plants. *J Virol.* 81: 6690-9.

Rinne P.L., van den Boogaard R., Mensink M.G., Kopperud C., Kormelink R., Goldbach R., van der Schoot C. 2005. Tobacco plants respond to the constitutive expression of the tospovirus movement protein NS with a heat-reversible sealing of plasmodesmata that impairs development. *Plant J.* 43: 688-707.

Rivkin MI, Deĭneko EV, Komarova ML, Kochetov AV, Shumnyĭ VK. 1993. [Assessment of virus resistance of transgenic tobacco and alfalfa plants bearing the human beta-interferon gene] *Dokl. Akad.Nauk.* 331: 652-4. Russian.

Rudolph C., Schreier P.H., Uhrig J.F. Peptide-mediated broad-spectrum plant resistance to tospoviruses. 2003. *Proc. Natl. Acad. Sci.* U S A. 100:.4429-34.

Sanford J.C., Johnson S.A. 1985. The concept of parasite derived resistance – deriving resistance genes from the parasite's own genome. *J. Theo. Biol.* 113: 395-405.

Sano T., Nagayama A., Ogawa T., Ishida I., Okada Y. 1997. Transgenic potato expressing a double-stranded RNA-specific ribonuclease is resistant to potato spindle tuber viroid. *Nat. Biotechnol.* 15: 1290-1294.

Sela I., Grafi G., Sher N., Edelbaum O., Yagev H., Gerassi E. 1987. Resistance systems related to the N gene and their comparison with interferon. Plant resistance to viruses. *Ciba foundation symposium.* 133: 109-119.

Sindelarova M., Sindelar L., Burketova L. 2002. Glucose-6-phosphate dehyfrogenase, ribonucleases and esterases upon tobacco mosaic virus infection and benzothiodiazole treatment in tobacco. *Biologia Plantarum.* 45:423-432.

Smirnov S.P., Krasheninnikova L.V., Pukhal'skiĭ V.A. 1991. [Stability to the tobacco mosaic virus in transgenic tobacco plants, producing human alpha-interferon] *Dokl Akad Nauk* SSSR. 317: 732-4. Russian.

Szittya G., Silhavy D., Molnár A., Havelda Z., Lovas A., Lakatos L., Bánfalvi Z., Burgyán J. 2003. Low temperature inhibits RNA silencing-mediated defence by the control of siRNA generation. *EMBO J.* 22: 633-40.

Tacke E., Salamini F., Rohde W. 1996. Genetic engineering of potato for broad-spectrum protection against virus infection. *Nat. Biotechnol.* 14: 1597-1601.

Trifonova E.A., Sapotsky M.V., Komarova M.L., Scherban A.B., Shumny V.K., Polyakova A.M., Lapshina L.A., Kochetov A.V., Malinovsky V.I. 2007. Protection of transgenic tobacco plants expressing bovine pancreatic ribonuclease against tobacco mosaic virus. *Plant Cell Rep.* 26: 1121-6.

Truve E., Aaspollu A., Honkanen J., Puska R., Mehto M., Hassi A. 1993. Transgenic potato plants expressing mammalian 2'-5' oligoadenylate synthetase are protected from potato virus X infection under field conditions. *Bio/Technology.* 11: 1048-1052.

Truve E., Kelve M., Aaspollu A., Kuusksalu A., Seppanen P., Saarma M. 1994. Principles and background for the construction of transgenic plants displaying multiple virus resistance. *Arch.virol.* (Suppl.), 941-950.

Vandenbussche F., Desmyter S., Ciani M., Proost P., Peumans W.J., Van Damme E.J. 2004. Analysis of the in planta antiviral activity of elderberry ribosome-inactivating proteins. *Eur J Biochem.* 271: 1508-15.

Vandenbussche F., Peumans W.J., Desmyter S., Proost P., Ciani M., Van Damme E.J. 2004. The type-1 and type-2 ribosome-inactivating proteins from Iris confer transgenic tobacco plants local but not systemic protection against viruses. *Planta.* 220: 211-221.

Voinnet O., Lederer C., Baulcombe D.C. 2000. A viral movement protein prevents spread of the gene silencing signal in *Nicotiana benthamiana. Cell.* 103: 157-167.

Wang P., Zoubenko O., Tumer N.E. 1998. Reduced toxicity and broad spectrum resistance to viral and fungal infection in transgenic plants expressing pokeweed antiviral protein II. *Plant Mol. Biol.* 38: 957-964.

Watanabe Y., Ogawa T., Takahashi H., Ishida I., Takeuchi Y., Yamamoto M., Okada Y. 1995. Resistance against multiple plant viruses in plants mediated by a double stranded-RNA specific ribonuclease. *FEBS Letters.* 372: 165-168.

Whitham S., McCormick S., Baker B. 1996. The N gene of tobacco confers resistance to tobacco mosaic virus in transgenic tomato. *Proc. Natl. Acad. Sci.* USA. 93: 8776-8781.

Xie Z., Johansen L.K., Gustafson A.M., Kasschau K.D., Lellis A.D., Zilberman D., Jacobsen S.E., Carrington J.C. 2004. Genetic and functional diversification of small RNA pathways in plants. *PLoS Biol.* 2: E104.

Ye Z.H., Droste D.L. 1996. Isolation and characterization of cDNAs encoding xylogenesis-associated and wounding-induced ribonucleases in Zinnia elegans. *Plant.Mol.Biol.* 30: 697-709.

Zhang L., French R., Langenberg W.G., Mitra A. 2001. Accumulation of barley stripe mosaic virus is significantly reduced in transgenic wheat plants expressing a bacterial ribonuclease. *Transgen. Res.* 10: 13-19.

In: Transgenic Plants: New Research
Editor: Oliver T. Chan

ISBN 978-1-60692-017-6
© 2009 Nova Science Publishers, Inc.

Chapter 8

BT CROPS UNDER INVASION OF BT-RESISTANT INSECTS: MODELING THE CROP PRODUCTION

Alexander B. Medvinsky[1,2,*], Maria M. Gonik[1,2], Bai-Lian Li[2], Yuri V. Tyutyunov[3], Alexey V. Rusakov[1,2] and Horst Malchow[4]

[1]Institute of Theoretical and Experimental Biophysics, Pushchino, Moscow Region, 142290, Russia

[2]Department of Botany and Plant Sciences, University of California, Riverside, California 92521-0124, USA

[3]Vorovich Research Institute of Mechanics and Applied Mathematics, Southern Federal University, Rostov on Don, 344090, Russia

[4]Institute of Environmental Systems Research, University of Osnabrueck, D-49069, Osnabrueck, Germany

ABSTRACT

Genetically modified crops are an effective management tool for worldwide growers. However, there is a concern that pests may develop resistance to Bt toxins produced by genetically-modified Bt plants. We study the impact of invasion of Bt-resistant pests on Bt crops. We show that the invasion leads to a fall in Bt crop production. The biomass of the Bt plants is shown to have only a weak dependence on the initial biomass of the insect invaders. Bt crop production is found to be essentially dependent on the initial Bt plant biomass. Furthermore, variations of the crop production under invasion of the Bt-resistant insects are shown to undergo badly predictable irregular oscillations. The complex character of the response of the Bt crop-pest system to the invasion of Bt-resistant insects can lead to significant complications in attempts to regulate the system dynamics.

---

* Corresponding author. Permanent address: Institute for Theoretical & Experimental Biophysics, Pushchino, Moscow Region, 142290 Russia. Tel.: +7 4967 739238; fax: +7 4967 330553. E-mail address: medvinsky@iteb.ru.

INTRODUCTON

The use of transgenic insecticidal Bt crops, which express insecticidal Bt toxins, has risen dramatically since their introduction in the mid-1990s [Shelton *et al.*, 2002]. Field performance of the Bt crops, which has been excellent against the insect pests, resulted in a positive impact on farm income and rapid adaptation of this technology by corn producers [James, 2004; Brookes, Barfoot, 2006]. However, widespread and long-term planting of Bt crops makes highly probable the development of pest resistance to Bt toxins produced by the Bt plants. Indeed, the increase in resistance to Bt- sprays in the field, as well as the greenhouse and laboratory selection of resistant strains of several major pests, demonstrates that the probability of appearance and expansion of Bt resistant insects can build up [Tabashnik *et al.*, 1990; Tabashnik, 1994; Alstad and Andow, 1995; Ostlie *et al.*, 1997; Gould, 1998; Tabashnik *et al.*, 2000; Janmaat and Meyers, 2003; Tabashnik *et al.*, 2003]. The invasion of Bt-resistant insects can impact the crop production [Medvinsky *et al.*, 2004; 2005] and lead to serious problems in using the genetically-modified plants in agriculture [Gould, 1998; Shelton *et al.*, 2002; Huang, 2006]. Mathematical simulations can reveal key factors and integrate them in order to evaluate an ecological effect of a Bt-resistant insect invasion.

Conceptual (minimal) mathematical modeling may assess the factors which influence both Bt-crop production and dynamics of the Bt-plant biomass under an invasion of Bt-resistant insects (see for example, Hillier and Birch, 2002; Medvinsky *et al.*, 2006; 2007; Zhadanovskaya *et al.*, 2006; Tyutyunov *et al.*, 2007). Historically, simple conceptual models have been demonstrated to be an appropriate tool for searching and understanding the nature of changes in the abundance of species [Lotka, 1925; Volterra, 1926; Kolmogorov *et al.*, 1937; Bazykin, 2004]. The conceptual models do provide a way of simulating and revealing mechanisms of complicated population dynamics [May, 1976; Medvinsky *et al.*, 2002; 2007].

In this paper, following the conceptual modeling approach, we develop a reaction-diffusion model of an agricultural ecosystem including transgenic Bt plants, insects susceptible to Bt toxins, and adapted Bt-resistant insects carrying a recessive mutation enabling them to grow on Bt plants. With the use of this model, we evaluate Bt crop production under invasion of the Bt-resistant insects. We show that the biomass of the Bt plants has only a weak dependence on the initial biomass of the insect invaders. Bt crop production is shown to be essentially dependent on the initial Bt plant biomass.

MATHEMATICAL MODEL

At any point X and time τ the total insect biomass I is given by the following equation:

$$I(X,\tau) = I_{rr}(X,\tau) + I_{rs}(X,\tau) + I_{ss}(X,\tau). \tag{1}$$

In (1), I_{rr}, I_{ss} and I_{rs} denote the biomass of Bt-resistant (*rr*), Bt-susceptible (*ss*) and heterozygote (*rs*) insects, respectively. We assume a constant 50/50 male-female ratio for all genotypes: *rr*, *rs*, and *ss*. Table 1 lists all possible outcomes of mating between different genotypes (in accordance with the Hardy – Weinberg – Castle theory [Castle, 1903; Hardy,

1908; Weinberg, 1908]. The probability of mating between males of genotype i and females of genotype j ($i, j = rr, rs, ss$) is $\frac{I_i I_j}{I^2}$. Similarly, the probability of mating between males of genotype j and females of genotype i is $\frac{I_j I_i}{I^2}$. We also assume that the effective insect fecundity η_{ij} depends on the plant biomass P. The total increase in the ss-insect biomass (cf. Table 1) can then be represented as $\frac{\eta_{ss}(P)}{2I}\left(I_{ss}+\frac{1}{2}I_{rs}\right)^2$ where the insect biomass I is defined in Eq. (1). Similarly, the total increase in the rr-insect biomass is $\frac{\eta_{rr}(P)}{2I}\left(I_{rr}+\frac{1}{2}I_{rs}\right)^2$. The total increase in the rs-insect biomass resulting from insect reproduction is given by $\frac{\eta_{rs}(P)}{I}\left(I_{ss}+\frac{1}{2}I_{rs}\right)\left(I_{rr}+\frac{1}{2}I_{rs}\right)$.

Taking into account all the foregoing results, we simulate the dynamics of the Bt plant and insect biomass in the following mathematical model:

$$\frac{\partial P}{\partial \tau} = rP\left(1-\frac{P}{K}\right)-\frac{1}{2I}\left[\eta_{ss}(P)\left(I_{ss}+\frac{1}{2}I_{rs}\right)^2+2\eta_{rs}(P)\left(I_{ss}+\frac{1}{2}I_{rs}\right)\left(I_{rr}+\frac{1}{2}I_{rs}\right)+\eta_{rr}(P)\left(I_{rr}+\frac{1}{2}I_{rs}\right)^2\right], \tag{2}$$

Table 1. Outcomes of mating between different genotypes

Female	Male	Outcome
ss	ss	ss
rr	rr	rr
ss	rr	rs
rr	ss	rs
ss	rs	$\frac{1}{2}ss+\frac{1}{2}rs$
rs	ss	$\frac{1}{2}ss+\frac{1}{2}rs$
rs	rr	$\frac{1}{2}rs+\frac{1}{2}rr$
rr	rs	$\frac{1}{2}rs+\frac{1}{2}rr$
rs	rs	$\frac{1}{4}ss+\frac{1}{4}rr+\frac{1}{2}rs$

$$\frac{\partial I_{ss}}{\partial \tau} = -\mu_{ss} I_{ss} + \frac{1}{2I}\delta_{ss}\eta_{ss}(P)\left(I_{ss} + \frac{1}{2}I_{rs}\right)^2 + D\frac{\partial^2 I_{ss}}{\partial X^2}, \tag{3}$$

$$\frac{\partial I_{rs}}{\partial \tau} = -\mu_{rs} I_{rs} + \frac{1}{2I}\delta_{rs}\eta_{rs}(P)\left[\left(I_{ss} + \frac{1}{2}I_{rs}\right)\left(I_{rr} + \frac{1}{2}I_{rs}\right)\right] + D\frac{\partial^2 I_{rs}}{\partial X^2}, \tag{4}$$

$$\frac{\partial I_{rr}}{\partial \tau} = -\mu_{rr} I_{rr} + \frac{1}{2I}\delta_{rr}\eta_{rr}(P)\left(I_{rr} + \frac{1}{2}I_{rs}\right)^2 + D\frac{\partial^2 I_{rr}}{\partial X^2}, \tag{5}$$

where $P(X,\tau)$ is the Bt plant biomass.

The parameters *r* and *K* in Eq. (2) denote the growth rate and the carrying capacity of the plant population. The parameters μ_{ss}, μ_{rs} and μ_{ss} in Eqs. (3) – (5) are the mortality rates per unit mass of the *ss*-, *rs*- and *rr*-insects, and *D* is the insect diffusion coefficient. The trophic functions $\eta_{ss}(P)$, $\eta_{rs}(P)$ and $\eta_{rr}(P)$ are in fact the effective fecundity rates of the *ss*-, *rs*- and *rr*-insects, correspondingly. The terms in Eqs. (2) – (5), which involve these functions, account for both the insect reproduction and maturation of the insect offspring. We allow for the fact that Bt plants selectively affect larvae [Rajamohan *et al.*, 1998], that is embodied in the form of the functions $\eta_{ss}(P)$ and $\eta_{rs}(P)$. Thus we suppose that the *ss*-, *rs*-insects are both liable to the action of Bt toxins, although not to the same extent. Consequently, their effective fecundity rates fall as soon as the Bt plant biomass is more than a critical value P_{cr}, so that

$$\eta_{ss}(P) = \eta_s^0 P \exp\left(-\frac{\lambda P}{K}\right), \tag{6}$$

$$\eta_{rs}(P) = \eta_{rs}^0 P\left(\exp\left(-\frac{\lambda P}{K}\right) + \omega_{rs}^*\right), \tag{7}$$

where η_s^0, η_{rs}^0 and $\lambda > 1$ are constants, and $P_{cr} = \dfrac{K}{\lambda}$. In contrast, the effective fecundity rate $\eta_{rr}(P)$ of Bt-resistant insects does not tend to decrease with an increase in the Bt plant biomass, so we suppose that

$$\eta_{rr}(P) = \frac{\eta_r^0 P}{C + P}, \tag{8}$$

where η_r^0 and *C* are constants. The parameters δ_{ss}, δ_{rs} and δ_{rr} are the respective yield coefficients of larvae of *ss*-, *rs*- and *rr*-insects to the plants. Our model similarly to the mathematical models in [Medvinsky *et al.* 2004; 2005; 2006; 2007] accounts for Bt plants being periodically sown, implying that at the beginning of each of the growing seasons T days in length the plant biomass

$$P = P_0 \text{ at } \tau = nT,\; n = 0,\, 1,\, 2,\, \ldots. \tag{9}$$

For later convenience, we introduce the dimensionless variables

$$p = \frac{P}{K},\; i_{ss} = \frac{I_{ss}}{K},\; i_{rs} = \frac{I_{rs}}{K},\; i_{rr} = \frac{I_{rr}}{K},\; t = \mathrm{A}\frac{\tau}{T},\; x = X\sqrt{\frac{\mathrm{A}}{DT}}$$

to render the mathematical model (1) – (9) in dimensionless form:

$$i(x,t) = i_{rr}(x,t) + i_{rs}(x,t) + i_{ss}(x,t), \tag{10}$$

$$\frac{\partial p}{\partial t} = \alpha p(1-p) - \frac{1}{i}\left[\omega_{ss}(p)\left(i_{ss} + \frac{1}{2}i_{rs}\right)^2 + \omega_{rs}(p)\left(i_{ss} + \frac{1}{2}i_{rs}\right)\left(i_{rr} + \frac{1}{2}i_{rs}\right) + \omega_{rr}(p)\left(i_{rr} + \frac{1}{2}i_{rs}\right)^2\right], \tag{11}$$

$$\frac{\partial i_{ss}}{\partial t} = -v_{ss} i_{ss} + \frac{1}{i}\delta_{ss}\omega_{ss}(p)\left(i_{ss} + \frac{1}{2}i_{rs}\right)^2 + \frac{\partial^2 i_{ss}}{\partial x^2}, \tag{12}$$

$$\frac{\partial i_{rs}}{\partial t} = -v_{rs} i_{rs} + \frac{1}{i}\delta_{rs}\omega_{rs}(p)\left[\left(i_{ss} + \frac{1}{2}i_{rs}\right)\left(i_{rr} + \frac{1}{2}i_{rs}\right)\right] + \frac{\partial^2 i_{rs}}{\partial x^2}, \tag{13}$$

$$\frac{\partial i_{rr}}{\partial t} = -v_{rr} i_{rr} + \frac{1}{i}\delta_{rr}\omega_{rr}(p)\left(i_{rr} + \frac{1}{2}i_{rs}\right)^2 + \frac{\partial^2 i_{rr}}{\partial x^2}, \tag{14}$$

$$\omega_{ss}(p) = \omega_s^0 p \exp(-\lambda p), \tag{15}$$

$$\omega_{rs}(p) = \omega_{rs}^0 p\left(\exp(-\lambda p) + \omega_{rs}^*\right), \tag{16}$$

$$\omega_{rr}(p) = \frac{\omega_r^0 p}{\gamma_r + p}, \tag{17}$$

$$p = p_0 \text{ at } t = n\mathrm{A},\; n = 0,\, 1,\, 2,\, \ldots. \tag{18}$$

In Eqs. (11) – (18), $\alpha = \frac{rT}{A}$, $\nu_{ss} = \frac{\mu_{ss}T}{A}$, $\nu_{rs} = \frac{\mu_{rs}T}{A}$, $\nu_{rr} = \frac{\mu_{rr}T}{A}$, $\omega_s^0 = \frac{KT}{2A}\eta_s^0$, $\omega_{rs}^0 = \frac{KT}{A}\eta_{rs}^0$, $\omega_r^0 = \frac{T}{2A}\eta_r^0$, $\gamma_r = \frac{C}{K}$. For our numerical work described below, we put $A = T$ so that $t = \tau$, $x = X\sqrt{\frac{1}{D}}$, $\alpha = r$, $\nu_{ss} = \mu_{ss}$, $\nu_{rs} = \mu_{rs}$, $\nu_{rr} = \mu_{rr}$, $\omega_s^0 = \frac{K}{2}\eta_s^0$, $\omega_{rs}^0 = K\eta_{rs}^0$, and $\omega_r^0 = \frac{1}{2}\eta_r^0$.

To investigate the dynamics of the plant-insect model system, we solved Eqs. (10) – (18) numerically by an explicit finite difference scheme. Neumann zero-flux boundary conditions were used. The initial conditions assumed at $t = 0$ were that plants and susceptible (*ss*) insects are homogeneously distributed in space, while heterozygote (*rs*) insects are concentrated in a small region at the center of the whole domain and Bt-resistant (*rr*) insects are absent. The mesh step sizes chosen were $\Delta x = 1$ and $\Delta t = 0.001$, sufficiently small to ensure the accuracy of the simulations.

RESULTS

One can see from Table 1 that frequencies of the Bt-susceptible *ss*- and *rs*-alleles should be seemingly higher than the frequency of the Bt-resistant *rr*-allele. However, the Bt-toxin action, which is simulated by Eqs. (15) and (16), induces the decrease of the growth rate of the Bt-susceptible *ss*- and *rs*-insects. These insects can go extinct resulting from the invasion of the Bt-resistant *rr*-insects, which in turn leads to dramatic decrease in the nutrient resource of the insects, Bt plant biomass. Hence, the Bt plant and insect populations dynamics are closely interrelated.

In order to characterize the impact of the *rr*-insects on the Bt plants let us take a look at the averaged harvest, *i.e.*, the time- and spatially averaged plant biomass at the expiration of growing seasons:

$$\langle p \rangle = \frac{1}{LN}\sum_{0}^{N}\int_{0}^{L} p(x, nA)dx. \tag{19}$$

Here, N is a number of the seasons. Table 2 shows the dependence of $\langle p \rangle$ defined by Eq. (19) on both the initial plant biomass p_0 defined by Eq. (18), and the initial biomass i_{rs}^0 of the heterozygote *rs*-insects under successful invasion of the Bt-resistant *rr*-insects, *i.e.*, when i_{ss} and i_{rs} become equal to zero. One can see from Table 2 that small values of the initial plant density p_0 can lead to the drop in the plant biomass $\langle p \rangle$ to zero. Likewise, $\langle p \rangle = 0$ under high values of the plant biomass p_0 (Table 2). In the first case, the Bt plants perish due to progressive growth of the Bt-resistant insect populations, which in turn leads to the posterior crash of the transgenic crops (Figure 1c). In the latter case, high values of the initial plant density p_0 cause the dramatic rise in the Bt-resistant insects grazing on the Bt plants, which results in extermination of the Bt crop (Figure 1a). For intermediate values of the initial plant density p_0 ($-6 \le \ln p_0 \le -2$ in Table 2), the Bt plant biomass does not perish under invasion of Bt-resistant *rr*-insects. In this case the plant biomass oscillates in time (Figure 1b), and $0.01 < \langle p \rangle < 0.05$ (Table 2). Notice that, as the data of Table 2 suggest, the dependence of the plant biomass $\langle p \rangle$ on p_0 depends only weakly on the initial biomass of the heterozygote *rs*-insects.

Comparison of Figure 1b and 1d shows that the invasion of the pests resistant to Bt toxins is followed by a drastic decrease in the Bt plant biomass. Another consequence of the invasion is the irregularity in the crop production. The irregularity can lead to a bad predictability of the crop production. In order to numerically evaluate the predictability of the crop production we used a modified algorithm from [Kaplan & Glass, 1995].

The changes in the annual crop production are characterized by the function

$$p_y(n) = \frac{1}{L}\int_0^L p(x, n\mathrm{A})dx, \tag{20}$$

where, like in (18), *n* specifies the number of a growing season: $n = 1, 2, 3, \ldots, N$. An example of the time series $p_y(n)$ is given in Figure 2. One can see the irregular character of this time series. The algorithm of evaluation of predictability of the crop production is based on the following steps:

i) constructing the vector $\vec{P}_y\left(\frac{N}{2}\right) = \left(p_y\left(\frac{N}{2}\right), p_y\left(\frac{N}{2}-1\right), p_y\left(\frac{N}{2}-2\right), \ldots, p_y\left(\frac{N}{2}-(p-1)\right)\right)$, where N is an even number, and *p* is the so-called embedding dimension [Ott, 2002];

ii) searching at the interval $\left[0, \frac{N}{2}\right]$ the *p*-dimensional vectors $\vec{\Pi}_y(n_i) = (\Pi_y(n_i), \Pi_y(n_i - 1), \ldots, \Pi_y(n_i - (p-1)))$, $i = 1, 2, \ldots, M$, such that $\left|\vec{P}_y\left(\frac{N}{2}\right) - \vec{\Pi}_y(n_i)\right| < \varepsilon$;

iii) giving a prediction of the value $\tilde{p}_y\left(\frac{N}{2}+1\right) = \frac{1}{M}\sum_{i=1}^{M} \Pi_y(n_i + 1)$;

iv) constructing the vector $\vec{P}\left(\frac{N}{2}+1\right)$;

v) starting the second iteration at the interval $\left[0, \frac{N}{2}+1\right]$ and then repeating the iterations until the point N is reached.

Table 2. The dependence of $\langle p \rangle$ defined by Eq. (19) on the initial plant biomass p_0 and the initial insect biomass i_{rs}^0

$\ln p_0$	i_{rs_0}	
	0.01	0.001
	$\langle p \rangle$	
-8	0	0
-7	0	0
-6	0.051	0.052
-5	0.021	0.024
-4	0.031	0.031
-3	0.015	0.015
-2	0.030	0.033
-1	0	0

The prediction error is

$$E(n) = \frac{1}{n}\sum_{k=1}^{n}\left(\frac{\widetilde{p}_y\left(\frac{N}{2}+k\right)}{p_y\left(\frac{N}{2}+k\right)} - 1\right)^2, n \le \frac{N}{2}. \tag{21}$$

Obviously, the closer the predicted values of $\widetilde{p}_y$ are to the real values p_y, the less is the prediction error. Figure 3a shows both real and predicted changes in the annual crop production. Figure 3c demonstrates the corresponding values of the prediction error $E(n)$ given by Eq. (21). One can see that the prediction error is growing in time and approaches values close to unity, which means the forecast becomes invalid very quickly. In fact, the predicted and real time series diverge after three growing seasons (Figure 3a).

Invasion of the Bt-resistant pests leads to high values of the prediction error in all instances when the crop production oscillated in a comparatively wide range. Otherwise, the values of $E(n)$ were close to the timely averaged crop production (Figure 3d), and the predicted time series did not exceed the amplitude of the crop production oscillations (Figure 3b). In this case, the algorithm above was fruitful in prediction of the crop production, which means that the difference between predicted and real values of the crop production remained close to each other. However, under these conditions, the predicted time series, as seen from

Figure 3b, does not reproduce the general run of the crop production curve. In this sense, the Bt-plant dynamics under invasion of the Bt-resistant insects remain poorly predictable.

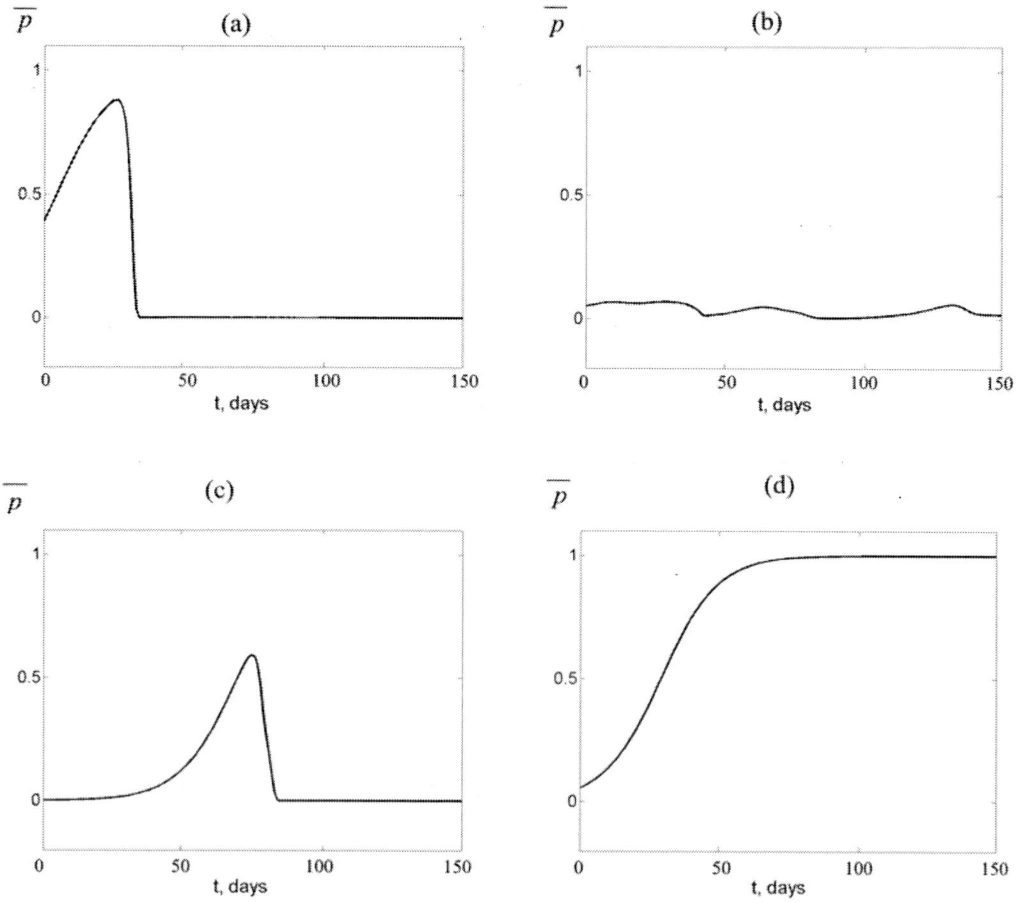

Figure 1. Annual changes in the spatially averaged Bt plant biomass $\bar{p}(t) = \frac{1}{L}\int_0^L p(x,t)dx$ resulting from the invasion of the Bt-resistant insect. (a) $p_0 = 10^{-1}$; (b) $p_0 = 10^{-3}$; (c) $p_0 = 10^{-7}$; (d) $p_0 = 10^{-3}$, no invasion. The simulations were carried out under the following set of the model parameters: $\alpha = 0.1$; $\omega_s^0 = \omega_{rs}^0 = \omega_r^0 = 1$; $\omega_{rs}^* = 0.1$; $\lambda = 100$; $\nu_{ss} = \nu_{rs} = \nu_{rr} = 0.2$; $\delta_{ss} = \delta_{rs} = \delta_{rr} = 1$; $\gamma_r = 0.1$. The parameter values are in agreement with observational and experimental data (see [Velkov et al., 2005] and the references therein). The only exception is the values of the parameters δ_{ss}, δ_{rs} and δ_{rr}, which in nature are less than 1; the parameter values $\delta_{ss} = \delta_{rs} = \delta_{rr} = 1$ imply the strongest impact of pests on the plant biomass. Notice that for smaller values of these parameters the results remain qualitatively the same, however, the duration of the transition period, when Bt-susceptible and Bt-resistant insects coexist, increases.

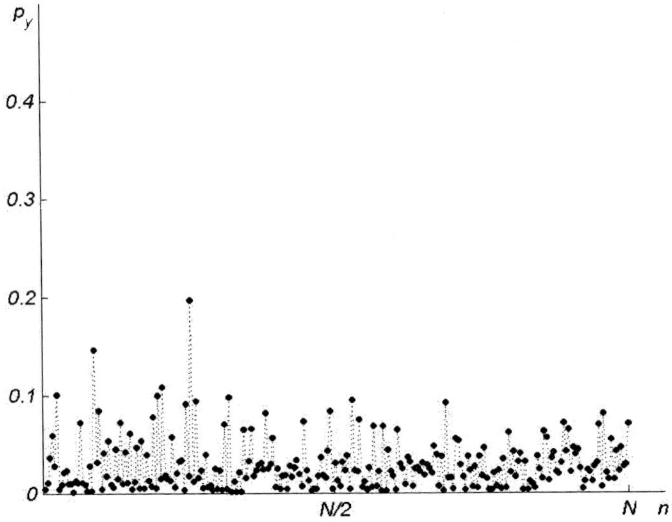

Figure 2. Variations of the crop production $p_y(n)$, which is given by Eq. 21; $p_0 = 10^{-3}$, other parameter values are the same as in Figure 1.

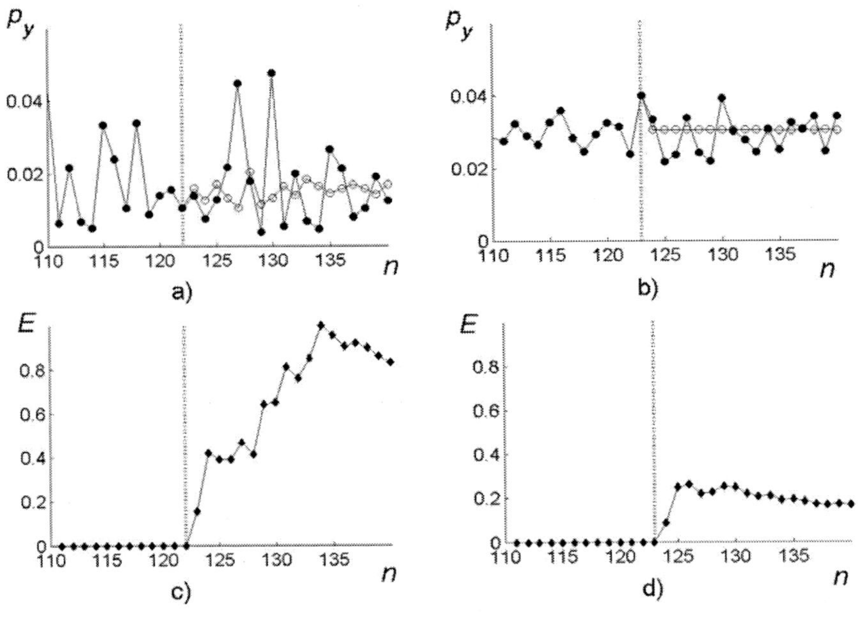

Figure 3. The results of prediction of $p_y(n)$. The vertical lines mark the initial point of the prediction. (a) Variations of the crop production (black circles) in comparison with the predicted time series (white circles); $p_0 = 10^{-3}$, other parameter values are the same as in Figure 1. (b) Variations of the crop production (black circles) in comparison with the predicted time series (white circles); $p_0 = 10^{-4}$, other parameter values are the same as in Figure 1. (c) The prediction error $E(n)$, $p_0 = 10^{-3}$. (d) The prediction error $E(n)$, $p_0 = 10^{-4}$.

CONCLUSION

In this work we use a conceptual reaction-diffusion model of the Bt crop – Bt-susceptible insect – Bt-resistant insect system to simulate the invasion of Bt-resistant insects and their impact on crop production. The obtained results imply that thorough measurements could allow us to estimate consequences of the Bt-resistant insect invasion for a given transgenic crop sown in a given geographical region – viz. of the constants ω_r^0 and γ_r, which parameterize the saturating effective fecundity rate of the Bt-resistant insects (Eq. 17), the insect mortality rates ν_{ss}, ν_{rs} ν_{rr} (Eq. 12, 13 and 14) and the other constants of the model above, which parameterize the growth rate of the Bt plants and effective fecundity rates of the Bt-susceptible insects: α, ω_s^0 and ω_{rs}^0 (Eq. 11, 15 and 16). To our knowledge, such measurements have not been carried out yet.

Since our model for the plant and insect dynamics depends on uncertain input parameters, the model output should not be interpreted as predictive of the plant-insect system behavior, but must be interpreted within the context of possible scenarios of population dynamics; see also [Medvinsky, 2006]. Recently, it has been shown that one of the scenarios of the Bt-plant biomass dynamics resulting from invasion of Bt-resistant insects involves coexistence of a chaotic attractor and a limit cycle, which means non-uniqueness of the plant-insect dynamics. The non-uniqueness has been shown [Medvinsky *et al.*, 2006] to lead to an essential decrease in the plant biomass under unfavorable environmental conditions, such as in the so-called zones of risky agriculture where environmental conditions, for example, temperature and light, can affect seed germination [Mayer and Poljakoff-Mayber, 1982]. In terms of the model (10)–(18) the impact of the environmental conditions on seed germination can appear as variations of the numerical value of the parameter p_0 in Eq. (18).

One can see from Table 2 that at both small and large values of the initial Bt-plant density p_0, the plant biomass $\langle p \rangle$ defined by Eq. (19) goes extinct, which is a result of the invasion of Bt-resistant pests (Figure 1). Surprisingly, the initial heterozygote pest biomass does not essentially affect the plant biomass $\langle p \rangle$ (Table 2). Notice, however, that the pest biomass may affect [Medvinsky *et al.*, 2008] the regularity of the annual oscillations of the plant biomass that take place for intermediate values of the parameter p_0 (Figure 1).

Figure 2 shows that the Bt crop production under invasion of the Bt-resistant insects is highly irregular. The analysis of predictability shows that the horizon of predictability [Kravtsov, Kadtke, 1996] of the Bt crop production variations does not exceed a few years (see Figure 3a as an example). Even in the special case that the prediction error is comparatively small (Figure 3d), the Bt crop production cannot be considered as predictable, since the predicted time series does not reproduce the time series obtained directly from computer simulations (Figure 3b).

All the figures above were obtained assuming that $\omega_s^0 = \omega_{rs}^0$, which implies that $\eta_s^0 > \eta_{rs}^0$ (see the chapter "Mathematical Model"). In other words, it means that the reproduction of the heterozygote *rs*-insects is characterized by lesser values of the fecundity rate. This can be due to the fitness cost of partial resistance of *rs*-pests to Bt toxins [Oppert *et al.*, 2000; Janmaat and Meyers, 2003; 2005; Carrière *et al.*, 2006; Wenes *et al.*, 2006]. Notice, however, that qualitatively the same results were obtained under assumption that there is freedom from the fitness cost, *i.e.*, $\eta_s^0 = \eta_{rs}^0$ (not shown). In this sense, the model (10) – (18) is structurally stable.

Within the framework of the given work, the appearance of pests resistant to Bt toxins is considered as a *post facto* phenomenon. The resistance development is a separate substantial problem, which has been considered in both theoretical and applied papers. In particular, using a demo-genetic model, it has been recently demonstrated [Zhadanovskaya *et al.*, 2006; Tyutyunov *et al.*, 2007] that the spatial organization of Bt and non-Bt fields, the so-called "high dose - refuge strategy" [Alstad and Andow, 1995], can delay the pest invasion by many hundreds of years. In the present paper, we address other problems, studying some dynamical consequences of the invasion of Bt resistant pests.

Thus, we show that the invasion of Bt-resistant insects may lead to both falls in the Bt crop production and to practically unpredictable variations of the production. The results obtained are particularly important for expanding transgenic cropping practice into zones of risky agriculture in many developing countries, as well as degraded agricultural regions in some industrialized nations. Consequences of biological invasions in a more general context remain a challenging problem.

ACKNOWLEDGMENTS

This work was partially supported by DFG, ISTC, RFBR grants, NSF Biocomplexity Program and the University of California Agricultural Experiment Station.

REFERENCES

Alstad, D.N. and Andow, D.A. (1995) Managing the evolution of insect resistance to transgenic plants. *Science* 268, 1894-1896.

Bazykin, A.D. (2004) *Nonlinear Dynamics of Interacting Populations*. ISC, Moscow, Izhevsk (in Russian).

Brookes, G. and Barfoot, P. (2006) Global impact of biotech crops: Socio-economic and environmental effects in the first ten years of commercial use. *AgBioForum* 9, 139-151.

Carrière, Y., Ellers-Kirk, C., Biggs, R.W., Nyboer, M.E., Unnithan, G.C., Dennehy, T.J. and Tabashnik, B.E. (2006) Cadherin-based resistance to *Bacillus thuringiensis* cotton in hybrid strains of pink bollworm: Fitness costs and incomplete resistance. *Journal of Economic Entomology* 99, 1925-1935.

Castle, W.E. (1903) The laws of Galton and Mendel and some laws governing race improvement by selection. *Proceedings of the American Academy of Arts and Science* 35, 233-242.

Gould, F. (1998) Sustainability of transgenic insecticidal cultivars: integrating pest genetics and ecology. *Annual Review of Entomology* 43, 701-726.

Hardy, G.H. (1908) Mendelian proportions in a mixed population. *Science* 28, 49-50.

Hillier, J.G. and Birch, A.N.E. (2002) A bi-trophic mathematical model for pest adaptation to a resistant crop. *Journal of Theoretical Biology* 215, 305-319.

Huang, F. (2006) Detection and monitoring of insect resistance to transgenic Bt crops. *Insect Science* 13, 73-84.

James, C. (2004) Global status of commercialized biotech/GM crops: 2004. *ISAAA Briefs*, 32, 1-12.

Janmaat, A.F. and Meyers, J. (2003) Rapid evolution and the cost of resistance to *Bacillus thuringiensis* in greenhouse populations of cabbage loopers, *Trichoplusia ni*. *Proceedings of the Royal Society of London B, Biological Sciences* 270, 2263-2270.

Janmaat, A.F. and Meyers, J. (2005) The cost of resistance to *Bacillus thuringiensis* varies with the host plant of *Trichoplusia ni*. *Proceedings of the Royal Society of London B, Biological Sciences* 272, 1031-1038.

Kaplan, D. and Glass, L. (1995) *Understanding Nonlinear Dynamics*. Springer, New York.

Kolmogorov, A., Petrovskii, I. and Piskunov, S. (1937) Étude de l'equation de la diffusion avec croissance de la quantité de matière et son application à un problème biologique. *Bull. Univ. Moscou Serie Int. A* 1, 1-25.

Kravtsov, Y.A., Kadtke, J.B. (eds.) (1996) *Predictability of Complex Dynamical Systems*. Springer, New York.

Lotka, A.J. (1925) *Elements of Physical Biology*. Williams and Wilkins, Baltimore.

May, R.M. (1976) Simple mathematical models with very complicated dynamics. *Nature* 261, 459-467.

Mayer, A.M., Poljakoff-Mayber, A. (1982) *The Germination of Seeds*. Pergamon. New York.

Medvinsky, A.B. (2006) Population dynamics: Limits of predictability. *Biophysics* 51, 908-916.

Medvinsky, A.B., Gonik, M.M., Li, B.-L. and Malchow, H. (2007) Beyond Bt resistance of pests in the context of population dynamical complexity. *Ecological Complexity* 4, 201-211.

Medvinsky, A.B., Gonik, M.M., Li, B.-L., Velkov, V.V. and Malchow, H. (2006) Invasion of pests resistant to Bt toxins can lead to inherent non-uniqueness in genetically modified Bt-plant dynamics: Mathematical modeling. *Journal of Theoretical Biology* 242, 539-546.

Medvinsky, A.B., Gonik, M.M., Tyutyunov Y.V., Li, B.-L., Rusakov, A.V. and Malchow, H. (2008) Insecticidal Bt crops under massive Bt-resistant pest invasion: Mathematical simulation. In: *Aspects of Mathematical Modelling* (ed. by R. Hosking and E. Venturino). Birkhäuser, Basel, pp. 81-94.

Medvinsky, A.B., Gonik, M.M., Velkov, V.V., Li, B.-L. and Malchow, H. (2005) Modeling invasion of pests resistant to Bt toxins produced by genetically modified plants: recessive vs. dominant invaders. *Natural Resource Modeling* 18, 347-362.

Medvinsky, A.B., Morozov, A.Y., Velkov, V.V., Li, B.-L., Sokolov, M.S. and Malchow, H. (2004) Modeling the invasion of recessive Bt-resistant insects: an impact on transgenic plants. *Journal of Theoretical Biology* 231, 121-127.

Medvinsky, A.B., Petrovskii, S.V., Tikhonova, I.A., Malchow, H. (2002) Spatiotemporal complexity of plankton and fish dynamics. *SIAM Review* 44, 311-370.

Oppert, B., Hammel, R. Throne, J.E. and Kramer, K.J. (2000) Fitness costs of resistance to *Bacillus thuringiensis* in the Indianmeal moth, *Plodia interpunctella*. *Entomologia Experimentalis et Applicata* 96, 281-287.

Ostlie, K.R., Hutchinson, W.D. and Hellmich, R.L. (1997) Bt-corn & Europen corn borer, long-term success through resistance management. *North Central Region Extension Publication 602*. University of Minnesota, St. Paul.

Ott, E. (2002) *Chaos in Dynamical Systems*. Cambridge University, Cambridge.

Rajamohan, F., Lee, M.K. and Dean, D.H. (1998) *Bacillus thuringiensis* insecticidal protein: molecular node of action. *Progress in Nucleic Acid Research and Molecular Biology* 60, 1-23.

Shelton, A.M., Zhao, J.-Z. and Roush, R.T. (2002) Economic, ecological, food safety, and social consequences of the deployment of Bt transgenic plants. *Annual Review of Entomology* 47, 845-881.

Tabashnik, B.E. (1994) Evolution of resistance to *Bacillus thuringiensis*. *Annual Review of Entomology* 39, 47-79.

Tabashnik, B.E., Carriere, Y., Dennehy, T.J., Morin, S., Sisterson, M.S., Roush, R.T., Shelton, A.M., Zhao, J.-Z. (2003) Insect resistance to transgenic Bt crops: Lessons from the laboratory and the field. *Journal of Economic Entomology* 96, 1031-1038.

Tabashnik, B.E., Cushing, N.L., Finson, N., Johnson, M.W. (1990) Field development of resistance to *Bacillus thuringiensis* in diamondback moth (*Lepidoptera: Plutellidae*). *Journal of Economic Entomology* 83, 1671-1676.

Tabashnik, B.E., Patin, A.L., Dennehy, T.J., Liu, Y.B., Carriere, Y., Sims, M.A., Antilla, L. (2000) Frequency of resistance to *Bacillus thuringiensis* in field populations of pink bollworm. *Proceedings of the National Academy of Sciences of the USA* 97, 12980-12984.

Tyutyunov, Yu.V., Zhadanovskaya, E.A., Arditi, R., Medvinsky, A.B. (2007) A spatial model of the development of pest resistance to a transgenic insecticidal crop: European corn borer on Bt maize. *Biophysics* 52, 52-67.

Velkov, V.V., Medvinsky, A.B., Sokolov, M.S., Marchenko, A.I. (2005) Will transgenic plants adversely affect environment? *Journal of Biosciences* 30, 515-548.

Volterra, V. (1926) Fluctuations in the abundance of a species considered mathematically. *Nature* 118, 558-560.

Weinberg, W. (1908) Über den Nachweis der Vererbung beim Menschen. *Jahreshefte des Vereins für vaterländische Naturkunde in Württemberg* 64, 368-382.

Wenes, A.-L., Bourguet, D., Andow, D.A., Courtin, C., Carré, G., Lorme, P., Sanchez, L. and Augustin, S. (2006) Frequency and fitness cost of resistance to *Bacillus thuringiensis* in *Chrysomela tremulae* (*Coleoptera*: *Chrysomelidae*). *Heredity* 97, 127-134.

Zhadanovskaya, E.A., Tyutyunov, Yu.V., Arditi, R. (2006) Modelling the "high dose - refuge" strategy when using genetically modified maize for suppression of the European corn borer. *Izvestiya VUZov, Severo-Kavkazskii region, Estestvennye nauki* 11, 7-13 (in Russian).

Reviewed by Dr. Ekaterina Zhadanovskaya.

Laboratory of Mathematical Modelling of Biological Processes, Department of Mathematical Methods in Economics and Ecology, Vorovich Research Institute of Mechanics and Applied Mathematics, Southern Federal University.

Stachki Street 200/1, 344090 Rostov-on-Don, Russia.

Phone: (+7 863) 297-52-53; Fax: (+7 863) 297-52-31.

E-mail: catherine@math.sfedu.ru, ejadan@mail.ru

In: Transgenic Plants: New Research
Editor: Oliver T. Chan

ISBN 978-1-60692-017-6
© 2009 Nova Science Publishers, Inc.

Chapter 9

THE USE OF TRANSGENIC PLANTS TO STUDY THE BIOSYNTHESIS AND REMODULATION OF CELL WALL POLYSACCHARIDES

Zoran Minic*
Department of Chemistry, University of Saskatchewan, 110 Science Place, Saskatoon, SK S7N 5C9, Canada

ABSTRACT

Plant cell wall contains various components such as polysaccharides, lignin and proteins whose composition and concentration change during plant development and growth. Polysaccharides are major components in plants cell walls and contain three major classes: cellulose, hemicelluloses, and pectins. During plant development a series of coordinated biochemical processes occurred in plant including biosynthesis and degradation of cell wall polymers. Biosynthesis and remodulation of cell wall polysaccharides are regulated by complex enzymatic machinery including various enzymes belonging to families of glycoside hydrolases and glycoside transferases. These enzymes play crucial roles in the structure and function of the plant cell wall polysaccharides. Various molecular, genetic and biochemical approaches have been used to study of these enzymes. Transgenic plants modified in the expression of biosynthetic or modifying enzymes have been intensively used to investigate the role of the particular activities and to determine alteration in cell wall composition. The present review described utilisation of mutant and transgenic plants related to glycosyl transferases and glucoside hydrolases in order to determine the molecular basis of the synthesis and remodulation of plant cell walls polysaccharides. The modulation of plant cell wall polysaccharides using transgenic plants could be important not only for fundamental research but also in plant biotechnology.

* E-mail: zom198@mail.usask.ca or zoranminic@yahoo.fr Phone: 1-306-717-4752; Fax: 1-306-966-4730

INTRODUCTION

The most widely used method for the introduction of new genes into plants is based on the ability of the natural DNA of *Agrobacterium tumefaciens*. These gram-negative soil bacteria are known to cause crown gall disease in plants. Its virulence is conferred by large tumour-inducing plasmids (Ti), of which a small specific part, named T-DNA (transferred DNA), possesses the capacity to be incorporated into host plant nuclear DNA. Inserted T-DNA (20 kb long) into the nuclear genome directs the plant cells to synthesize the hormones which cause the crown gall symptoms (Chilton et al.,1977). In addition, infected plants synthesize novel metabolites, opines that are not found in normal plant tissues. Two short regions on the ends of the T-DNA, known as the left border (LB) and right border (RB), are necessary for the transfer of the T-DNA into the nucleus and for subsequent integration (Gordon, 1998; Zambryski et al.1989). The precise molecular mechanism of how the bacterial DNA moves from the cytoplasm to the nucleus of the plant cell or how the T-DNA becomes integrated into the plant chromosome is not clear (Gheysen et al., 1991). The T-DNA is inserted randomly within the genome (Thomas et al., 1994; Azpiroz-Leehan and Feldmann, 1997) and it appears that host DNA repair mechanisms might have been recruited to assist in integration (Tinland, 1996, Nacry et al., 1998).

T-DNA vector from *Agrobacterium* was used for making transgenic plants. Except for the border repeats (LB and RB), none of the T-DNA sequence is required for transfer and integration; consequently, the T-DNA genes can be replaced by any other DNA of interest and transferred to the plant genome. In addition, change of the length of the T-DNA is not crucial for transfer because small as well as large T-DNAs (Hamilton et al., 1996) can be transferred by the *Agrobacterium* into the plant cell. Moreover, the removal of the plant hormone biosynthetic T-DNA gene has as a consequence that the transformed plant cells do not proliferate into tumorous tissues but may regenerate into normal plants. In general, these transgenic plants have phenotypes indinguishable from untransformed plants. Numerous vector systems are now available containing the T-DNA borders, selectable marker genes and various reporters. The selectable markers include genes that provide antibiotic or herbicide resistance. For example, one of the most well-known is the neomycin phosphotransferase gene which provides resistance to the kanamycin. Selectable marker genes are intentionally introduced into cells as developmental tools in recombinant DNA technology. They act as "genetic tags" and are very important to easily identify successfully modified cells. Several reporter genes are used in plants, including β-glucuronidase, luciferase, genes involved in anthocyanin biosynthesis and the green fluorescent protein. These reporter genes are important for determination of sequences and for analysis of expression of genes under different conditions.

When the plant transformation methods are developed large production of the transgenic plants have been realized in order to study gene expression and gene function. Several hundred mutants have been characterized and the disrupted genes isolated using forward genetic strategies based on T-DNA mutagenesis (Koncz et al., 1992; Azpiroz-Leehan and Feldmann, 1997). Further modifications of T-DNA have allowed additional approaches for assessing gene function, such as enhancer, promoter trapping and activation tagging (Topping and Lindsey, 1997; Campisi et al., 1999; Weigel et al., 2000). Forward genetics, which uses the T-DNA mutant phenotype coupled with genomics molecular analyses, has been important for helping to understand many biological functions. In addition to forward genetics, gene function has been investigated using reverse genetics which is a complement to forward genetics. In reverse genetics, the functional study of a gene starts with the gene sequence rather than with a mutant phenotype (Krysan et al., 1999).

Arabidopsis thaliana has been used as model in the study of many molecular aspects of plant biology. Since this plant has a small genome size (125 Mb) it was chosen as the subject for genome sequencing. The availability of the genomic sequence information of *Arabidopsis* allowed for easier analysis of the function of genes. In addition, several approaches were developed for isolating and determining the function of specific genes such as T-DNA and activation tagging, transposon tagging, and various gene silencing methods (Feldmann, 1991; Tissier et al., 1999; Winkler et al., 1998; Dalmay et al., 2000; Chuang et al., 2000 ; Smith et al., 2000 ; Mysore et al., 2000). Moreover, numerous T-DNA or transposon insertion collections are available in public databases that can be exploited to link DNA sequences to phenotypes (Eckardt, 2001). The results obtained for the *Arabidopsis* experimental system are important for understanding the function of genes in economically important plants. These new approaches are likely to speed up our understanding of various aspects of plant physiology.

The application of new genetic technologies has helped and revolutionized research in the plant cell wall, as well as cell wall polysaccharide biosynthesis and the mechanisms that control the assembly of these polysaccharides into a functional wall. This review describes recent advances in the use of transgenic plants to study the biosynthesis and remodulation of cell wall polysaccharides. The investigation is focused particularly on enzymes glycosyl transferases (GTs) and glycoside hydrolases (GHs) which are known to be crucial for the synthesis and assembly of cell walls polysaccharides.

Plant Cell Wall

Plant cell wall represents most abundant organic carbon in the biosphere. During plant life, cell wall is involved in various physiological functions such as growth, intercellular communications, defense against pathogen attack, mechanical resistance, and interaction with the environment (Cosgrove, 1997; 1999; Fry, 1995; 2004; Minic and Jouanin, 2006; Minic, 2008).

Plant cells can have two types of cell walls: the primary and secondary cell walls (Figure 1). Primary walls surround the growing and dividing plant cells and are composed of cellulose, hemicelluloses, pectic polysaccharides and many proteins. The polysaccharide composition of the primary cell wall can vary; approximately equal amounts of pectin and hemicellulose are present in dicot, whereas hemicellulose is more abundant in grasses. The secondary cell wall is deposited in fully expanded and specialized cells (xylem and fibers). The secondary walls of woody tissue and grasses are composed predominantly of cellulose, lignin, and hemicellulose (xylan, glucuronoxylan, arabinoxylan, or glucomannan). Compared to primary walls, secondary walls contain more cellulose where the cellulose fibrils are embedded in a network of hemicellulose and lignin. Cross-linking of this network results in the elimination of water from the wall and the formation of a hydrophobic composite that limits accessibility of hydrolytic enzymes and is a major contributor to the structural characteristics of secondary walls. Secondary walls also, as a raw material and renewable biomass, have important economical roles. Various polysaccharide constituents of plant cell walls have different uses in agriculture, food industry, energy production and materials for various applications. However, numerous fundamental and technical challenges must be overcome in order to efficiently utilize these polysaccharides to a greater extent.

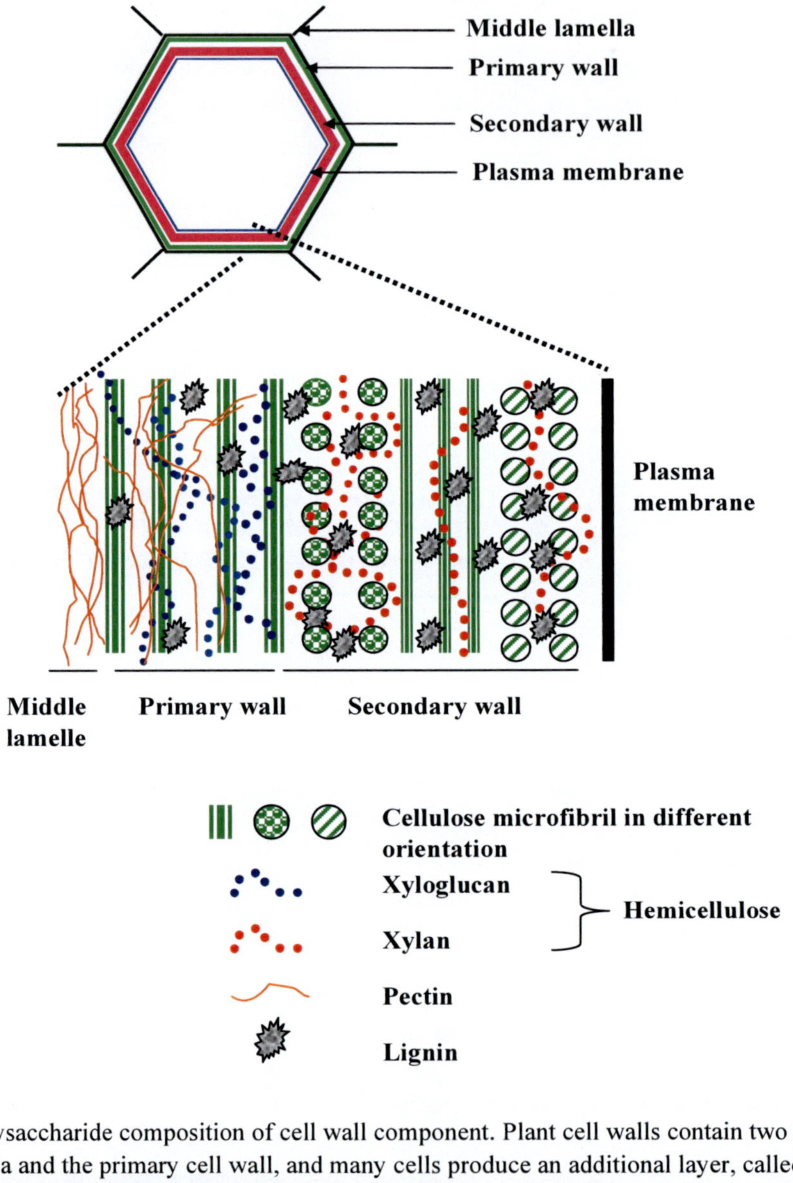

Figure 1. Polysaccharide composition of cell wall component. Plant cell walls contain two layers, the middle lamella and the primary cell wall, and many cells produce an additional layer, called the secondary wall. The middle lamella forms the interface between the primary walls of neighbouring cells. The primary wall is formed during the early stage of growth. Secondary cell wall can form between the primary cell wall and plasma membrane and consist of three layers, which differ in the orientation of their cellulose microfibrils.

Cell Wall Polysaccharides

Plant cell wall polysaccharides are the most abundant organic compounds found in nature. They make up to 90% of plant cell wall and can be divided into three groups: cellulose, hemicelluloses, and pectins. Their composition varies from one plant species to another (dicot/monocot), as well as within different tissues of the same plant (Cosgrove, 1997; Heredia et al., 1995; Popper and Fry, 2003). Structural proteins and enzymes constitute the remaining 10% of the plant cell wall. They play a crucial role in cell wall structure and architecture, cell wall metabolism, cell enlargement, signalling, response to abiotic and biotic stresses, and many other physiological processes (Cosgrove, 1997; 1999; Fry, 1995; 2004; Carpita, 1996; Carpita and Gibeaut, 1993).

Cellulose represents the major cell wall polysaccharide and is composed of (1→4) β-D-glucan chains assembled together by hydrogen bonding into very long crystalline microfibrils. Cellulose is synthesized at the plasma membrane level and then deposited into the wall (Kohorn, 2000).

Hemicelluloses constitute up to 30-35% of the dry weight of the plant cell wall (Heredia et al., 1995) and are mainly composed of xyloglucan and xylan with minor amounts of several mannose-containing polysaccharides (Bewley et al., 1997; Handford et al., 2003). The hemicelluloses are synthesized in the Golgi apparatus and are then excreted into the wall. The major hemicellulose polymer in cereals and wood is xylan (Prade, 1996; Kulkarni et al., 1999). Xylan, which accounts for up to 30% of the mass of the secondary walls in wood and grasses, contributes to the recalcitration of these walls to enzymatic degradation. Xylan is composed of a backbone of β-(1→4)-linked D-xylopyranosyl units substituted with L-arabinofuranose, glucuronic acid, 4-O-methylglucuronic acid, and acetyl side-groups to varying extent (Prade, 1996; Kulkarni et al., 1999). Arabinoxylan has a β-(1→4)-D-xylopyranoside backbone carrying O-2, O-3 or O-2,3 α-L-arabinofuranosyl substitutions with an overall L-Ara/D-Xyl ratio of about 0.7 (Izydorczyk and Biliaderis, 1995). Xyloglucan possesses a (1→4) β-D-glucan backbone like cellulose substituted with xylosyl, galactosyl and L-fucosyl residues (Madson et al., 2003). Xyloglucans interact with cellulose microfibrils by formation of hydrogen bonds (Lima and Buckeridge, 2001), thus contributing to the structural integrity of the cellulose network (McCann and Roberts, 1991). Binding capacity between cellulose and xyloglucans depends on the number of side-chains branching fucose and galactose residues (Madson et al., 2003; Lima and Buckeridge, 2001). A minor amount of the hemicellulose is composed of galacto(gluco)mannan, which consists of a backbone of β-(1→4)-linked D-mannose and D-glucose residues with occasional side chains of single units of D-galactose (Dey, 1978).

Pectins are synthesized in the Golgi apparatus and excreted to the wall (Willats et al., 2001). Pectins are heteropolysaccharides composed of homogalacturonan (HG), rhamnogalacturonan I (RG-I) and substituted homogalacturonans named rhamnogalacturonan II (RG-II) and xylogalacturonan (XGA) (Vincken et al., 2003). Homogalacturonan is composed of long chains of (1→4) α-D-galacturonic acid, and is initially highly methyl-esterified. Rhamnogalacturonan I (RG I) is made of alternating *α*-D-rhamnose and α-D-galacturonic acid residues, with long side-chains attached to the rhamnose residues of either unbranched (1→4) β-D-galactan or branched α-L-arabinans or type I arabinogalactans. Xylogalacturonan (XGA) is a branched galacturonan with β-(1→3)-D-xylopyranoside chains

(Schols et al., 1995; Visser and Voragen, 1996). The α-D-galacturonide (GalA) residues of XGA can be methyl-esterified like HG (Schols et al., 1995). Rhamnogalacturonan II (RG II) is composed of a backbone of (1→4) α-D-galacturonic acid like homogalacturonan, but with complex side chains of several types of neutral sugars.

All these polysaccharides (hemicelluloses, pectins) and the aromatic lignin polymer interact with the cellulose fibrils to create a rigid structure strengthening the plant cell wall (Obembe et al., 2006; Hrmova et al., 2007). The polymers among them also form covalent cross-links, which are thought to be involved in limiting cell growth and reducing cell wall biodegradability.

Enzymes Involved in Biosynthesis of Cell Wall Polysaccharides

Plant cell wall polysaccharides are synthesized by glycosyltransferases (GTs), and most of them are located in the Golgi vesicles (Scheible and Pauly, 2004; Farrokhi et al., 2006). These enzymes use nucleotide sugars as donor substrates and transfer the sugar moiety to an acceptor substrate. In *Arabidopsis* 415 putative GT genes have been identified on the basis of their gene sequence (http://afmb.cnrs-mrs.fr/CAZY/) and classified into various gene families according to their sequence and structure (Coutinho and Henrissat, 1999). The biological activity of only a few of these putative GTs have been demonstrated. For some of them the isolation, characterization and mapping of wall mutants, together with biochemical approaches, have been shown to be involved in wall polysaccharide synthesis. Investigations about polysaccharides biosynthesis related with new genetic technology, particularly using transgenic plants, is described below.

Cellulose

Cellulose micrifibrils constitute the major cell wall polysaccharides and play an important role in the organization of plant cell walls (Dublin et al., 2002; Somerville, 2006). Although the chemical structure of cellulose is simple its biosynthesis and regulation appears to be a complex process. Many reviews described different aspect on cellulose structure and biosynthesis; however, many important questions regarding regulation and biosynthesis of cellulose remain unanswered (Lerouxel et al. 2006; Someville, 2006; Doblin et al., 2002; Robert et al., 2004; Taylor 2008; Hayashi et al., 2005; Scheible and Pauly, 2004). For example, the isolation and measurement of cellulose synthase activity *in vitro* remain challenging. Till now cellulose synthase activity was obtained only by using blackberry well suspension culture (Lai-Kee-Him et al., 2002). An important progress in the description of mechanisms involved in the biosynthesis of cellulose was made while exploring a large number of mutants from different plants. These mutants are derived from mutagenesis using ethylmethanesulfonate, insertional T-DNA, an insertional transponsable element, RNA interference and antisense alleles (Scheible and Pauly, 2004).

The enzymes that catalyze glucan-chain elongation in cellulose belong to the glycosidetransferase family 2 (GT2). In plants this family of GTs, designated as the cellulose synthase catalytic subunits (CESA), has multiple members (Lerouxel et al. 2006). *Arabidopsis* contains 10 *CesA* genes (named *AtCesA1-10*) while rice and maize contain 12. Mutations of these genes have been studied and for most of them phenotypes has been observed. For example, nulls mutants of *CESA1* from *Arabidopsis* cause embryo lethality and root swelling (Beeckman et al., 2002; Gillmor et al., 2002). The *AtCESA2* knockout mutant, *atcesa2*, had severe defects in the cell wall formation and microtubule orientation, resulting in abnormal plant growth and development (Chu et al., 2007). A mutation in the *CESA3* gene showed reduced cellulose levels (Caño-Delgado et al., 2003). Null mutation in *CESA6* cause a strong growth defect and reduced cellulose content in roots and dark-grown hypocotyls (Desprez et al., 2002). Mutation in *CESA4*, *CESA7* and *CESA8* enhanced resistance to the soil-borne bacterium *Ralstonia solanacearum* and the necrotrophic fungus *Plectosphaerella cucumerina* (Hernández-Blanco et al., 2007). The *CESA7* and *CESA8* mutants have identical phenotypes and exhibited irregular xylem (Taylor et al., 2000). A resent study showed that that *CESA3* and *CESA6* are expressed in the same cells and at the same time in the hypocotyl of etiolated seedlings and migrate with comparable velocities along linear trajectories at the cell surface (Desprez et al., 2007). In addition, using bimolecular fluorescence complementation experiments has shown that CESA1, CESA3, and CESA6 can interact in vivo. Moreover, the cellulose synthase activity was related with three components, CESA1, CESA3, and CESA6 (Persson et al., 2007). Taken together, all these data suggest that cellulose subunits are essential components of the cellulose synthesizing complex which is composed of three types of CESA subunits (CESA1, CESA3, and CESA6) (Taylor et al., 2003; Desprez et al., 2007; Persson et al., 2007).

In addition to genetic analyses, structure and properties of cellulose synthase have been also studied by electron microscopy in freeze-fractured plasma membrane (Delmar, 1999, Tsekos 1999; Doblin et al., 2002). Both electron microscopy and genetic studies are allowed to propose a model of a structure of the cellulose synthase. The cellulose synthase complex, known as rosette, is composed of a total 36 CESA proteins arranged in six groups of six individual proteins (Figure 2) (Scheible et al. 2001; Doblin et al., 2002; Taylor, 2007).

Aside from CESA, other proteins are also required for cellulose synthesis in plants (Lerouxel et al., 2006). It has been suggested that the β-1,4-endocellulase KORRIGAN is involved in biosynthesis of cellulose (Nicol et al., 1998). Various other proteins are also involved such as KOBITO1 (Pagant et al., 2002), a protein with an unknown function, and COBRA protein which is essential for microfibrial organization (Roudier et al., 2005). Finally, the role of the cytoskeleton, proteins that are localized to the cell wall (e.g. kinesins and glycosylphosphatidylinositol-anchored proteins), has been suggested to regulate cellulose deposition and cell wall ultrastructure (Paradez et al., 2006; Wasteneys and Fujita, 2006).

Although significant progress was made in the identification of proteins involved in cellulose synthase complex as well some structural aspect of enzyme resolved, the precise mechanism and identification of other genes involved in the regulation of biosynthesis of cellulose are not yet understood. In addition, it appears that the cellulose biosynthesis requires cellular components such as the cytoskeleton, which play an important role in the orientation of microfibrils (Somerville, 2006, 2006; Zhong and Ye 2007). Further challenges will include investigation of the relationship between these cellular components and cellulose synthase complex.

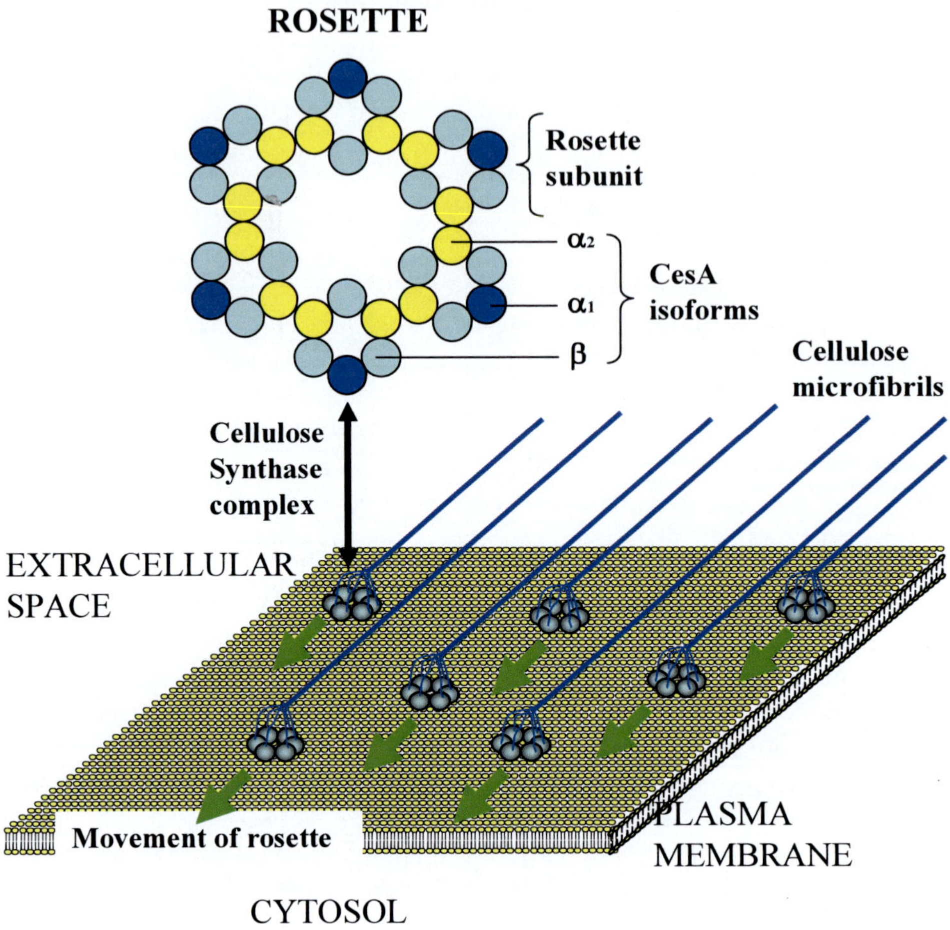

Figure 2. The cellulose synthase rosette complex is located at the plasma membrane and contains six subunits, each subunit containing six CesA polypeptides (36 per rosette) (Doblin et al., 2002; Scheible et al. 2001). In this rosette structure three types of CesA isoforms, α1, α2 and β are required for spontaneous rosette assembly. These protein complexes move through the membrane during the synthesis of glucan chains that aggregate to form cellulose microfibrils (Taylor, 2007).

HEMICELLULOSE

The cellulose synthase like (*CSL*) family of genes that have sequence similarity with *CESA* genes has been hypothesed to be involved in the backbone biosynthesis of glucomannan, mannan, xyloglucan and mixed-linkage glucans (Dhugga et al., 2004; Lerouxel et al., 2006).

Two glycosyltransferases are involved together in biosynthesis of galactomannans: (1→4)- β-D-mannosyltransferase (ManS), which makes the beta-1,4-mannan backbone of galactomannan and a mannan-specific (1→6)- α-D-galactosyltransferase ("galactomannan galactosyltransferase" [GMGT]) which adds galactosyl residues to the mannan backbone. The complementary DNA (cDNA) clone encoding ManS has been isolated from guar seed endosperm (Dhugga et al., 2004). The soybean somatic embryos expressing *ManS* cDNA contained high levels of ManS activities. Furthermore, the mechanism and regulation of galactomannan biosynthesis have been studied using sense, antisense, and sense/antisense ("hairpin loop") constructs of the galactomannan galactosyltransferase (GMGT) cDNA in *L. japonicus* endosperms (Edwards et al., 2004). Some of these transgenic lines exhibited variations in galactomannan structure and amount in the endosperms of *L. japonicus*in seeds. In addition, microsomal membranes prepared from the developing T_2 generation endosperms of transgenic lines showed reduced GMGT activity of mannan synthase in comparison to the wild type.

Recently, the genes encoding enzymes involve in biosynthesis of xyloglucan (XyG) have been identified and characterized (Figure 3). The gene(s) responsible for synthesizing the β-1,4-glucan backbone of XyG were analyzed from nasturtium (*Tropaeolum majus*) seed (Cocuron et al. 2007). These genes belonging to the *CSLC* family encode enzymes that produce β-1,4-linked glucans. A single member of this subfamily was found to be over represented in the cDNA library. Heterologous expression of this gene in the yeast *Pichia pastoris* resulted in the production of a β-1,4 glucan, confirming that the CSLC protein is involved in synthesis of the XyG backbone (Cocuron et al. 2007). In addition to the backbone biosynthesis several other genes involved in side chain addition were identified. Isolated *mur2* mutants from *Arabidopsis* corresponding to fucosyltransferase were shown to be deficient in fucose, a component of xyloglucan molecules (Vanzin et al., 2002). Another, mur3 *Arabidopsis* mutant, also showed deficiency in fucose and lacks galactose on the xylose moiety of xyloglucan. This mutant has led to the identification of a xyloglucan-galactosyltransferase that specifically adds galactose onto the xylose residue of xyloglucan polymer (Madson et al., 2003). This encoding xyloglucan-galactosyltransferase enzyme belongs to family 47 of GTs. Two *Arabidopsis* genes from CAZy family GT34 were expressed in insect cells and demonstrated that both can add multiple xylosyl residues to cellohexaose cellopentaose and cellohexaose acceptor substrates (Cavalier and Keegstra, 2006). This xyloglucan xylosyltransferase belong to family 34 of GTs.

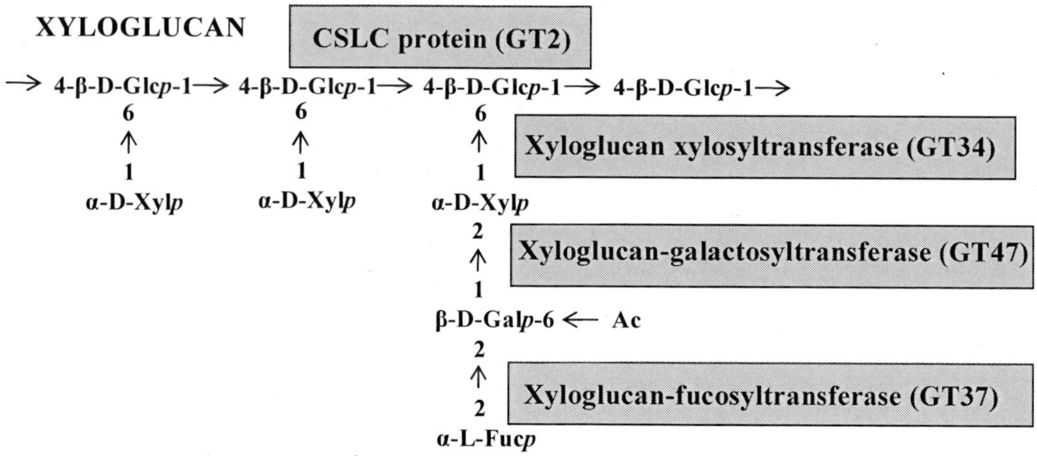

Figure 3. Representation of glycosyl transferases involved in the biosynthesis of plant xyloglucans.

Glucuronoxylans (GXs) together with cellulose and lignin are the three major components of secondary cell walls in woody plant tissues (Filonova et al., 2007). GX is composed of a linear backbone of β-(1,4)-linked D-xylosyl (Xyl) residues substituted by a single α-D-glucuronic acid (GlcA) or 4-*O*-methyl-α-D-glucuronic acid (MeGlcA) residue at O2 (Shimizu et al., 1976; Johansson and Samuelson, 1977; Andersson et al., 1983). Mutations of *Arabidopsis thaliana IRREGULAR XYLEM8* (*IRX8*) and *IRX9* were previously shown to be required for normal vessel morphology and wall thickness and for normal amounts of xylose and cellulose in cell walls (Brown et al., 2005; Persson et al., 2005; Zhong et al., 2005). The *IRX8* gene encodes a putative GT belonging to family GT8 (Brown et al., 2005; Persson et al., 2005) and the *IRX9* gene, which encodes a putative GT belonging to family GT43 (Coutinho et al., 2003). *Arabidopsis irx9* mutant plants showed increased number of GX chains and decreased the GX chain length (Lee et al., 2007). *Arabidopsis thaliana FRAGILE FIBER8* (*FRA8*) gene, which encodes a putative GT in family GT47 (Coutinho et al., 2003), is specifically expressed in cells undergoing secondary wall thickening. Plants carrying mutations in this gene have reduced amounts of wall GX and a decreased ratio of GlcA to MeGlcA residues in the GX (Zhong et al., 2005). Taken together, these investigations suggest that the coordinated action of *FRA8*, *IRX8*, and *IRX9* is necessary for the regulation of GX chain length and for the normal assembly and morphology of secondary cell walls in *Arabidopsis*. *FRA8* and *IRX8* (or their homologs) are likely to be required for the synthesis of the glycosyl sequence at the reducing end of the GX, whereas *IRX9* is involved in elongation of the GX chain.

Arabinoxylans (AXs) are major components of graminaceous plant cell walls, such as rice (*Oryza sativa*), wheat (*Triticum aestivum*), and barley (*Hordeum vulgare*). Using the bioinformatics approach, based on estimation of differential expression of orthologous genes between taxonomic divisions of species, has predicted that genes in the GT43, GT47, and GT61 families are the most likely candidates to encode the enzymes that synthesize AX and its side chains (Mitchell et al., 2007). These analyses together with known activities of some enzymes suggest that the GT43 families encode β-1,4-xylan synthases, genes in the GT47 family encode xylan α-1,2- or α-1,3-arabinosyl transferases, and genes in the GT61 family encode feruloyl-AX.

PECTIN

It is a fact that pectin has a complex structure; consequently, numerous enzymes are involved in their biosynthesis. It is predicted that biosynthesis of pectins require at least 53 different GTs (Ridley et al., 2001) but very little information is available about them. Some genes involved in biosynthesis of pectin have been identified using genetic strategies. These include *QUA1* (Bouton et al., 2002), *PARVUS/GLZ1* (Lao et al.,Shao et al., 2004) and *NpGUT1* (Iwai et al., 2002).

Two allelic *Arabidopsis* cell wall mutants namely *quasimodo1* (*qua1-1* and *qua1-2*) have been identified and characterized (Bouton et al., 2002). Mutant cell wall line showed a 25% reduction in galacturonic acid levels compared with the wild type, indicating reduced pectin content, whereas neutral sugars remained unchanged. Moreover, *Arabidopsis thaliana* wild type and mutant *qua1-1* has been used to compare cell wall biosynthesis in stem tissue. In comparison to the wild type, α-1-4-D-galacturonosyltransferase (GalAT) activity was consistently reduced in *qua1-1* stems (Orfila et al., 2005). The *QUA1* gene is predicted to encode the homogalacturonan GalAT, which belongs to CAZy family GT8 (Orfila et al., 2005). Isolated transposon-tagged mutant from *Arabidopsis thaliana*, designated *parvus*, has been shown to exhibit humidity-sensitive semi-sterile dwarf phenotype (Lao et al., 2003). In addition, analyses of cell-wall carbohydrate of the *parvus* mutant indicated reduced levels of rhamnogalacturonan I branching and alterations in the abundance of some xyloglucan linkages that may be indirect consequences of the mutation. Another *Arabidopsis* mutant, designated *gaolaozhuangren1* (*glz1*), which is allelic to *parvus* exhibited a reduced plant stature, reduced size of organs in the shoot and dark-green leaves, indicating an important role of *GLZ1* gene in normal development (Iwai et al., 2002). *GLZ1* is expressed predominantly in stem and silique, and moderately in the flower. The carbohydrate analyses revealed that the composition and distribution were aberrant in the *glz1* mutant. On the basis of the results obtained using biochemical and genomics strategies Sterling et al., (2006) 24 genes of CAZy family GT8 have been classified which are closely related to galactonosyltransferase 1 (GAUT1). These studies provide the genetic and biochemical tools required to study other putative pectin biosynthetic genes.

Analyses of two independent mutant lines with T-DNA insertion in the *ARABINAN DEFICIENT 1* (*ARAD1*) genes from *Arabidopsis* revealed that these genes were expressed in all tissues but particularly in the vascular tissues of leaves and stems (Harholt et al., 2006). Analysis of cell wall polysaccharides isolated from leaves and stems showed that arabinose content was reduced to about 75% and 46% respectively, of wild-type levels. Isolated rhamnogalacturonan I from mutant tissues contained only about 30% of the wild-type amount of arabinose, confirming the specific deficiency in arabinan. These data suggest that *ARAD1* encode an arabinan alpha-1,5-arabinosyltransferase which belongs to members of CAZy family GT47.

The involvement of gene *NpGUT1 from Nicotiana tabacum*, which contains a putative glycosyltransferase catalytic domain characteristic for glycosyltransferase family 47, has been observed in the biosynthesis of rhamnogalacturonan II (RG-II) sugar chains (Iwai et al., 2002). *NpGUT1* expression in plants was higher during the reproductive stage than in the vegetative stage, and the suppression of *NpGUT1* expression results in flowers that are completely sterile. Mutation of the *NpGUT1* gene causes defects in the glucuronic acid of RG-II and drastically reduced the formation of borate cross-linking of RG-II (Iwai et al., 2006).

Enzymes Involved in Remodulation of Cell Wall Polysaccharides

During plant growth, the size of cells changes and a coordinated series of biochemical events occur resulting in the biosynthesis and degradation of cell wall components (Minic, 2008; Minic and Jouanin, 2006). Consequently, numerous enzymes must be implicated in these processes to allow the biosynthesis and remodelling of polysaccharide components. In addition to GTs which involve in the synthesis, a vast number of carbohydrate-active enzymes (CAZymes) play a key role in the modification and degradation of cell wall oligosaccharides and polysaccharides. The substantial role in the breakdown of glycosidic linkages is performed by glycoside hydrolases (GHs). Table 1 present some families of GTs and GHs which are involved or are supposed to be involved in biosynthesis and remodulation of cell wall polysaccharides in plants. However, other enzymes and proteins are also involved in biosynthesis, modification and assembly of cell wall polysaccharides such as carbohydrate esterases, polysaccharide lyases, expansins and various structural proteins (Girke et al., 2004).

Previous analyses showed that a total of 200 genes in the *Arabidopis* genome that encode GHs are predicted to be potentially involved in polysaccharides remodulation (Minic and Jouanin, 2006). These enzymes belong to 13 different families. Enzymatic activities have never been determined for most of these predicted enzymes. Very little information is available concerning the biochemical properties and mode of action of these proteins. In addition, listed GHs may have other functions than hydrolysis of cell wall components such as defence against pathogen attack and hydrolysis of different metabolites present in cell wall (Minic, 2008).

Recent reports have described various approaches which are used in order to identify and characterize these enzymes (Minic, 2008; Minic and Jouanin, 2006). As for GTs, the collection of T-DNA insertion mutants were used in order to characterize GHs which are involved in cell wall polysaccharides remodulation. The results of this research allowed for the identification of genes involved in the remodulation of cell wall polysaccharides (Minic and Jouanin, 2006). For some transgenic plants, phenotypes were observed (Goujon et al., 2003; Minic et al., 2006). However, as plants possess multiple genes of GHs which have the same or similar function, the determination of the function of a particular encoding enzyme is a very difficult problem. For example, biochemical characterization of some plant enzymes belonging to families 3 and 51 of GHs were performed by in vitro experiment. Purified enzymes from both familes showed similar enzymatic activities using natural or artificial substrates (Lee et al., 2001; Kotake et al., 2006; Minic et al., 2004, Minic et al., 2006 Tateishi et al., 2005). To understand their precise physiological roles, application of a different experimental approach is needed (Chavez Montes et al., 2008). In this way, interesting progress was made to study the functions of GHs by using biochemical analysis of mutant or transgenic plants. This strategy consists the extraction of proteins from wild and transgenic plants and analyses of enzymatic activities using various substrates after fractionations of proteins by classical chromatographic techniques. Principle of this strategy is present in Figure 4. This methodology is very useful for the identification of substrate specificities of enzymes without employing purification of the enzyme to homogeneity. Several recent papers describe the analysis of enzymes activity using this strategy (Goujon et al., 2003; Minic et al., 2006; Macquet et al., 2007; Chavez Montes et al., 2008). Finally, this strategy could be used in the characterization of GTs which are involved in cell wall polysaccharides biosynthesis.

Table 1. List of the *Arabidopsis thaliana* different CAZy families of GHs and GTs which are supposed to be involved in cell wall biosynthesis and reorganisation (Scheible and Pauly, 2004; Farrokhi et al., 2004; 2006; Minic and Lise, 2006; Minic, 2008)

CAZy Family GHs	GTs	Cell wall polysaccharide substrate
GH1, GH9,	GT2	Glucan/cellulose
GH5, GH27	GT2, GT34,	Galactomannan, mannan
GH10, GH3, GH43, GH51	GT8, GT43, GT47, GT61	Xylan, arabinoxylan, arabinan
GH1, GH16, GH31,	GT2, GT34, GT37, GT47	Xyloglucan
GH28, GH35	GT8	Homogalacturonan, galactan
	GT47	RG-II

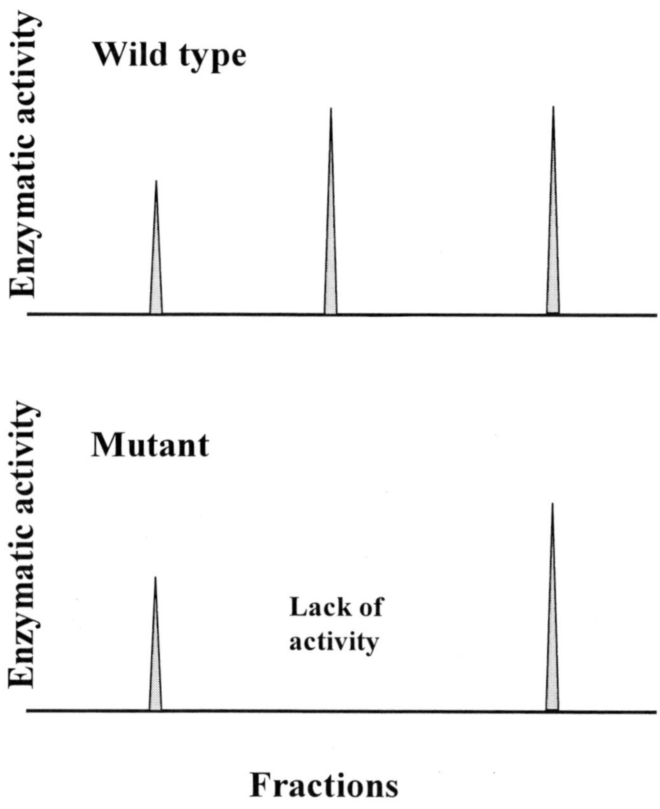

Figure 4. General principle of biochemical analysis of enzymatic activity using protein extract obtained from wild-type and transgenic plant containing null mutations in the gene encoding enzymes involved in cell wall polysaccharide remodulation. In the firstinstance, protein extracts obtained from wild-type (WT) and mutant lines were fractionated using the classic chromatographic technique for purification of proteins. The enzymatic activity of each fraction was determined and chromatographic profile between WT and mutant lines is compared. The absence of activity in the mutant line is a consequence of a mutated gene.

CONCLUSION AND PERSPECTIVES

Plants produce various cell wall polysaccharides and as a consequence a large array of enzymes involved in their biosynthesis and remodulations. Superfamilies of GT and GH enzymes have crucial role in these processes (Table 1); however, little information is available concerning them. These enzymes have been studied using different molecular, genetic and biochemical approaches. Many genes encoding these enzymes have been identified and functional analyses of some of them have been performed. A wide variety of transgenic plants with altered expression of one or more polysaccharides genes have been exploited in order to identify and characterize these encoding enzymes. Here is summarized the most recent advances in the study of transgenic plants related to the cell wall, focusing particularly on GTs and GHs. An important advance in this research area is possible through the exploration of different collections of the mutants. However, although these collections of the mutants are available very few of these mutants are characterized (Fagard et al., 2000).Therefore, further exploration of these mutants would be helpful in order to identify genes involved in polysaccharides metabolisms.

One important challenge is the identification of candidate genes for further analyses because this could facilitate the isolation of mutants. Bioinformatics, microarray and proteomic technology could produce supplementary information for the selection of candidate genes. The application of bioinformatics analysis might help to predict enzyme family members with similar functional properties. Proteomic and cDNA microarray analyses of different plant organs in various stages of development can be used to evaluate eventual enzyme/gene candidates for further investigation (Minic and Jouanin 2006; Hilson et al. 2004; Aspeborg et al. 2005). The second important problem in the analysis of mutant and transgenic plants is related to the development of a tool to screen mutants which alert polysaccharides content and structure (Mouille et al., 2006). Recently, Fourier transform infrared (FTIR) microspectroscopy was used for screening the mutants which altered cell wall composition or architecture (Mouille et al., 2006). In addition, antibodies against various polysaccharide epitopes also can be used in mutant screens (Willats et al. 2000).

Although the molecular genetics approach led to an increase in our knowledge about the identification of the genes which play key roles in cell wall biosynthesis, the precise functions of identified genes are not well documented. Purification of the enzymes or heterologous expression would be most useful to identify sequence, determine substrate specificity and perform structural analysis. However, the purification of protein may be difficult to achieve. For this reason, there will be a need to develop new biochemical tools for the evaluation of enzymatic functions and properties of corresponding genes using transgenic and mutant plant. Some effort has been developed in order to determine biochemical functions of enzymes using transgenic plants. However, new simple analytical and enzymatic test array might be developed for studies GHa and GTs since substrates or potential substrates of these enzymes are mostly known.

Research on the polysaccharides metabolism could allow for modification of cell wall structure and composition through genetics. In future, it is possible to design synthesis of polysaccharides in the cell wall for specific purposes. The modulation of plant cell wall polysaccharides using transgenic plants could be important for the improvement of architecture, fruit texture, fiber and wood quality, as well as for the properties of processed food products.

REFERENCES

Andersson, S.I., Samuelson, O., Ishihara, M. & Shimizu, K. (1983) Structure of the reducing end-groups in spruce xylan. *Carbohydr Res* 111: 283-288.

Aspeborg, H., Schrader, J., Coutinho, P.M., Stam, M., Kallas, A., Djerbi, S., Nilsson, P., Denman, S., Amini, B., Sterky, F., Master, E., Sandberg, G., Mellerowicz, E., Sundberg, B., Henrissat, B. & Teeri, T.T. (2005) Carbohydrate-active enzymes involved in the secondary cell wall biogenesis in hybrid aspen. *Plant Physiol* 137: 983–997.

Azpiroz-Leehan, R. & Feldmann, K.A. (1997) T-DNA insertion mutagenesis in Arabidopsis: going back and forth. *Trends Genet* 13: 152-156.

Beeckman, T., Przemeck, G.K., Stamatiou, G., Lau, R., Terryn, N., De Rycke, R., Inzé, D. & Berleth, T. (2002) Genetic complexity of cellulose synthase A gene function in Arabidopsis embryogenesis. *Plant Physiol* 130: 1883–1893.

Bewley, J.D., Burton, R.A., Morohashi, Y. & Fincher, G.B. (1997) Molecular cloning of a cDNA encoding a (1→4)-beta-mannan endohydrolase from the seeds of germinated tomato (*Lycopersicon esculentum*). *Planta* 203: 454–459.

Bouton, S., Leboeuf, E., Mouille, G., Leydecker, M.T., Talbotec, J., Granier, F., Lahaye, M., Hofte, H. & Truong, H.N. (2002). QUASIMODO1 encodes a putative membrane-bound glycosyltransferase required for normal pectin synthesis and cell adhesion in *Arabidopsis. Plant Cell* 14: 2577–2590.

Brown, D.M., Zeef, L.A., Ellis, J., Goodacre, R. & Turner, S.R. (2005) Identification of novel genes in *Arabidopsis* involved in secondary cell wall formation using expression profiling and reverse genetics. *Plant Ce*ll 17: 2281–2295.

Campisi, L., Yang, Y., Yi, Y., Heilig, E., Herman, B., Cassista, A.J., Allen, D.W., Xiang, H. & Jack, T. (1999) Generation of enhancer trap lines in *Arabidopsis* and characterization of expression patterns in the inflorescence. *Plant J* 17: 699–707.

Caño-Delgado, A., Penfield, S., Smith, C., Catley, M. & Bevan, M. (2003) Reduced cellulose synthesis invokes lignification and defense responses in *Arabidopsis thaliana. Plant* J 34: 351–362.

Carpita, N.C. (1996) Structure and biogenesis of the cell walls of grasses. *Annu. Rev. Plant Physiol. Plant Mol Biol* 47: 445–476.

Carpita, N.C. & Gibeaut, D.M. (1993) Structural models of primary cell walls in flowering plants: consistency of molecular structure with the physical properties of the walls during growth. *Plant* J 3:1–30.

Cavalier, D.M. & Keegstra, K. (2006) Two xyloglucan xylosyltransferases catalyze the addition of multiple xylosyl residues to cellohexaose. *J Biol Chem* 281: 34197-34207.

Chilton, M.D., Drummond, M.H., Merio, D.J., Sciaky, D., Montoya, A.L., Gordon, M.P. & Nester, E.W. (1977) Stable incorporation of plasmid DNA into higher plant cells: the molecular basis of crown gall tumorigenesis. *Cell* 11: 263-271.

Chavez Montes, R.A., Ranocha, P., Martinez, Y., Minic, Z., Jouanin, L., Marquis, M., Saulnier, L., Fulton, L.M., Cobbett, C.S., Bitton, F., Renou, J.P., Jauneau, A. & Goffner,

D.(2008) Cell wall modifications in Arabidopsis thaliana plants with altered {alpha}-L-arabinofuranosidase activity. *Plant Physiol.* 147:63-77.

Chuang, C.F. & Meyerowitz, E.M. (2000) Specific and heritable genetic interference by double-stranded RNA in *Arabidopsis thaliana.* Proc Natl Acad Sci USA. 97: 4985–4990.

Chu, Z., Chen, H., Zhang, Y., Zhang, Z., Zheng, N., Yin, B., Yan, H., Zhu, L., Zhao, X., Yuan, M., Zhang, X. & Xie, Q. (2007) Knockout of the AtCESA2 Gene Affects microtubule orientation and causes abnormal cell expansion in *Arabidopsis. Plant Physiol* 143: 213–224.

Cocuron, J.C., Lerouxel, O., Drakakaki, G., Alonso, A.P., Liepman, A.H., Keegstra, K., Raikhel, N., Wilkerson, C.G. (2007) A gene from the cellulose synthase-like C family encodes a β-1,4 glucan synthase. *Proc. Natl Acad Sci* USA104:8550–8555.

Cosgrove, D.J. (1997) Assembly and enlargement of the primary cell wall in plants. *Annu Rev Cell Dev Biol.* 13: 171–201.

Cosgrove, D.J. (1999) Enzymes and other agents that enhance cell wall extensibility. *Annu Rev Plant Physiol Plant Mol Biol* 50: 391–417.

Coutinho, P.M. & Henrissat, B. (1999) Carbohydrate-active enzymes server at URL: http://afmb.cnrs-mrs.fr/_cazy/CAZY/index. html

Coutinho, P.M., Deleury, E., Davies, G.J. & Henrissat B. (2003) An evolving hierarchical family classification for glycosyltransferases. *J Mol Biol* 328: 307–317.

Dalmay, T., Hamilton, A., Mueller, E. & Baulcombe, D.C. (2000) Potato virus X amplicons in *Arabidopsis* mediate genetic and epigenetic gene silencing. *Plant Cell* 12: 369–379.

Delmer, D. (1999). Cellulose biosynthesis: Exciting times for a difficult field of study. Annu Rev Plant Physiol Plant Mol Biol 50: 245–276.

Dey, P.M. (1978) Biochemistry of plant galactomannans. *Adv Carbohydr Chem Biochem* 35: 3411–3476.

Desprez, T., Juraniec, M., Crowell, E.F., Jouy, H., Pochylova, Z., Parcy, F., Höfte, H., Gonneau, M. & Vernhettes, S. (2007) Organization of cellulose synthase complexes involved in primary cell wall synthesis in *Arabidopsis thaliana. Proc Natl Acad Sci* U S A. 104:15572-15577.

Desprez, T., Vernhettes, S., Fagard, M., Refrégier, G., Desnos, T., Aletti, E., Py, N., Pelletier, S. & Höfte, H. (2002) Resistance against herbicide isoxaben and cellulose deficiency caused by distinct mutations in same cellulose synthase isoform CESA6. *Plant Physiol* 128: 482-490.

Dhugga, K.S., Barreiro, R., Whitten, B., Stecca, K., Hazebroek, J., Randhawa, G.S., Dolan, M., Kinney, A.J., Tomes, D., Nichols, S. & Anderson, P. (2004) Guar seed beta-mannan synthase is a member of the cellulose synthase super gene family. *Science.* 303: 363-366.

Doblin, M.S., Kurek, I., Jacob-Wilk, D. & Delmer, D.P. (2002) Cellulose biosynthesis in plants: from genes to rosettes. *Plant Cell Physiol* 43: 1407–1420.

Eckardt, N.A. (2001) The new biology. Genomics fosters a "systems approach" and collaborations between academic, government, and industry scientists. *Plant Cell* 13:725-732.

Edwards, M.E., Choo, T.-S., Dickson, C.A., Scott, C., Gidley, M.J. & Reid, J.S.G. (2004) The seeds of *Lotus japonicus* lines transformed with sense, antisense, and sense/antisense galactomannan galactosyltransferase constructs have structurally altered galactomannans in their endosperm cell walls. *Plant Physiol* 134: 1153–1162.

Farrokhi, N., Burton, R.A., Brownfield, L., Hrmova, M., Wilson, S.M., Bacic, A. & Fincher, G.B. (2006) Plant cell wall biosynthesis: genetic, biochemical and functional genomics approaches to the identification of key genes. *Plant Biotechnol* J 4: 145-67.

Feldmann, K.A. (1991) T-DNA insertion mutagenesis in *Arabidopsis*: mutational spectrum. *Plant J* 1: 71–82.

Filonova, L., Gunnarsson, L.C., Daniel, G. & Ohlin, M. (2007) Synthetic xylan-binding modules for mapping of pulp fibres and wood sections. *BMC Plant Biol* 7: 54.

Fry, S.C. (1995) Polysaccharide-modifying enzymes in the plant cell wall. *Annu Rev Plant Physiol Plant Mol Biol* 46: 497–520.

Fry, S.C. (2004) Primary cell wall metabolism: tracking the careers of wall polymers in living plant cells. *New Phytol* 161: 641–675.

Gheysen, G., Villarroel, R. & Van Montagu, M. (1991) Illegitimate recombination in plants: a model for T-DNA integration. *Genes Dev* 5: 287-297

Gillmor, C.S., Poindexter, P., Lorieau, J., Palcic, M.M. & Somerville, C. (2002) α-Glucosidase I is required for cellulose biosynthesis and morphogenesis in *Arabidopsis. J Cell Biol* 156: 1003–1013.

Girke, T., Lauricha, J., Tran, H., Keegstra, K. & Raikhel, N.(2004) The Cell Wall Navigator database. A systems-based approach to organism-unrestricted mining of protein families involved in cell wall metabolism. *Plant Physiol.* 136: 3003-3008.

Gordon, M.P. (1998) Discovery of the T-DNA of *Agrobacterium tumefaciens.* In *Discoveries in Plant Biology,* (S Kung, S Yang, eds), Vol. 1. World Scientific, Singapore, pp 111-115.

Goujon, T., Minic, Z., El Amrani, A., Lerouxel, O., Aletti, E., Lapierre, C., Joseleau, J.P. & Jouanin, L. (2003) AtBXL1, a novel higher plant (*Arabidopsis thaliana*) putative beta-xylosidase gene, is involved in secondary cell wall metabolism and plant development. *Plant J* 33: 677–690.

Hamilton, C.M., Frary, A., Lewis, C. & Tanksley, S.D. (1996) Stable transfer of intact high molecular weight DNA into plant chromosomes. *Proc Natl Acad Sci* USA 93: 9975–9979.

Heredia A., Jimenez A.and Guillen, R. (1995) Composition of plant cell walls. *Z. Lebensm Unters Forsch* 200: 24–31.

Handford, M.G., Baldwin, T.C., Goubet, F., Prime, T.A., Miles, J., Yu, X. & Dupree, P. (2003) Localisation and characterisation of cell wall mannan polysaccharides in *Arabidopsis thaliana. Planta* 218: 27–36.

Harholt, J., Jensen, J.K., Sørensen, S.O., Orfila, C., Pauly, M., & Scheller, H.V. (2006). Arabinan deficient 1 is a putative arabinosyltransferase involved in biosynthesis of pectic arabinan in *Arabidopsis. Plant Physiol* 140: 49–58.

Hayashi, T., Yoshida, K., Park, Y.W., Konishi, T. & Baba, K. (2005) Cellulose metabolism in plants. *In Survey of cell biology* (K.W. Jeon ed.), Academic Press, San Diego, CA. Vol. 247: pp. 1–34.

Hernández-Blanco, C., Feng, D.X., Hu, J. Sánchez-Vallet, A., Deslandes, L., Llorente, F., Berrocal-Lobo, M., Keller, H., Barlet, X., Sánchez-Rodríguez, C., Anderson, L.K., Somerville, S., Marco, Y. & Molina, A. (2006) Impairment of cellulose synthases required for Arabidopsis secondary cell wall formation enhances disease resistance. *Plant Cell* 19: 890–903.

Hilson, P., Allemeersch, J., Altmann, T., Aubourg, S., Avon, A., Beynon, J., Bhalerao, R.P., Bitton, F., Caboche, M., Cannoot, B., Chardakov, V., Cognet-Holliger, C., Colot, V., Crowe, M., Darimont, C., Durinck, S., Eickhoff, H., de Longevialle, A.F., Farmer, E.E., Grant, M., Kuiper, M.T., Lehrach, H., Leon, C., Leyva, A., Lundeberg, J., Lurin, C., Moreau, Y., Nietfeld, W., Paz-Ares, J., Reymond, P., Rouze, P., Sandberg, G., Segura, M.D., Serizet, C., Tabrett, A., Taconnat, L., Thareau, V., Van Hummelen, P., Vercruysse, S., Vuylsteke, M., Weingartner, M., Weisbeek, P.J., Wirta, V., Wittink, F.R., Zabeau, M. & Small, I. (2004) Versatile gene-specific sequence tags for *Arabidopsis* functional genomics: transcript profiling and reverse genetics applications. *Genome Research* 14: 2176-2189.

Hrmova, M., Farkas, V., Lahnstein, J. & Fincher, G.B. (2007) A barley xyloglucan xyloglucosyl transferase covalently links xyloglucan, cellulosic substrates, and (1,3-1,4)-beta-D-glucans. *J Biol Chem.* 282: 12951-12962.

Iwai, H., Hokura, A., Oishi, M., Chida, H., Ishii, T., Sakai, S. & Satoh, S. (2006) The gene responsible for borate cross-linking of pectin Rhamnogalacturonan-II is required for plant reproductive tissue development and fertilization. *Proc Natl Acad Sci* U S A 103: 16592-16597.

Iwai, H., Masaoka, N., Ishii, T. & Satoh, S. (2002). A pectin glucuronosyltransferase gene is essential for intercellular attachment in the plant meristem. *Proc Natl Acad Sci* USA 99: 16319–16324.

Izydorczyk, M.S. & Biliaderis, C.G. (1995) Cereal arabinoxylans: advances in structure and physicochemical properties. *Carbohydr Polym* 28: 33–48.

Johansson, M.H. & Samuelson, O. (1977) Reducing end groups in birch xylan and their alkaline degradation. *Wood Sci Technol* 11: 251-263.

Kohorn, B.D. (2000) Plasma membrane-cell wall contacts. *Plant Physiol* 124: 31-38.

Kotake, T., Tsuchiya, K., Aohara, T., Konishi, T., Kaneko, S., Igarashi, K., Samejima, M. & Tsumuraya, Y. (2006) An alpha-L-arabinofuranosidase/beta-D-xylosidase from immature seeds of radish (*Raphanus sativus* L.). *J Exp Bot* 57: 2353-2362.

Kulkarni, N., Shendye, A. & Rao, M. (1999) Molecular and biotechnological aspect of xylanases. *FEMS Microbiol Rev* 23: 411–456.

Koncz, C., Nemeth, K., Redei, G.P. & Schell, J. (1992) T-DNA insertional mutagenesis in *Arabidopsis*. *Plant Mol Biol* 20: 963-976.

Krysan, P.J., Young, J.C. & Sussman, M.R. (1999) T-DNA as an insertional mutagen in Arabidopsis. *Plant Cell* 11: 2283–2290.

Lai Kee Him, J., Chanzy, H., Mueller, M., Putaux, JL, Imai, T. & Bulone, V. (2002) In vitro versus in vivo cellulose microfibrils from plant primary wall synthases: structural differences. *J Biol Chem* 277: 36931-36939

Lao, N.T., Long, D., Kiang, S., Coupland, G., Shoue, D.A., Carpita, N.C. & Kavanagh, T.A. (2003). Mutation of a family 8 glycosyltransferase gene alters cell wall carbohydrate composition and causes a humidity-sensitive semi-sterile dwarf phenotype in *Arabidopsis. Plant Mol Biol* 53: 687–701.

Lee, C., O'Neill, M.A., Tsumuraya, Y., Darvill, A.G. & Ye, Z.H. (2007) The irregular xylem9 mutant is deficient in xylan xylosyltransferase activity. *Plant Cell Physiol* 48: 1624-1634.

Lee, R.C., Hrmova, M., Burton, R.A., Lahnstein, J. & Fincher, G.B. (2003) Bifunctional family 3 glycoside hydrolases from barley with alpha -L-arabinofuranosidase and beta-D-xylosidase activity. Characterization, primary structures, and COOH-terminal processing. *J Biol Chem* 278: 5377-5387.

Lerouxel, O., Cavalier, D.M., Liepman, A.H. & Keegstra, K. (2006) Biosynthesis of plant cell wall polysaccharides - a complex process. *Curr Opin Plant Biol* 9: 621-630.

Lima, D.U. & Buckeridge, M.S. (2001) Interaction between cellulose and storage xyloglucans: the influence of the degree of galactosylation. *Carbohydr Polym* 46: 157–163.

Macquet, A., Ralet, M.C., Loudet, O., Kronenberger, J., Mouille, G., Marion-Poll, A. & North, H.M. (2007) A naturally occurring mutation in an Arabidopsis accession affects a beta-D-galactosidase that increases the hydrophilic potential of rhamnogalacturonan I in seed mucilage. *Plant Cell* 19: 3990-4006.

Madson, M., Dunand, C., Li, X., Verma, R., Vanzin, G.F., Caplan, J., Shoue, D.A., Carpita, N.C. & Reiter, W.D. (2003) The MUR3 gene of *Arabidopsis* encodes a xyloglucan galactosyltransferase that is evolutionarily related to animal exostosins. *Plant Cell* 15: 1662–1670.

McCann, M.C. & Roberts, K. (1991) Architecture of the primary cell wall. *In*: *The cytoskeletal basis of plant growth and form* (C.W. Lloyd, Ed), Academic Press, Inc, New York, N.Y., pp. 109–129.

Minic, Z., Rihouey, C., Do, C.T., Lerouge, P. & Jouanin, L. (2004) Purification and characterization of enzymes exhibiting beta-D-xylosidase activities in stem tissues of *Arabidopsis*. *Plant Physiol* 135: 867-878.

Minic, Z., Do, C-T., Rihouey, C., Morin, H., Lerouge, P. & Jouanin, L. (2006) Purification, functional characterization, cloning and identification of mutants of a seed specific arabinan hydrolase in *Arabidopsis*. *J Exp Bot* 57: 2339–2351.

Minic, Z. & Jouanin, L. (2006) Plant glycoside hydrolases involved in cell wall polysaccharides degradation. *Plant Plysiol Biochem* 44: 435–449.

Minic, Z. (2008) Physiological roles of plant glycoside hydrolases. *Planta.* 227: 723-740.

Mitchell, R.A., Dupree, P. & Shewry, P.R. (2007) A novel bioinformatics approach identifies candidate genes for the synthesis and feruloylation of arabinoxylan. *Plant Physiol* 144: 43-53.

Mysore, K.S., Kumar, C.T. & Gelvin, S.B. (2000) Arabidopsis ecotypes and mutants that are recalcitrant to *Agrobacterium* root transformation are susceptible to germ-line transformation. *Plant J* 21: 9-16.

Nacry, P., Camilleri, C., Courtial, B., Caboche, M. & Bouchez, D. (1998) Major chromosomal rearrangements induced by T-DNA transformation in *Arabidopsis*. *Genetics* 149: 641-650.

Nicol, F., His, I., Jauneau, A., Vernhettes, S., Canut, H. & Höfte, H. (1998) A plasma membrane-bound putative endo-1,4-beta-D-glucanase is required for normal wall assembly and cell elongation in *Arabidopsis*. *EMBO J* 17: 5563-5576.

Obembe, O.O., Jacobsen, E., Visser, R.G.F. & Vincken, J-P. (2006) Cellulose-hemicellulose networks as target for in *planta* modification of the properties of natural fibres. *Biotechnol Mol Biol Rev* 1: 76-86.

Orfila, C., Sørensen, S.O., Harholt, J., Geshi, N., Crombie, H., Truong, H.-N., Reid, J.S.G., Knox, J.P. & Scheller, H.V. (2005). Quasimodo1 is expressed in vascular tissue of *Arabidopsis thaliana* inflorescence stems, and affects homogalacturonan and xylan biosynthesis. *Planta* 222: 613–622.

Pagant, S., Bichet, A., Sugimoto, K., Lerouxel, O., Desprez, T., McCann, M., Lerouge, P., Vernhettes, S. & Höfte, H. (2002) KOBITO1 encodes a novel plasma membrane protein necessary for normal synthesis of cellulose during cell expansion in *Arabidopsis*. *Plant Cell* 14: 2001-2013.

Paradez, A., Wright, A. & Ehrhardt, D.W. (2006) Microtubule cortical array organization and plant cell morphogenesis. *Curr Opin Plant* Biol 9: 571–578

Persson, S., Paredez, A., Carroll, A., Palsdottir, H., Doblin, M., Poindexter, P., Khitrov, N., Auer, M. & Somerville, C.R. (2007) Genetic evidence for three unique components in primary cell-wall cellulose synthase complexes in *Arabidopsis*. *Proc Natl Acad Sci* USA USA 104: 15566–15571.

Persson, S., Wei, H., Milne, J., Page, G.P. & Somerville, C.R. (2005) Identification of genes required for cellulose synthesis by regression analysis of public microarray data sets. *Proc Natl Acad Sci* USA 102: 8633–8638.

Popper, Z.A. & Fry, S.C. (2003) Primary cell wall composition of bryophytes and charophytes. *Ann Bot* (Lond) 91: 1–12.

Prade, R.A. (1996) Xylanases: from biology to biotechnology. *Biotechnol Genet Eng Rev* 13: 101–131.

Ridley, B.L., O'Neill, M.A. & Mohnen, D. (2001). Pectins: Structure, biosynthesis, and oligogalacturonide-related signaling. *Phytochemistry* 57: 929–967.

Robert S., Mouille G. & Höfte, H. (2004) The mechanism and regulation of cellulose synthesis in primary walls: lessons from cellulose-deficient *Arabidopsis* mutants. *Cellulose* 11: 351–364.

Roudier, F., Fernandez, A.G., Fujita, M., Himmelspach, R., Borner, G.H., Schindelman, G., Song, S., Baskin, T.I., Dupree, P., Wasteneys, G.O. & Benfey, P.N. (2005) COBRA, an *Arabidopsis* extracellular glycosyl-phosphatidyl inositol-anchored protein, specifically controls highly anisotropic expansion through its involvement in cellulose microfibril orientation. *Plant Cell* 17: 1749-1763.

Scheible, W.R., Eshed, R., Richmond, T., Delmer, D. & Somerville, C. (2001) Modifications of cellulose synthase confer resistance to isoxaben and thiazolidinone herbicides in *Arabidopsis Ixr1* mutants. *Proc Natl Acad Sci* USA 98: 10079-10084.

Scheible, W.R. & Pauly, M. (2004) Glycosyltransferases and cell wall biosynthesis: novel players and insights. *Curr Opin Plant Biol* 7: 285–295.

Schols, H.A., Vierhuis, E., Bakx, E.J. & Voragen, A.G. (1995) Different populations of pectic hairy regions occur in apple cell walls. *Carbohydr Res* 275: 343–360.

Shao, M., Zheng, H., Hu, Y., Liu, D., Jang, J.C., Ma, H. & Huang, H. (2004) The *GAOLAOZHUANGREN1* gene encodes a putative glycosyltransferase that is critical for normal development and carbohydrate metabolism. *Plant Cell Physiol* 45: 1453-1460.

Shimizu, K., Ishihara, M. & Ishihara, T. (1976) Hemicellulases of brown rotting fungus, *Tyromyces palustris*. II. The oligosaccharides from the hydrolysate of a hardwood xylan by the intracellular xylanase. *Mokuzai Gaikkashi* 22: 618-625.

Smith, N.A., Singh, S.P., Wang, M.B., Stoutjesdijk, P.A., Green, A.G. & Waterhouse, P.M. (2000) Total splicing by intron-spliced hairpin RNAs. *Nature* 407: 319–320.

Somerville, C.R. (2006) Cellulose synthesis in higher plants. *Annu Rev Cell Dev Biol* 22: 53-78.

Somerville, C.R. (2006) Cellulose synthesis in higher plants. *Annu Rev Cell Dev Biol* 22: 53–78.

Sterling, J.D., Atmodjo, M.A., Inwood, S.E., Kolli, V.S.K., Quigley, H.F., Hahn, M.G. & Mohnen, D. (2006). Functional identification of an *Arabidopsis* pectin biosynthetic homogalacturonan galacturonosyltransferase. *Proc Natl Acad Sci* USA 103: 5236–5241.

Tateishi A, Mori H, Watari J, Nagashima K, Yamaki S, Inoue H (2005) Isolation, characterization, and cloning of {alpha}-l-arabinofuranosidase expressed during fruit ripening of Japanese pear. *Plant Physiol* 138:1653-1664.

Taylor, N.G. (2008) Cellulose biosynthesis and deposition in higher plants. *New Phytol.* 178: 239-252.

Taylor, N.G. (2007) Identification of cellulose synthase AtCesA7 (IRX3) in vivo phosphorylation sites-a potential role in regulating protein degradation. *Plant Mol Biol* 64: 161-171.

Taylor, N.G., Gardiner, J.C., Whiteman, R. & Turner, S.R.. (2004) Cellulose synthesis in the *Arabidopsis* secondary cell wall. *Cellulose* 11: 329–338.

Taylor, N.G., Howells, R.M., Huttly, A.K., Vickers, K. & Turner, S.R. (2003) Interactions among three distinct CesA proteins essential for cellulose synthesis. *Proc Natl Acad Sci* USA 100: 1450–1455.

Taylor, N.G., Laurie, S. & Turner, S.R. (2000) Multiple cellulose synthase catalytic subunits are required for cellulose synthesis in *Arabidopsis*. *Plant Cell* 12: 2529–2539.

Thomas, C.M., Jones, D.A., English, J.J., Carroll, B.J., Bennetzen, J.L., Harrison, K., Burbidge, A., Bishop, G.J. & Jones, J.D. (1994) Analysis of the chromosomal distribution of transposon-carrying T-DNAs in tomato using the inverse polymerase chain reaction. *Mol Gen Genet* 242: 573-585.

Tinland, B. (1996) The integration of T-DNA into plant genomes. *Trends Plant Sci* 1: 178-183.

Tissier, A.F., Marillonnet, S., Klimyuk, V., Patel, K., Torres, M.A., Murphy, G. & Jones, J.D.G. (1999) Multiple independent defective suppressor-mutator transposon insertions in *Arabidopsis*: A tool for functional genomics. *Plant Cell* 11: 1841–1852.

Topping, J.F. & Lindsey, K. (1997) Promoter trap markers differentiate structural and positional components of polar development in *Arabidopsis*. *Plant Cell* 9: 1713–1725.

Tsekos, I. (1999) The sites of cellulose synthesis in algae: Diver-sity and evolution of cellulose-synthesizing enzyme complexes. *J Phycol* 35: 635-655.

Vanzin, G.F., Madson, M., Carpita, N.C., Raikhel, N.V., Keegstra, K. & Reiter, W.D. (2002) The mur2 mutant of *Arabidopsis* thaliana lacks fucosylated xyloglucan because of a lesion in fucosyltransferase AtFUT1. *Proc Natl Acad Sci* U S A. 99: 3340-3345.

Vincken, J.P., Schols, H.A., Oomen, R.J., McCann, M.C., Ulvskov, P., Voragen, A.G. & Visser, R.G. (2003) If homogalacturonan were a side chain of rhamnogalacturonan I. Implications for cell wall architecture. *Plant Physiol* 132: 1781–1789.

Visser, J. & Voragen, A.G.J. (1996) *Progress in Biotechnology 14: Pectins and Pectinases.* Proceedings of an International Symposium, Wageningen. Elsevier, Amsterdam, Science BV, Amsterdam, 990 pp.

Wasteneys, G.O. & Fujita, M. (2006) Establishing and maintaining axial growth: wall mechanical properties and the cytoskeleton. *J Plant Res 1*99: 5–10.

Weigel, D., Ahn, J., Blazquez, M.A., Borevitz, J.O., Christensen, S.K., Fankhauser, C., Ferrandiz, C., Kardailsky, I., Malancharuvil, E.J. & Neff, M.M. (2000) Activation tagging in *Arabidopsis*. P*lant Physiol1*22: 1003–1014.

Willats, W.G., McCartney, L., Mackie, W. & Knox, J.P. (2001) Pectin: cell biology and prospects for functional analysis. *Plant Mol Biol* 47: 9–27.

Willats, W.G.T., Steele-King, C.G., McCartney, L., Orfila, C., Marcus, S.E. & Knox, J.P. (2000) Making and using antibody probes to study plant cell walls. *Plant Physiol Biochem* 38: 27–36.

Winkler, R.G., Frank, M.R., Galbraith, D.W., Feyereisen, R., Feldmann, K.A. (1998) Systematic reverse genetics of transfer-DNA-tagged lines of *Arabidopsis*. *Plant Physiol* 118: 743–750.

Zambryski, P., Tempe, J. & Schell, J. (1989) Transfer and function of T-DNA genes from *Agrobacterium* Ti and Ri plasmids in plants. *Cell* 56: 193-201.

Zhong, R., Pena, M.J., Zhou, G.-K., Nairn, C.J., Wood-Jones, A., Richardson, E.A., Morrison, W.H., Darvill, A.G., York, W.S. & Ye, Z.H. (2005) *Arabidopsis Fragile Fiber8*, which encodes a putative glucuronyltransferase, is essential for normal secondary wall synthesis. *Plant Cell* 17: 3390–3408.

Zhong, R. & Ye, Z.H. (2007) Regulation of cell wall biosynthesis. *Curr Opin Plant Biol* 10: 564-572.

INDEX

B

C

D

E

H

I

J

K

L

M

N

O

P

Q

R

S

T

U

V

W

X

Y

Z